AF443358

BIOLOGICAL AND MEDICAL PHYSICS, BIOMEDICAL ENGINEERING

For further volumes:
http://www.springer.com/series/3740

BIOLOGICAL AND MEDICAL PHYSICS, BIOMEDICAL ENGINEERING

The fields of biological and medical physics and biomedical engineering are broad, multidisciplinary and dynamic. They lie at the crossroads of frontier research in physics, biology, chemistry, and medicine. The Biological and Medical Physics, Biomedical Engineering Series is intended to be comprehensive, covering a broad range of topics important to the study of the physical, chemical and biological sciences. Its goal is to provide scientists and engineers with textbooks, monographs, and reference works to address the growing need for information.

Books in the series emphasize established and emergent areas of science including molecular, membrane, and mathematical biophysics; photosynthetic energy harvesting and conversion; information processing; physical principles of genetics; sensory communications; automata networks, neural networks, and cellular automata. Equally important will be coverage of applied aspects of biological and medical physics and biomedical engineering such as molecular electronic components and devices, biosensors, medicine, imaging, physical principles of renewable energy production, advanced prostheses, and environmental control and engineering.

P.O.J. Scherer
Sighart F. Fischer

Theoretical Molecular Biophysics

With 178 Figures

 Springer

Professor Dr. Philipp Scherer
Professor Dr. Sighart F. Fischer
Technical University München, Physics Department T38
85748 München, Germany
E-mail: philipp.scherer@ph.tum.de, sighart.fischer@ph.tum.de

Biological and Medical Physics, Biomedical Engineering   ISSN 1618-7210
ISBN 978-3-540-85609-2   e-ISBN 978-3-540-85610-8
DOI 10.1007/978-3-540-85610-8
Springer Heidelberg Dordrecht London New York

Library of Congress Control Number: 2010930050

Cover design: eStudio Calamar Steinen

Printed on acid-free paper

Springer is part of Springer Science+Business Media (www.springer.com)

Preface

Biophysics deals with biological systems, such as proteins, which fulfill a variety of functions in establishing living systems. While the biologist uses mostly a phenomenological description, the physicist tries to find the general concepts to classify the materials and dynamics which underly specific processes. The phenomena span a wide range, from elementary processes, which can be induced by light excitation of a molecule, to communication of living systems. Thus, different methods are appropriate to describe these phenomena. From the point of view of the physicist, this may be Continuum Mechanics to deal with membranes, Hydrodynamics to deal with transport through vessels, Bioinformatics to describe evolution, Electrostatics to deal with aspects of binding, Statistical Mechanics to account for temperature and to learn about the role of the entropy, and last but not least Quantum Mechanics to understand the electronic structure of the molecular systems involved. As can be seen from the title, Molecular Biophysics, this book will focus on systems for which sufficient information on the molecular level is available. Compared to crystallized standard materials studied in solid-state physics, the biological systems are characterized by very big unit cells containing proteins with thousands of atoms. In addition, there is always a certain amount of disorder, so that the systems can be classified as complex. Surprisingly, the functions like a photocycle or the folding of a protein are highly reproducible, indicating a paradox situation in relation to the concept of maximum entropy production. It may seem that a proper selection in view of the large diversity of phenomena is difficult, but exactly this is also the challenge taken up within this book. We try to provide basic concepts, applicable to biological systems or soft matter in general. These include entropic forces, phase separation, cooperativity and transport in complex systems, like molecular motors. We also provide a detailed description for the understanding of elementary processes like electron, proton, and energy transfer and show how nature is making use of them for instance in photosynthesis. Prerequisites for the reader are a basic understanding in the fields of Mechanics, Electrostatics, Quantum Mechanics, and Statistics. This means the book is for graduate students, who want to specialize in the field of Biophysics. As we try to derive all equations in detail, the book may also be useful to physicists or chemists who are interested in applications of Statistical Mechanics or Quantum Chemistry to biological systems. This book is the outcome of a course presented by the authors as a basic element of the newly established graduation branch "Biophysics" in the Physics Department of the Technische Universität Muenchen.

The authors thank Dr. Florian Dufey and Dr. Robert Raupp-Kossmann for their contributions during the early stages of the evolving manuscript.

Garching,
May 2010

Sighart Fischer
Philipp Scherer

Contents

Part VII Elementary Photoinduced Processes

Part IX Appendix

Part I

Statistical Mechanics of Biopolymers

1

Random Walk Models for the Conformation

In this chapter, we study simple statistical models for the entropic forces which are due to the large number of conformations characteristic for biopolymers like DNA or proteins (Fig. 1.1).

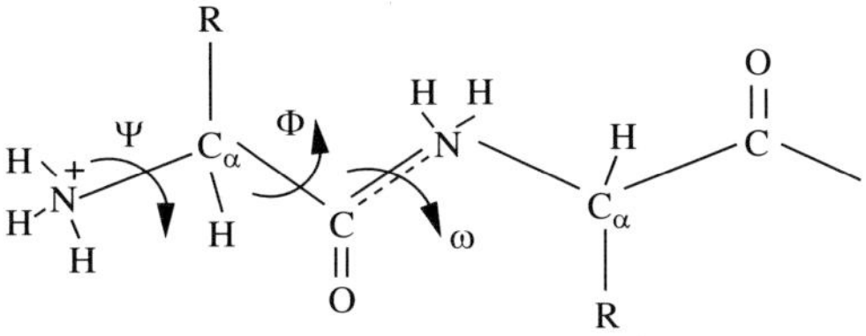

Fig. 1.1. Conformation of a protein. The relative orientation of two successive protein residues can be described by three angles (Ψ, Φ, ω). For a real protein, the ranges of these angles are restricted by steric interactions, which are neglected in simple models

1.1 The Freely Jointed Chain

We consider a three-dimensional chain consisting of M units. The configuration can be described by a point in a $3(M + 1)$-dimensional space (Fig. 1.2):

$$(\mathbf{r}_0, \mathbf{r}_2, \ldots, \mathbf{r}_M).$$ (1.1)

The M bond vectors

$$\mathbf{b}_i = \mathbf{r}_i - \mathbf{r}_{i-1}$$ (1.2)

have a fixed length $|\mathbf{b}_i| = b$ and are randomly oriented. This can be described by a distribution function:

$$P(\mathbf{b}_i) = \frac{1}{4\pi b^2}\delta(|b_i| - b).$$ (1.3)

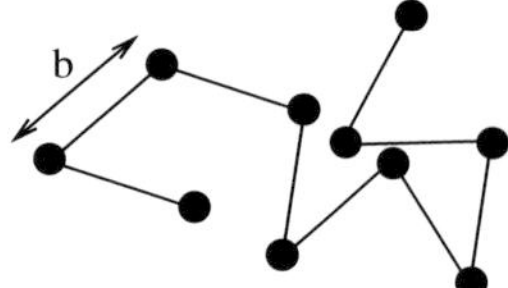

Fig. 1.2. Freely jointed chain with bond length b

Since the different units are independent, the joint probability distribution factorizes

$$P(\mathbf{b}_1, \ldots, \mathbf{b}_M) = \prod_{i=1}^{M} P(\mathbf{b}_i). \tag{1.4}$$

There is no excluded volume interaction between any two monomers. Obviously, the end-to-end distance

$$\mathbf{R} = \sum_{i=1}^{N} \mathbf{b}_i \tag{1.5}$$

has an average value of $\overline{\mathbf{R}} = 0$ since

$$\overline{\mathbf{R}} = \sum \overline{\mathbf{b}_i} = M \int \mathbf{b}_i P(\mathbf{b}_i) = 0. \tag{1.6}$$

The second moment is

$$\overline{R^2} = \overline{\left(\sum_i \mathbf{b}_i \sum_j \mathbf{b}_j \right)} = \sum_{i,j} \overline{\mathbf{b}_i \mathbf{b}_j} \tag{1.7}$$

$$= \sum_i \overline{b_i^2} + \sum_{i \neq j} \overline{\mathbf{b}_i \mathbf{b}_j} = Mb^2.$$

1.1.1 Entropic Elasticity

The distribution of the end-to-end vector is

$$P(\mathbf{R}) = \int P(\mathbf{b}_1, \ldots, \mathbf{b}_M) \delta \left(\mathbf{R} - \sum \mathbf{b}_i \right) d^3 b_1, \ldots, d^3 b_M. \tag{1.8}$$

This integral can be evaluated by replacing the δ-function by the Fourier integral:

$$\delta(\mathbf{R}) = \frac{1}{(2\pi)^3} \int e^{-i\mathbf{k}\mathbf{R}} d^3 k \tag{1.9}$$

which gives

$$P(\mathbf{R}) = \int d^3 k \, e^{-i\mathbf{k}\mathbf{R}} \prod_{i=1}^{M} \left(\int \frac{1}{4\pi b^2} \delta(|\mathbf{b}_i| - b) e^{i\mathbf{k}\mathbf{b}_i} d^3 b_i \right). \tag{1.10}$$

The inner integral can be evaluated in polar coordinates:

$$\int \frac{1}{4\pi b^2}\delta(|\mathbf{b}_i| - b)e^{i\mathbf{k}\mathbf{b}_i}\,d^3 b_i \tag{1.11}$$

$$= \int_0^{2\pi} d\phi \int_0^\infty b_i^2\,db_i\,\frac{1}{4\pi b^2}\delta(|b_i| - b)\int_0^\pi \sin\theta\,d\theta\,e^{ikb_i\cos\theta}.$$

The integral over θ gives

$$\int_0^\pi \sin\theta\,d\theta\,e^{ikb_i\cos\theta} = \frac{2\sin kb_i}{kb_i} \tag{1.12}$$

and hence

$$\int \frac{1}{4\pi b^2}\delta(|\mathbf{b}_i| - b)e^{i\mathbf{k}\mathbf{b}_i}\,d^3 b_i = 2\pi\int_0^\infty db_i\,\frac{1}{4\pi b^2}\delta(b_i - b)b_i^2\,\frac{2\sin kb_i}{kb_i}$$

$$= \frac{\sin kb}{kb}, \tag{1.13}$$

and finally we have

$$P(\mathbf{R}) = \frac{1}{(2\pi)^3}\int d^3k\,e^{-i\mathbf{k}\mathbf{R}}\left(\frac{\sin kb}{kb}\right)^M. \tag{1.14}$$

The function

$$\left(\frac{\sin kb}{kb}\right)^M \tag{1.15}$$

has a very sharp maximum at $kb = 0$. For large M, it can be approximated quite accurately by a Gaussian:

$$\left(\frac{\sin kb}{kb}\right)^M \approx e^{-M(1/6)k^2 b^2} \tag{1.16}$$

which gives

$$P(\mathbf{R}) \approx \frac{1}{(2\pi)^3}\int d^3k\,e^{-i\mathbf{k}\mathbf{R}}e^{-(M/6)k^2 b^2} = \left(\frac{3}{2\pi b^2 M}\right)^{3/2}e^{-3R^2/(2b^2 M)}. \tag{1.17}$$

We consider $\mathbf{R}$ as a macroscopic variable. The free energy is (no internal degrees of freedom, $E = 0$)

$$F = -TS = -k_\mathrm{B}T\ln P(\mathbf{R}) = \frac{3R^2}{2b^2 M}k_\mathrm{B}T + \text{const.} \tag{1.18}$$

The quadratic dependence on L is very similar to a Hookean spring. For a potential energy

$$V = \frac{k_{\mathrm{s}}}{2}x^2,$$ (1.19)

the probability distribution of the coordinate is

$$P(x) = \sqrt{\frac{k_{\mathrm{s}}}{2\pi k_{\mathrm{B}}T}}\,e^{-k_{\mathrm{s}}x^2/2k_{\mathrm{B}}T}$$ (1.20)

which gives a free energy of

$$F = -k_{\mathrm{B}}T\ln P = \mathrm{const} + \frac{k_{\mathrm{s}}x^2}{2}.$$ (1.21)

By comparison, the apparent spring constant is

$$k_{\mathrm{s}} = \frac{3k_{\mathrm{B}}T}{Mb^2}.$$ (1.22)

1.1.2 Force–Extension Relation

We consider now a chain with one fixed end and an external force acting in x-direction at the other end [1] (Fig. 1.3).

The projection of the ith segment onto the x-axis has a length of (Fig. 1.4)

$$b_i = -b\cos\theta \in [-b, b].$$ (1.23)

We discretize the continuous range of b_j by dividing the interval $[-b, b]$ into n bins of width $\Delta b = 2b/n$ corresponding to the discrete values $l_i, i = 1, \ldots, n$.

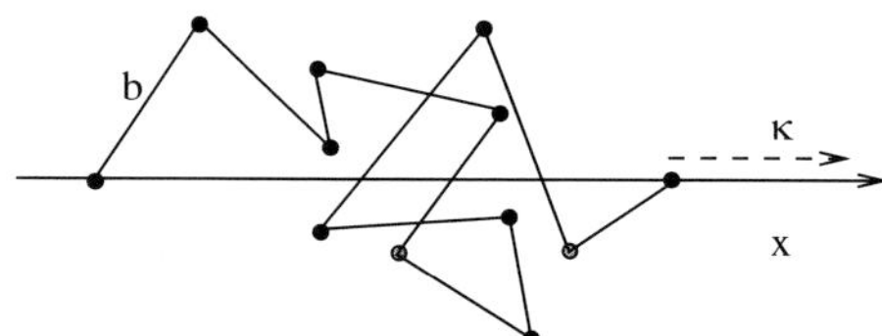

Fig. 1.3. Freely jointed chain with external force

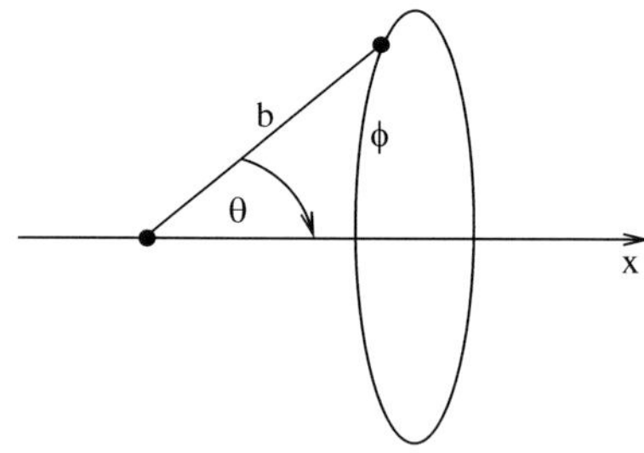

Fig. 1.4. Projection of the bond vector

The chain members are divided into n groups according to their bond projections b_j. The number of units in each group is denoted by M_i so that

$$\sum_{i=1}^{n} M_i = M \tag{1.24}$$

and the end-to-end length is

$$\sum_{i=1}^{n} l_i M_i = L. \tag{1.25}$$

The probability distribution is

$$P(\theta, \phi)\mathrm{d}\theta\mathrm{d}\phi = \frac{\sin(\theta)\mathrm{d}\theta\mathrm{d}\phi}{4\pi}. \tag{1.26}$$

Since we are only interested in the probability of the l_i, we integrate over ϕ

$$P(\theta)\mathrm{d}\theta = \frac{\sin(\theta)\mathrm{d}\theta}{2} \tag{1.27}$$

and transform variables to have

$$P(l)\mathrm{d}l = P(-b\cos\theta)\mathrm{d}(-b\cos\theta) = \frac{1}{2b}\mathrm{d}(-b\cos\theta) = \frac{1}{2b}\mathrm{d}l. \tag{1.28}$$

The canonical partition function is

$$Z(L, M, T) = \sum_{\{M_i\} \sum M_i l_i = L} \frac{M!}{\prod_j M_j!} \prod_i z_i^{M_i} = \sum_{\{M_i\}} M! \prod_i \frac{z_i^{M_i}}{M_i!}. \tag{1.29}$$

The $z_i = z$ are the independent partition functions of the single units which we assume as independent of i. The degeneracy factor $M!/\prod M_i!$ counts the number of microstates for a certain configuration $\{M_i\}$. The summation is only over configurations with a fixed end-to-end length. This makes the evaluation rather complicated. Instead, we introduce a new partition function by considering an ensemble with fixed force and fluctuating length

$$\Delta(\kappa, M, T) = \sum_L Z(L, M, T)\mathrm{e}^{\kappa L/k_\mathrm{B}T}. \tag{1.30}$$

Approximating the logarithm of the sum by the logarithm of the maximum term, we see that

$$-k_\mathrm{B}T\ln\Delta = -k_\mathrm{B}T\ln Z - \kappa L \tag{1.31}$$

gives the Gibbs free enthalpy[1]

---

[1] $-\kappa L$ corresponds to $+pV$.

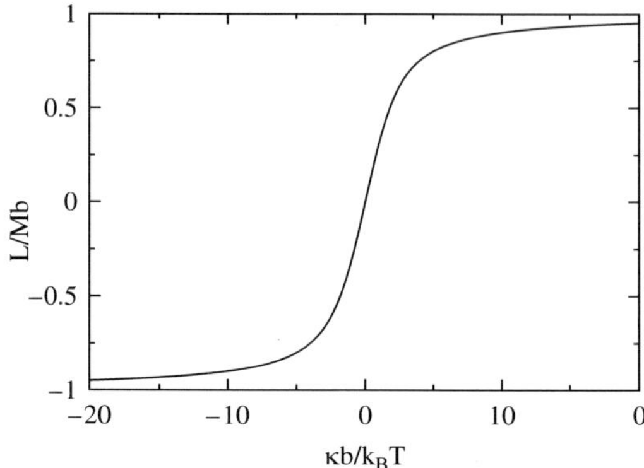

Fig. 1.5. Force–extension relation

$$G(\kappa, M, T) = F - \kappa L. \tag{1.32}$$

In this new ensemble, the summation over L simplifies the partition function:

$$\Delta = \sum_{\{M_i\}} e^{\kappa \sum M_i l_i / k_B T} M! \prod \frac{z^{M_i}}{M_i!} = \sum_{\{M_i\}} M! \prod \frac{\left(z e^{\kappa l_i / k_B T}\right)^{M_i}}{M_i!}$$

$$= \left(\sum z e^{\kappa l_i / k_B T}\right)^M = \xi(\kappa, T)^M. \tag{1.33}$$

Now returning to a continuous distribution of $l_i = -b \cos\theta$, we have to evaluate

$$\xi = \int_{-b}^{b} P(l)\,\mathrm{d}l, \quad z e^{\kappa l / k_B T} = z \frac{\sinh t}{t} \tag{1.34}$$

with (Fig. 1.5)

$$t = \frac{\kappa b}{k_B T}. \tag{1.35}$$

From

$$\mathrm{d}G = -S\mathrm{d}T - L\mathrm{d}\kappa, \tag{1.36}$$

we find

$$L = -\left.\frac{\partial G}{\partial \kappa}\right|_T = \frac{\partial}{\partial \kappa}\left(M k_B T \ln\left(\frac{z}{\kappa b / k_B T} \sinh \frac{\kappa b}{k_B T}\right)\right)$$

$$= M k_B T \left(-\frac{1}{\kappa} + \frac{b}{k_B T} \coth\left(\frac{\kappa b}{k_B T}\right)\right) = M b \mathfrak{L}\left(\frac{\kappa b}{k_B T}\right),$$

with the Langevin function

$$\mathfrak{L}(x) = \coth(x) - \frac{1}{x}. \tag{1.37}$$

1.2 Two-Component Model

A one-dimensional random walk model can be applied to a polymer chain which is composed of two types of units (named α and β), which may interconvert. This model is for instance applicable to the dsDNA $\rightarrow$ SDNA transition or the α-Helix $\rightarrow$ random coil transition of proteins which show up as a plateau in the force–extension curve [1, 2].

We follow the treatment given in [1] which is based on an explicit evaluation of the partition function. Alternatively, the two-component model can be mapped onto a one-dimensional Ising model, which can be solved by the transfer matrix method [3, 4]. We assume that of the overall M units, M_α are in the α-configuration and $M - M_\alpha$ are in the β-configuration. The lengths of the two conformers are l_α and l_β, respectively (Fig. 1.6).

1.2.1 Force–Extension Relation

The total length is given by

$$L = M_\alpha l_\alpha + (M - M_\alpha)l_\beta = M_\alpha(l_\alpha - l_\beta) + Ml_\beta. \tag{1.38}$$

The number of configurations with length L is given by

$$\Omega(L) = \Omega\left(M_\alpha = \frac{L - Ml_\beta}{l_\alpha - l_\beta}\right) = \frac{M!}{\left(\frac{L-Ml_\beta}{l_\alpha-l_\beta}\right)!\left(\frac{Ml_\alpha-L}{l_\alpha-l_\beta}\right)!}. \tag{1.39}$$

From the partition function

$$Z = z_\alpha^{M_\alpha} z_\beta^{M-M_\alpha}\Omega = z_\alpha^{(L-Ml_\beta)/(l_\alpha-l_\beta)} z_\beta^{(Ml_\alpha-L)/(l_\alpha-l_\beta)}\Omega, \tag{1.40}$$

application of Stirling's approximation gives for the free energy

$$\begin{aligned}
F &= -k_{\mathrm{B}}T\ln Z \\
&= -k_{\mathrm{B}}T\frac{L - Ml_\beta}{l_\alpha - l_\beta}\ln z_\alpha - k_{\mathrm{B}}T\frac{Ml_\alpha - L}{l_\alpha - l_\beta}\ln z_\beta - k_{\mathrm{B}}T\left(M\ln M - M\right) \\
&= -k_{\mathrm{B}}T\left\{-\left(\frac{L - Ml_\beta}{l_\alpha - l_\beta}\right)\ln\left(\frac{L - Ml_\beta}{l_\alpha - l_\beta}\right) + \left(\frac{L - Ml_\beta}{l_\alpha - l_\beta}\right)\right. \\
&\quad \left. -\left(\frac{Ml_\alpha - L}{l_\alpha - l_\beta}\right)\ln\left(\frac{Ml_\alpha - L}{l_\alpha - l_\beta}\right) + \left(\frac{Ml_\alpha - L}{l_\alpha - l_\beta}\right)\right\}.
\end{aligned} \tag{1.41}$$

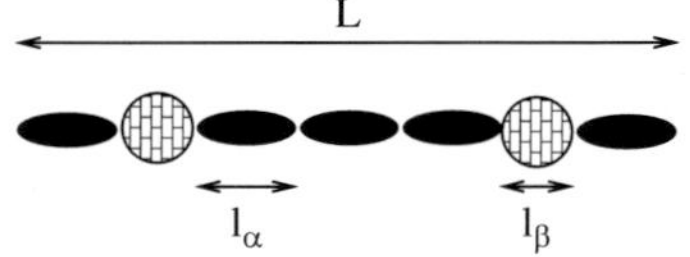

Fig. 1.6. Two-component model

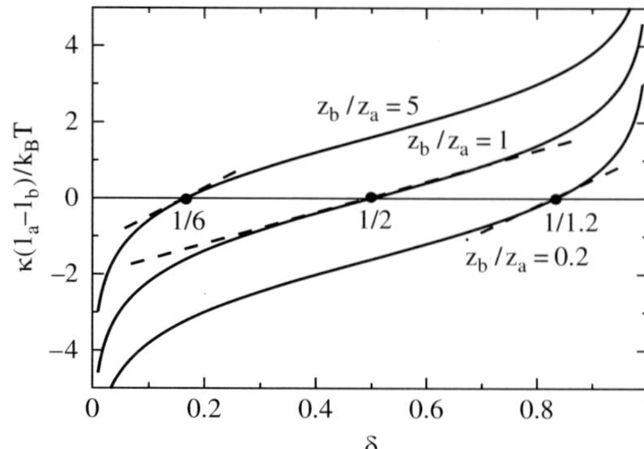

Fig. 1.7. Force–extension relation for the two-component model

The derivative of the free energy gives the force–extension relation

$$\kappa = \frac{\partial F}{\partial L} = \frac{k_B T}{l_\alpha - l_\beta} \ln\left(\frac{M l_\beta - L}{L - M l_\alpha}\right) + \frac{k_B T}{l_\alpha - l_\beta} \ln\frac{z_\beta}{z_\alpha}. \tag{1.42}$$

This can be written as a function of the fraction of segments in the α-configuration

$$\delta = \frac{M_\alpha}{M} = \frac{L - M l_\beta}{M(l_\alpha - l_\beta)} \tag{1.43}$$

in the somewhat simpler form (Fig. 1.7)

$$\kappa \frac{l_\alpha - l_\beta}{k_B T} = \ln\frac{\delta}{1 - \delta} + \ln\frac{z_\beta}{z_\alpha}. \tag{1.44}$$

The mean extension for zero force is obtained by solving $\kappa(L) = 0$:

$$\overline{L}_0 = M\left(\frac{z_\alpha l_\alpha + z_\beta l_\beta}{z_\alpha + z_\beta}\right), \tag{1.45}$$

$$\delta_0 = \frac{\overline{L}_0 - M l_\beta}{M(l_\alpha - l_\beta)} = \frac{z_\alpha}{z_\alpha + z_\beta}. \tag{1.46}$$

Taylor series expansion around $\overline{L}_0$ gives the linearized force–extension relation:

$$\begin{aligned}
\kappa = \frac{\partial F}{\partial L} &= \frac{k_B T}{M(l_\alpha - l_\beta)^2}\frac{(z_\alpha + z_\beta)^2}{z_\alpha z_\beta}(L - \overline{L}_0) + \cdots \\
&\approx \frac{k_B T}{l_\alpha - l_\beta}\frac{(z_\alpha + z_\beta)^2}{z_\alpha z_\beta}(\delta - \delta_0) \\
&= \frac{k_B T}{l_\alpha - l_\beta}\frac{1}{\delta_0(1 - \delta_0)}(\delta - \delta_0).
\end{aligned} \tag{1.47}$$

1.2.2 Two-Component Model with Interactions

We now consider additional interaction between neighboring units. We introduce the interaction energies $w_{\alpha\alpha}, w_{\alpha\beta}, w_{\beta\beta}$ for the different pairs of neighbors and the numbers $N_{\alpha\alpha}, N_{\alpha,\beta}, N_{\beta\beta}$ of such interaction terms. The total interaction energy is then

$$W = N_{\alpha\alpha}w_{\alpha\alpha} + N_{\alpha\beta}w_{\alpha\beta} + N_{\beta\beta}w_{\beta\beta}. \tag{1.48}$$

The numbers of pair interactions are not independent from the numbers of units M_α, M_β. Consider insertion of an additional α-segment into a chain. Figure 1.8 counts the possible changes in interaction terms. In any case by insertion of an α-segments, the expression $2N_{\alpha\alpha} + N_{\alpha\beta}$ increases by 2.

Similarly, insertion of an extra β-segment increases $2N_{\beta\beta} + N_{\alpha\beta}$ by 2 (Fig. 1.9).

This shows that there are linear relationships of the form

$$2N_{\alpha\alpha} + N_{\alpha\beta} = 2M_\alpha + c_1,$$
$$2N_{\beta\beta} + N_{\alpha\beta} = 2M_\beta + c_2. \tag{1.49}$$

The two constants depend on the boundary conditions as can be seen from an inspection of the shortest possible chains with two segments. They are zero for periodic boundaries and will therefore be neglected in the following since the numbers M_α and M_β are much larger (Fig. 1.10).

We substitute

$$N_{\alpha\alpha} = M_\alpha - \frac{1}{2}N_{\alpha\beta}, \tag{1.50}$$

$\alpha\downarrow$	M_α	M_β	$N_{\alpha\alpha}$	$N_{\alpha\beta}$	$N_{\beta\beta}$	$2N_{\alpha\alpha}+N_{\alpha\beta}$
αα	+1		+1			+2
αβ	+1		+1			+2
βα	+1		+1			+2
ββ	+1			+2	−1	+2

Fig. 1.8. Insertion of an α-segment. $2N_{\alpha\alpha} + N_{\alpha\beta}$ increases by 2

$\beta\downarrow$	M_α	M_β	$N_{\alpha\alpha}$	$N_{\alpha\beta}$	$N_{\beta\beta}$	$2N_{\beta\beta}+N_{\alpha\beta}$
αα		+1	−1	+2		+2
αβ		+1			+1	+2
βα		+1			+1	+2
ββ		+1			+1	+2

Fig. 1.9. Insertion of a β-segment. $2N_{\beta\beta} + N_{\alpha\beta}$ increases by 2

	N_α	N_β	$N_{\alpha\alpha}$	$N_{\alpha\beta}$	$N_{\beta\beta}$	$2N_{\alpha\alpha} + N_{\alpha\beta} - 2N_\alpha$	$2N_{\beta\beta} + N_{\alpha\beta} - 2N_\beta$
$\alpha\ \alpha$	2	0	1	0	0	−2	0
$\beta\ \beta$	0	2	0	0	1	0	−2
$\alpha\ \beta$	1	1	0	1	0	−1	−1
$\overset{\frown}{\alpha\ \alpha}$	2	0	2	0	0	0	0
$\overset{\frown}{\beta\ \beta}$	0	2	0	0	2	0	0
$\overset{\frown}{\alpha\ \beta}$	1	1	0	2	0	0	0

Fig. 1.10. Determination of the constants. c_1 and c_2 are zero for periodic boundary conditions and in the range $-2, \ldots, 0$ for open boundaries

$$N_{\beta\beta} = M_\beta - \frac{1}{2} N_{\alpha\beta}, \tag{1.51}$$

$$w = w_{\alpha\alpha} + w_{\beta\beta} - 2w_{\alpha\beta} \tag{1.52}$$

to have the interaction energy

$$W = w_{\alpha\alpha}\left(M_\alpha - \frac{1}{2}N_{\alpha\beta}\right) + w_{\beta\beta}\left(M_\beta - \frac{1}{2}N_{\alpha\beta}\right) + w_{\alpha\beta}N_{\alpha\beta}$$

$$= w_{\alpha\alpha}M_\alpha + w_{\beta\beta}(M - M_\alpha) - \frac{w}{2}N_{\alpha\beta}. \tag{1.53}$$

The canonical partition function is

$$Z(M_\alpha, T) = z_\alpha^{M_\alpha} z_\beta^{M_\beta} \sum_{N_{\alpha\beta}} g(M_\alpha, N_{\alpha\beta}) e^{-W(N_{\alpha\beta})/k_B T}$$

$$= \left(z_\alpha e^{-w_{\alpha\alpha}/k_B T}\right)^{M_\alpha} \left(z_\beta e^{-w_{\beta\beta}/k_B T}\right)^{(M-M_\alpha)}$$

$$\times \sum_{N_{\alpha\beta}} g(M_\alpha, N_{\alpha\beta}) e^{N_{\alpha\beta} w / 2k_B T}. \tag{1.54}$$

The degeneracy factor g will be evaluated in the following.

The chain can be divided into blocks containing only α-segments (α-blocks) or only β-segments (β-blocks). The number of boundaries between α-blocks and β-blocks obviously is given by $N_{\alpha\beta}$. Let $N_{\alpha\beta}$ be an odd number. Then there are $(N_{\alpha\beta}+1)/2$ blocks of each type. (We assume that $N_{\alpha\beta}, M_\alpha, M_\beta$ are large numbers and neglect small differences of order 1 for even $N_{\alpha\beta}$.) In each α-block, there is at least one α-segment. The remaining $M_\alpha - (N_{\alpha\beta}+1)/2$ α-segments have to be distributed over the $(N_{\alpha\beta} + 1)/2$ α-blocks (Fig. 1.11).

Therefore, we need the number of possible ways to arrange $M_\alpha - (N_{\alpha\beta} + 1)/2$ segments and $(N_{\alpha\beta} - 1)/2$ walls which is given by the number of ways

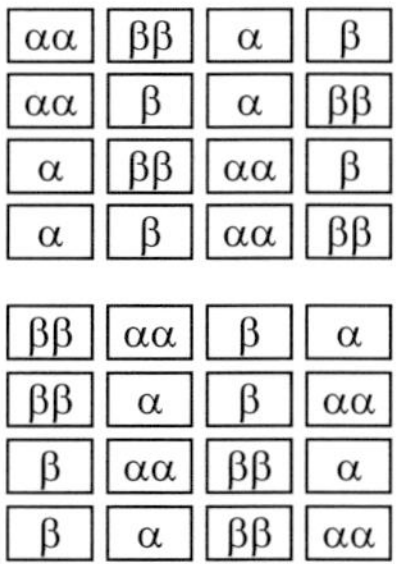

Fig. 1.11. Degeneracy factor g. The possible eight configurations are shown for $M = 4$, $M_\alpha = 3$, $M_\beta = 3$, and $N_{\alpha\beta} = 3$

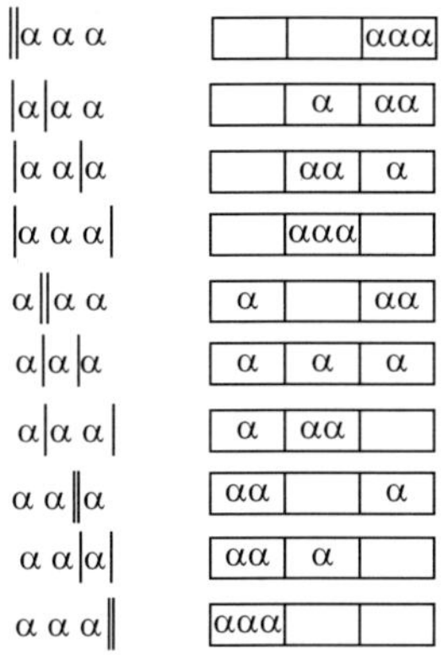

Fig. 1.12. Distribution of three segments over three blocks. This is equivalent to the arrangement of three segments and two border lines

to distribute the $(N_{\alpha\beta} - 1)/2$ walls over the total of $M_\alpha - 1$ objects which is given by (Fig. 1.12)

$$\frac{(M_\alpha - 1)!}{(\frac{N_{\alpha\beta}-1}{2})!(M_\alpha - \frac{N_{\alpha\beta}+1}{2})!} \approx \frac{M_\alpha!}{(\frac{N_{\alpha\beta}}{2})!(M_\alpha - \frac{N_{\alpha\beta}}{2})!}. \tag{1.55}$$

The same consideration for the β-segments gives another factor of

$$\frac{(M - M_\alpha)!}{(\frac{N_{\alpha\beta}}{2})!(M - M_\alpha - \frac{N_{\alpha\beta}}{2})!}. \tag{1.56}$$

Finally, there is an additional factor of 2 because the first block can be of either type. Hence for large numbers, we find

$$g(M_\alpha, N_{\alpha\beta}) = 2\frac{(M_\alpha)!}{(\frac{N_{\alpha\beta}}{2})!(M_\alpha - \frac{N_{\alpha\beta}}{2})!}\frac{(M - M_\alpha)!}{(\frac{N_{\alpha\beta}}{2})!(M - M_\alpha - \frac{N_{\alpha\beta}}{2})!}. \tag{1.57}$$

We look for the maximum summand of Z as a function of $N_{\alpha\beta}$. The corresponding number will be denoted as $N_{\alpha\beta}^*$ and is determined from the condition

$$0 = \frac{\partial}{\partial N_{\alpha\beta}} \ln\left(g(M_\alpha, N_{\alpha\beta})e^{wN_{\alpha\beta}/2k_{\mathrm B}T}\right) = \frac{w}{2k_{\mathrm B}T} + \frac{\partial}{\partial N_{\alpha\beta}} \ln g(M_\alpha, N_{\alpha\beta}).$$
(1.58)

Stirling's approximation gives

$$0 = \frac{w}{2k_{\mathrm B}T} + \frac{1}{2}\ln\left(M_\alpha - \frac{N^*_{\alpha\beta}}{2}\right) + \frac{1}{2}\ln\left(M - M_\alpha - \frac{N^*_{\alpha\beta}}{2}\right) - \ln\left(\frac{N^*_{\alpha\beta}}{2}\right)$$
(1.59)

or

$$0 = \frac{w}{k_{\mathrm B}T} + \ln\left(\frac{(M_\alpha - \frac{N^*_{\alpha\beta}}{2})(M - M_\alpha - \frac{N^*_{\alpha\beta}}{2})}{(\frac{N^*_{\alpha\beta}}{2})^2}\right).$$
(1.60)

Taking the exponential gives

$$\left(M_\alpha - \frac{N^*_{\alpha\beta}}{2}\right)\left(M - M_\alpha - \frac{N^*_{\alpha\beta}}{2}\right) = e^{-w/k_{\mathrm B}T}\left(\frac{N_{\alpha\beta}}{2}\right)^2.$$
(1.61)

Introducing the relative quantities

$$\delta = \frac{M_\alpha}{M}, \quad \gamma = \frac{N^*_{\alpha\beta}}{2M},$$
(1.62)

we have to solve the quadratic equation

$$(\delta - \gamma)(1 - \delta - \gamma) = \gamma^2 e^{-w/k_{\mathrm B}T}.$$
(1.63)

The solutions are

$$\gamma = \frac{-1 \pm \sqrt{(1 - 2\delta)^2 + 4e^{-w/k_{\mathrm B}T}\delta(1 - \delta)}}{2(e^{-w/k_{\mathrm B}T} - 1)}.$$
(1.64)

Series expansion in $w/k_{\mathrm B}T$ gives

$$\gamma = \frac{k_{\mathrm B}T}{2w} + \frac{1}{4} + \frac{w}{24k_{\mathrm B}T} + \cdots$$
$$\cdots \pm \left(-\frac{k_{\mathrm B}T}{2w} + \delta(1 - \delta) - \frac{1}{4} + \left((\delta - \delta^2)^2 - \frac{1}{24}\right)\frac{w}{k_{\mathrm B}T}\right) + \cdots .$$
(1.65)

The $-$ alternative diverges for $w \to 0$ whereas the $+$ alternative

$$\gamma = \delta(1 - \delta) + (\delta - \delta^2)^2\frac{w}{k_{\mathrm B}T} - \delta^2\left(\frac{1}{2}(1 - \delta)^2 + 2\delta(\delta - 1)^3\right)\frac{w}{k_{\mathrm B}T}\cdots$$
(1.66)

approaches the value

$$\gamma_0 = \delta - \delta^2$$
(1.67)

which is the only solution of the one case

$$(\delta - \gamma_0)(1 - \delta - \gamma_0) = \gamma_0^2 \to \delta(1 - \delta) - \gamma = 0.$$
(1.68)

For $N_{\alpha\beta}^*$, we obtain approximately

$$N_{\alpha\beta}^* = 2M \left(\delta(1-\delta) + (\delta - \delta^2)^2 \frac{w}{k_{\rm B}T} + \cdots \right). \qquad (1.69)$$

Let us now apply the maximum term method, which approximates the logarithm of a sum by the logarithm of the maximum summand:

$$\begin{aligned}
F &= -k_{\rm B}T \ln Z(M_\alpha, T) \\
&\approx -k_{\rm B}TM_\alpha \ln z_\alpha - k_{\rm B}T(M - M_\alpha) \ln z_\beta + M_\alpha w_{\alpha\alpha} + (M - M_\alpha)w_{\beta\beta} \\
&\quad -k_{\rm B}T \ln g(M_\alpha, N_{\alpha\beta}^*) - \frac{wN_{\alpha\beta}^*}{2}.
\end{aligned} \qquad (1.70)$$

The force–length relation is now obtained from

$$\begin{aligned}
\kappa &= \frac{\partial F}{\partial L} = \frac{\partial F}{\partial M_\alpha} \frac{\partial \left(\frac{L - Ml_\beta}{l_\alpha - l_\beta} \right)}{\partial L} = \frac{1}{l_\alpha - l_\beta} \frac{\partial F}{\partial M_\alpha} \\
&= \frac{1}{l_\alpha - l_\beta} \left(-k_{\rm B}T \ln z_\alpha + k_{\rm B}T \ln z_\beta + w_{\alpha\alpha} - w_{\beta\beta} - k_{\rm B}T \frac{\partial}{\partial M_\alpha} \ln g \right) \\
&\quad + \frac{1}{l_\alpha - l_\beta} \frac{\partial N_{\alpha\beta}^*}{\partial M_\alpha} \frac{\partial}{\partial N_{\alpha\beta}^*} \left(-k_{\rm B}T \ln g - \frac{wN_{\alpha\beta}^*}{2} \right).
\end{aligned} \qquad (1.71)$$

The last part vanishes due to the definition of $N_{\alpha\beta}^*$. Now using Stirling's formula, we find

$$\frac{\partial}{\partial M_\alpha} \ln g = \ln \frac{M_\alpha}{M - M_\alpha} + \ln \frac{M - M_\alpha - \frac{N_{\alpha\beta}^*}{2}}{M_\alpha - \frac{N_{\alpha\beta}^*}{2}} = \ln \frac{\delta}{1 - \delta} + \ln \frac{1 - \delta - \gamma}{\delta - \gamma} \qquad (1.72)$$

and substituting γ, we have finally

$$\begin{aligned}
\kappa \frac{(l_\alpha - l_\beta)}{k_{\rm B}T} &= \ln \frac{z_\beta e^{-w_{\beta\beta}/k_{\rm B}T}}{z_\alpha e^{-w_{\alpha\alpha}/k_{\rm B}T}} - \ln \frac{(1 - \delta)}{\delta} \\
&\quad + (2\delta - 1) \frac{w}{k_{\rm B}T} + \delta(3\delta - 2\delta^2 - 1) \left(\frac{w}{k_{\rm B}T} \right)^2.
\end{aligned} \qquad (1.73)$$

Linearization now gives

$$\delta_0 = \frac{z_\alpha e^{-w_{\alpha\alpha}/k_{\rm B}T}}{z_\alpha e^{-w_{\alpha\alpha}/k_{\rm B}T} + z_\beta e^{-w_{\beta\beta}/k_{\rm B}T}},$$

$$\begin{aligned}
\kappa &= \frac{k_{\rm B}T}{l_\alpha - l_\beta} \frac{1}{\delta_0(1 - \delta_0)} (\delta - \delta_0) \\
&\quad + \frac{w}{k_{\rm B}T}(2\delta_0 - 1) + \left(\frac{w}{k_{\rm B}T} \right)^2 (3\delta_0^2 - \delta_0 - 2\delta_0^3) \\
&\quad + \left(2\frac{w}{k_{\rm B}T} + \left(\frac{w}{k_{\rm B}T} \right)^2 (6\delta_0 - 6\delta_0^2 - 1) \right) (\delta - \delta_0).
\end{aligned} \qquad (1.74)$$

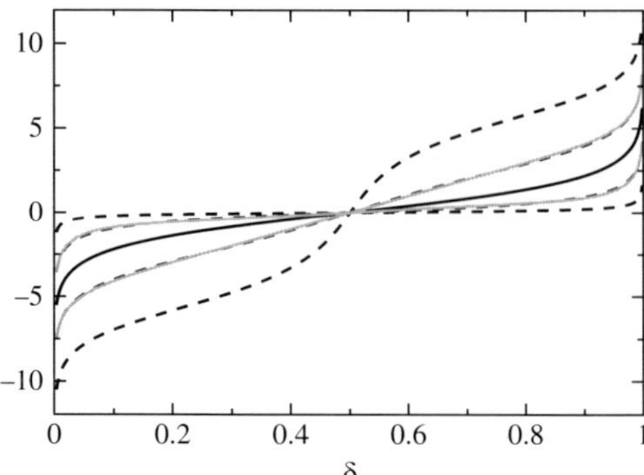

Fig. 1.13. Force–length relation for the interacting two-component model. *Dashed curves*: exact results for $w/k_BT = 0, \pm 2, \pm 5$. *Solid curves*: series expansion (1.73) which gives a good approximation for $|w/k_BT| < 2$

For negative w, a small force may lead to much larger changes in length than with no interaction. This explains, for example, how in proteins huge channels may open although the acting forces are quite small. In the case of Myoglobin, this is how the penetration of oxygen in the protein becomes possible (Fig. 1.13).

Problems

1.1. Gaussian Polymer Model

The simplest description of a polymer is the Gaussian polymer model which considers a polymer to be a series of particles joined by Hookean springs

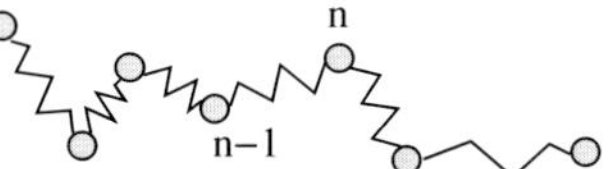

(a) The vector connecting monomers $n-1$ and n obeys a Gaussian distribution with average zero and variance
$$\left\langle (\mathbf{r}_n - \mathbf{r}_{n-1})^2 \right\rangle = b^2.$$

Determine the distribution function $P(\mathbf{r}_n - \mathbf{r}_{n-1})$ explicitly.

(b) Assume that the distance vectors $\mathbf{r}_n - \mathbf{r}_{n-1}$ are independent and calculate the distribution of end-to-end vectors $P(\mathbf{r}_N - \mathbf{r}_0)$.

(c) Consider now a polymer under the action of a constant force κ in x-direction. The potential energy of a conformation is given by
$$V = \sum_{n=1}^{N} \frac{f}{2}(\mathbf{r}_n - \mathbf{r}_{n-1})^2 - \kappa(x_N - x_0)$$

and the probability of this conformation is

$$P \sim e^{-V/k_{\mathrm{B}}T}.$$

Determine the effective spring constant. f

(d) Find the most probable configuration by searching for the minimum of the energy

$$\frac{\partial V}{\partial x_n} = \frac{\partial V}{\partial y_n} = \frac{\partial V}{\partial z_n} = 0.$$

Calculate the length of the polymer for the most probable configuration (according to the maximum term method, the average value coincides with the most probable value in the thermodynamic limit)

1.2. Three-Dimensional Polymer Model

Consider a model of a polymer in three-dimensional space consisting of N-links of length b. The connection of the links i and $i+1$ is characterized by the two angles ϕ_i and θ_i. The vector $\mathbf{r}_i$ can be obtained from the vector $\mathbf{r}_1$ through application of a series of rotation matrices[2]

$$\mathbf{r}_i = R_1 R_2 \cdots R_{i-1} \mathbf{r}_1$$

with

$$\mathbf{r}_1 = \begin{pmatrix} 0 \\ 0 \\ b \end{pmatrix}$$

$$R_i = R_z(\phi_i) R_y(\theta_i) = \begin{pmatrix} \cos\phi_i & \sin\phi_i & 0 \\ -\sin\phi_i & \cos\phi_i & 0 \\ 0 & 0 & 1 \end{pmatrix} \begin{pmatrix} \cos\theta_i & 0 & -\sin\theta_i \\ 0 & 1 & 0 \\ \sin\theta_i & 0 & \cos\theta_i \end{pmatrix}.$$

Calculate the mean square of the end-to-end distance $\left\langle \left(\sum_{i=1}^{N} \mathbf{r}_i \right)^2 \right\rangle$ for the following cases:

(a) The averaging $\langle \cdots \rangle$ includes averaging over all angles ϕ_i and θ_i.

(b) The averaging $\langle \cdots \rangle$ includes averaging over all angles ϕ_i while the angles θ_i are held fixed at a common value θ.

(c) The averaging $\langle \cdots \rangle$ includes averaging over all angles ϕ_i while the angles θ_i are held fixed at either $\theta_{2i+1} = \theta_a$ or $\theta_{2i} = \theta_b$ depending on whether the number of the link is odd or even.

---

[2] The rotation matrices act in the laboratory fixed system (xyz). Transformation into the coordinate system of the segment $(x'y'z')$ changes the order of the matrices. For instance, $\mathbf{r}_2 = R(y,\theta_1) R(z,\phi_1) \mathbf{r}_1 = R(z,\phi_1) R(y',\theta_1) R^{-1}(z,\phi_1) R(z,\phi_1) \mathbf{r}_1 = R(z,\phi_1) R(y',\theta_1) \mathbf{r}_1$ (this is sometimes discussed in terms of active and passive rotations).

(d) How large must N be, so that it is a good approximation to keep only terms which are proportional to N?

(e) What happens in the second case if θ is chosen as very small (wormlike chain)?

Hint. Show first that after averaging over the ϕ_i, the only terms of the matrix which have to be taken into account are the elements $(R_i)_{33}$. The appearing summations can be expressed as geometric series.

1.3. Two-Component Model

We consider the two-component model of a polymer chain which consists of M-segments of two different types α, β (internal degrees of freedom are neglected). The number of configurations with length L is given by the Binomial distribution:

$$\Omega(L, M, T) = \frac{M!}{M_\alpha!(M - M_\alpha)!}, \quad L = M_\alpha l_\alpha + (M - M_\alpha)l_\beta.$$

(a) Make use of the asymptotic expansion[3] of the logarithm of the Γ-function

$$\ln(\Gamma(z + 1)) = (\ln z - 1)z + \ln(\sqrt{2\pi}) + \frac{1}{2}\ln z + \frac{1}{12z} + O(z^{-3}),$$

$$N! = \Gamma(N + 1)$$

to calculate the leading terms of the force–extension relation which is obtained from

$$\kappa = \frac{\partial}{\partial L}(-k_{\mathrm{B}}T \ln \Omega(L, M, T)).$$

Discuss the error of Stirling's approximation for $M = 1{,}000$ and $l_\beta/l_\alpha = 2$.

(b) Now switch to an ensemble with constant force κ. The corresponding partition function is

$$Z(\kappa, M, T) = \sum_L e^{\kappa L/k_{\mathrm{B}}T}\Omega(L, M, T).$$

Calculate the first two moments of the length

$$\overline{L} = -\frac{\partial}{\partial \kappa}(-k_{\mathrm{B}}T \ln Z) = Z^{-1}k_{\mathrm{B}}T\frac{\partial}{\partial \kappa}Z,$$

$$\overline{L^2} = Z^{-1}(k_{\mathrm{B}}T)^2\frac{\partial^2}{\partial \kappa^2}Z,$$

and discuss the relative uncertainty $\sigma = (\sqrt{\overline{L^2}} - \overline{L}^2)/\overline{L}$. Determine the maximum of σ.

---

[3] Several asymptotic expansions can be found in [5].

2

Flory–Huggins Theory for Biopolymer Solutions

In the early 1940s, Flory [6] and Huggins [7], both working independently, developed a theory based upon a simple lattice model that could be used to understand the nonideal nature of polymer solutions. We consider a lattice model where the lattice sites are chosen to be the size of a solvent molecule and where all lattice sites are occupied by one molecule [1].

2.1 Monomeric Solution

As the simplest example, consider the mixing of a low-molecular-weight solvent (component α) with a low-molecular-weight solute (component β). The solute molecule is assumed to be the same size as a solvent molecule and therefore every lattice site is occupied by one solvent molecule or by one solute molecule at a given time (Fig. 2.1).

The increase in entropy ΔS_m due to mixing of solvent and solute is given by

$$\Delta S_\mathrm{m} = k_\mathrm{B} \ln \Omega = k_\mathrm{B} \ln \left(\frac{N!}{N_\alpha! N_\beta!} \right), \tag{2.1}$$

where $N = N_\alpha + N_\beta$ is the total number of lattice sites. Using Stirling's approximation leads to

$$\begin{aligned}
\Delta S_\mathrm{m} &= k_\mathrm{B}(N \ln N - N - N_\alpha \ln N_\alpha + N_\alpha - N_\beta \ln N_\beta + N_\beta) \\
&= k_\mathrm{B}(N \ln N - N_\alpha \ln N_\alpha - N_\beta \ln N_\beta) \\
&= -k_\mathrm{B} N_\alpha \ln \frac{N_\alpha}{N} - k_\mathrm{B} N_\beta \ln \frac{N_\beta}{N}.
\end{aligned} \tag{2.2}$$

Inserting the volume fractions

$$\phi_\alpha = \frac{N_\alpha}{N_\alpha + N_\beta}, \quad \phi_\beta = \frac{N_\beta}{N_\alpha + N_\beta}, \tag{2.3}$$

the mixing entropy can be written in the well-known form

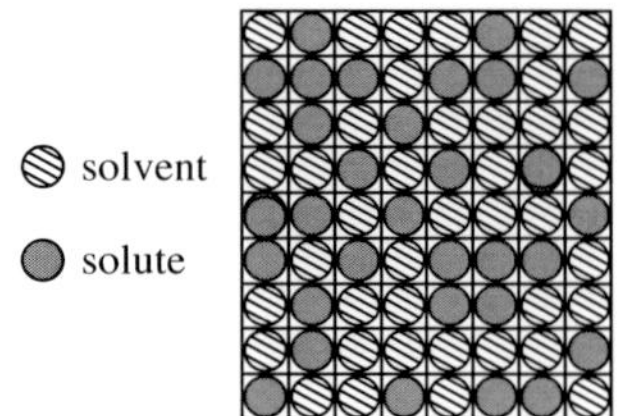

Fig. 2.1. Two-dimensional Flory–Huggins lattice

$$\Delta S_{\mathrm{m}} = -N k_{\mathrm{B}}(\phi_\alpha \ln \phi_\alpha + \phi_\beta \ln \phi_\beta). \tag{2.4}$$

Neglecting boundary effects (or using periodic b.c.), the number of nearest neighbor pairs is (c is the coordination number)

$$N_{nn} = N\frac{c}{2}. \tag{2.5}$$

These are divided into

$$N_{\alpha\alpha} = \frac{N_\alpha c}{2}\phi_\alpha = \frac{N\phi_\alpha^2 c}{2}, \quad N_{\beta\beta} = \frac{N_\beta c}{2}\phi_\beta = \frac{N\phi_\beta^2 c}{2},$$
$$N_{\alpha\beta} = N\phi_\alpha\phi_\beta c. \tag{2.6}$$

The average interaction energy is

$$\overline{w} = \frac{1}{2}Nc\phi_\alpha^2 w_{\alpha\alpha} + \frac{1}{2}Nc\phi_\beta^2 w_{\beta\beta} + Nc\phi_\alpha\phi_\beta w_{\alpha\beta} \tag{2.7}$$

which after the substitution

$$w_{\alpha\beta} = \frac{1}{2}(w_{\alpha\alpha} + w_{\beta\beta} - w) \tag{2.8}$$

becomes

$$\overline{w} = -\frac{1}{2}Nc\phi_\alpha\phi_\beta w$$
$$= +\frac{1}{2}Nc\phi_\alpha(\phi_\alpha w_{\alpha\alpha} + \phi_\beta w_{\alpha\alpha}) + \frac{1}{2}Nc\phi_\beta(\phi_\beta w_{\beta\beta} + \phi_\alpha w_{\beta\beta}) \tag{2.9}$$

and since $\phi_\alpha + \phi_\beta = 1$

$$\overline{w} = -\frac{1}{2}Nc\phi_\alpha\phi_\beta w + \frac{1}{2}Nc(\phi_\alpha w_{\alpha\alpha} + \phi_\beta w_{\beta\beta}). \tag{2.10}$$

Now the partition function is

$$Z = z_\alpha^{N_\alpha} z_\beta^{N_\beta} \mathrm{e}^{-N_\alpha c w_{\alpha\alpha}/2k_{\mathrm{B}}T} \mathrm{e}^{-N_\beta c w_{\beta\beta}/2k_{\mathrm{B}}T} \mathrm{e}^{N_\alpha N_\beta c w/2Nk_{\mathrm{B}}T} \frac{N!}{N_\alpha! N_\beta!}$$
$$= \left(z_\alpha \mathrm{e}^{-c w_{\alpha\alpha}/2k_{\mathrm{B}}T}\right)^{N_\alpha} \left(z_\beta \mathrm{e}^{-c w_{\beta\beta}/2k_{\mathrm{B}}T}\right)^{N_\beta} \mathrm{e}^{N_\alpha N_\beta c w/2Nk_{\mathrm{B}}T} \frac{N!}{N_\alpha! N_\beta!}.$$
$$\tag{2.11}$$

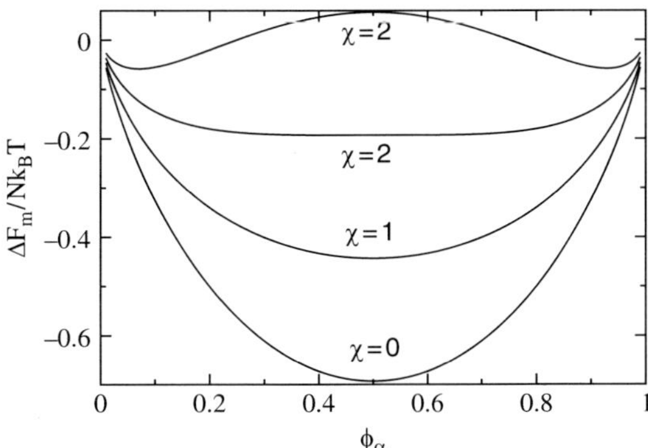

Fig. 2.2. Free energy change $\Delta F/Nk_{\mathrm{B}}T$ of a binary mixture with interaction

The free energy is

$$F = -k_{\mathrm{B}}T \ln Z = -N_\alpha k_{\mathrm{B}}T \ln z_\alpha - N_\beta k_{\mathrm{B}}T \ln z_\beta$$
$$= +N_\alpha \frac{c}{2} w_{\alpha\alpha} + N_\beta \frac{c}{2} w_{\beta\beta} - \frac{N_\alpha N_\beta c w}{2N} + Nk_{\mathrm{B}}T(\phi_\alpha \ln \phi_\alpha + \phi_\beta \ln \phi_\beta). \tag{2.12}$$

For the pure solvent, the free energy is

$$F(N_\alpha = N, N_\beta = 0) = -N_\alpha k_{\mathrm{B}}T \ln z_\alpha + N_\alpha \frac{c}{2} w_{\alpha\alpha} \tag{2.13}$$

and for the pure solute

$$F(N_\alpha = 0, N_\beta = N) = -N_\beta k_{\mathrm{B}}T \ln z_\beta + N_\beta \frac{c}{2} w_{\beta\beta}. \tag{2.14}$$

Hence the change in free energy is

$$\Delta F_{\mathrm{m}} = -\frac{N_\alpha N_\beta c w}{2N} + Nk_{\mathrm{B}}T(\phi_\alpha \ln \phi_\alpha + \phi_\beta \ln \phi_\beta) \tag{2.15}$$

with the energy change (van Laar heat of mixing)

$$\Delta E_{\mathrm{m}} = -\frac{N_\alpha N_\beta c w}{2N} = -N \frac{c w}{2} \phi_\alpha \phi_\beta = Nk_{\mathrm{B}}T\chi\phi_\alpha\phi_\beta. \tag{2.16}$$

The last equation defines the Flory interaction parameter (Fig. 2.2):

$$\chi = -\frac{c w}{2k_{\mathrm{B}}T}. \tag{2.17}$$

For $\chi > 2$, the free energy has two minima and two stable phases exist. This is seen from solving

$$0 = \frac{\partial \Delta F}{\partial \phi_\alpha} = N k_B T \frac{\partial}{\partial \phi_\alpha} \left(\chi \phi_\alpha (1 - \phi_\alpha) + \phi_\alpha \ln \phi_\alpha + (1 - \phi_\alpha) \ln(1 - \phi_\alpha) \right)$$

$$= N k_B T \left(\chi(1 - 2\phi_\alpha) + \ln \frac{\phi_\alpha}{1 - \phi_\alpha} \right). \tag{2.18}$$

This equation has as one solution $\phi_\alpha = 1/2$. This solution becomes instable for $\chi > 2$ as can be seen from the sign change of the second derivative

$$\frac{\partial^2 \Delta F}{\partial \phi_\alpha^2} = N k_B T \left(\frac{1 - 2\chi \phi_\alpha + 2\chi \phi_\alpha^2}{\phi_\alpha^2 - \phi_\alpha} \right) = N k_B T (4 - 2\chi). \tag{2.19}$$

2.2 Polymeric Solution

Now, consider N_β polymer molecules which consist of M units and hence occupy a total of $M N_\beta$ lattice sites. The volume fractions are (Fig. 2.3)

$$\phi_\alpha = \frac{N_\alpha}{N_\alpha + M N_\beta}, \qquad \phi_\beta = \frac{M N_\beta}{N_\alpha + M N_\beta} \tag{2.20}$$

and the number of lattice sites is

$$N = N_\alpha + M N_\beta. \tag{2.21}$$

The entropy change is given by

$$\Delta S = \Delta S_m + \Delta S_d = k_B \ln \Omega(N_\alpha, N_\beta). \tag{2.22}$$

It consists of the mixing entropy and a contribution due to the different conformations of the polymers (disorder entropy). The latter can be eliminated by subtracting the entropy for $N_\alpha = 0$:

$$\Delta S_m = \Delta S - \Delta S_d = k_B \frac{\ln \Omega(N_\alpha, N_\beta)}{\ln \Omega(0, N_\beta)}. \tag{2.23}$$

In the following, we will calculate $\Omega(N_\alpha, N_\beta)$ in an approximate way. We use a mean-field method where one polymer after the other is distributed over the

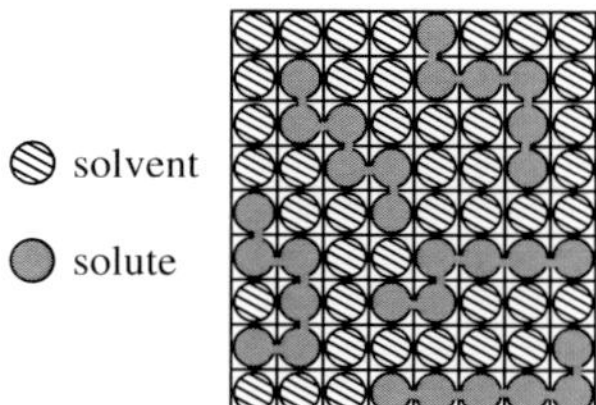

Fig. 2.3. Lattice model for a polymer

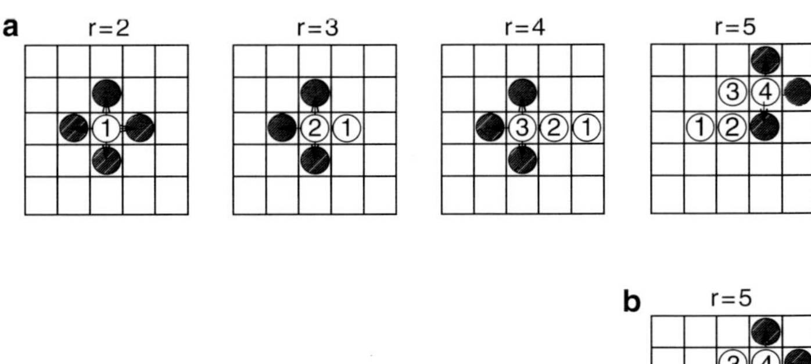

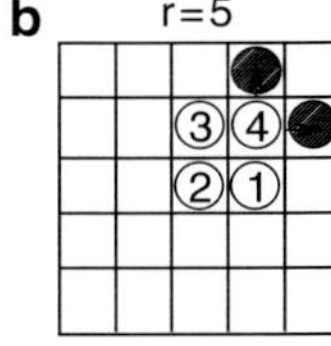

Fig. 2.4. Available positions. The second segment can be placed onto four positions, the third segment onto three positions. For the following segments, it is assumed that they can be also placed onto three positions (**a**). Configurations like (**b**) are neglected

lattice, taking into account only the available volume but not the configuration of all the other polymers. Under that conditions, Ω factorizes

$$\Omega = \frac{1}{N_\beta!} \prod_{i=1}^{N_\beta} \nu_i, \tag{2.24}$$

where ν_i counts the number of possibilities to put the ith polymer onto the lattice. It will be calculated by adding one segment after the other and counting the number of possible ways:

$$\nu_{i+1} = \prod_{s=1}^{M} n_s^{i+1}. \tag{2.25}$$

The first segment of the $(i+1)$th polymer molecule can be placed onto

$$n_1^{i+1} = N - iM \tag{2.26}$$

lattice sites (Fig. 2.4).

The second segment has to be placed on a neighboring position. Depending on the coordination number of the lattice, there are c possible neighboring sites. But only a fraction of

$$f = 1 - \frac{iM}{N} \tag{2.27}$$

of these is unoccupied. Hence for the second segment, we have

$$n_2^{i+1} = c\left(1 - \frac{iM}{N}\right). \tag{2.28}$$

For the third segment only $c - 1$ neighboring positions are available

$$n_3^{i+1} = (c - 1) \left(1 - \frac{iM}{N} \right). \tag{2.29}$$

For the following segments $r = 4 \ldots M$, we assume that the number of possible sites is the same as for the third segment. This introduces some error since for some configurations the number is reduced due to the excluded volume. This error, however, is small compared to the crudeness of the whole model. Multiplying all the factors, we have

$$\nu_{i+1} = (N - iM)c(c - 1)^{M-2} \left(1 - \frac{iM}{N} \right)^{M-1}$$

$$\approx (N - iM)^M \left(\frac{c - 1}{N} \right)^{M-1} \tag{2.30}$$

and the entropy is

$$\Delta S = k_{\mathrm{B}} \ln \left[\frac{1}{N_\beta!} \prod_{i=1}^{N_\beta} (N - iM)^M \left(\frac{c - 1}{N} \right)^{M-1} \right]$$

$$= -k N_{B\beta} \ln N_\beta + k_{\mathrm{B}} N_\beta + k_{\mathrm{B}} N_\beta (M - 1) \ln \left(\frac{c - 1}{N} \right)$$

$$+ k_{\mathrm{B}} M \sum_{i=1}^{N_\beta} \ln(N - iM). \tag{2.31}$$

The sum will be approximated by an integral

$$\sum_{i=1}^{N_\beta} \ln(N - iM) \approx \int_0^{N_\beta} \ln(N - Mx)\mathrm{d}x = \left(x - \frac{N}{M} \right) (\ln(N - Mx) - 1) \Big|_0^{N_\beta}$$

$$= -\frac{N_\alpha}{M}(\ln N_\alpha - 1) + \frac{N}{M}(\ln N - 1). \tag{2.32}$$

Finally, we get

$$\Delta S = -k_{\mathrm{B}} N_\beta \ln N_\beta + k_{\mathrm{B}} N_\beta + k_{\mathrm{B}} N_\beta (M - 1) \ln \left(\frac{c - 1}{N} \right)$$

$$+ k_{\mathrm{B}} (N \ln N - N + N_\alpha - N_\alpha \ln N_\alpha). \tag{2.33}$$

The disorder entropy is obtained by substituting $N_\alpha = 0$ and $N = MN_\beta$:

$$\Delta S_{\mathrm{d}} = \Delta S(N_\alpha = 0)$$

$$= -k_{\mathrm{B}} N_\beta \ln N_\beta + k_{\mathrm{B}} N_\beta + k_{\mathrm{B}} N_\beta (M - 1) \ln \left(\frac{c - 1}{MN_\beta} \right)$$

$$+ k_{\mathrm{B}} (MN_\beta \ln MN_\beta - MN_\beta). \tag{2.34}$$

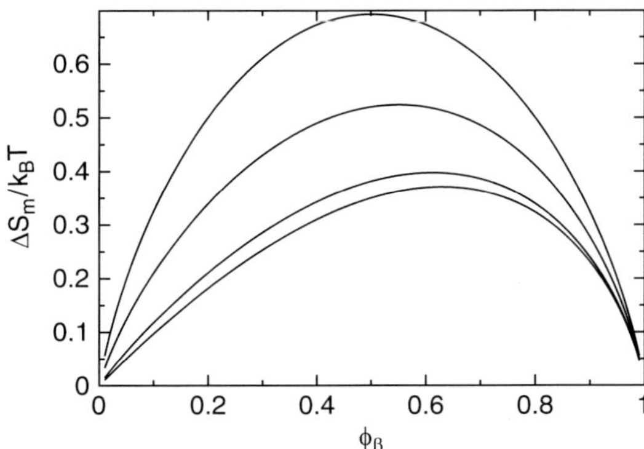

Fig. 2.5. Mixing entropy. $\Delta S_\mathrm{m}/Nk_\mathrm{B}$ is shown as a function of ϕ_β for $M = 1,\ 2,$ 10, 100

The mixing entropy is given by the difference (Fig. 2.5):

$$
\begin{aligned}
\Delta S_\mathrm{m} &= \Delta S - \Delta S_\mathrm{d} \\
&= k_\mathrm{B}(N \ln N - N + N_\alpha - N_\alpha \ln N_\alpha - MN_\beta \ln MN_\beta + MN_\beta) \\
&\quad + k_\mathrm{B}N_\beta(M - 1)(\ln MN_\beta - \ln N) \\
&= k_\mathrm{B}\left(N \ln N - N_\alpha \ln N_\alpha - MN_\beta \ln MN_\beta + MN_\beta \ln MN_\beta \right. \\
&\quad \left. - MN_\beta \ln N - N_\beta \ln MN_\beta + N_\beta \ln N \right) \\
&= -k_\mathrm{B}\left(N_\alpha \ln \frac{N_\alpha}{N} + N_\beta \ln \frac{MN_\beta}{N}\right) = -k_\mathrm{B}N_\alpha \ln \phi_\alpha - k_\mathrm{B}N_\beta \ln \phi_\beta \\
&= -Nk_\mathrm{B}\left(\phi_\alpha \ln \phi_\alpha + \frac{\phi_\beta}{M} \ln \phi_\beta\right).
\end{aligned}
\tag{2.35}
$$

In comparison with the expression for a solution of molecules without internal flexibility, we obtain an extra contribution to the entropy of

$$
- N_\beta k_\mathrm{B} \ln \frac{MN_\beta}{N} + N_\beta k_\mathrm{B} \ln \frac{N_\beta}{N} = -N_\beta k_\mathrm{B} \ln M.
\tag{2.36}
$$

Next, we calculate the change of energy due to mixing, ΔE_m. $w_{\alpha\alpha}$ is the interaction energy between nearest-neighbor solvent molecules, $w_{\beta\beta}$ between nearest-neighbor polymer units (not chemically bonded), and $w_{\alpha\beta}$ between one solvent molecule and one polymer unit. The probability that any site is occupied by a solvent molecule is ϕ_α and by a polymer unit is ϕ_β. We introduce an effective coordination number $\bar{c}$ which takes into account that a solvent molecule has c neighbors, whereas a polymer segment interacts only with c-2 other molecules. Then

$$
N_{\alpha\alpha} = \bar{c}\phi_\alpha \frac{N_\alpha}{2}, \quad N_{\beta\beta} = M\bar{c}\phi_\beta \frac{N_\beta}{2},
\tag{2.37}
$$

$$
N_{\alpha\beta} = \bar{c}\phi_\alpha N_\beta.
\tag{2.38}
$$

In the pure polymer $\phi_\beta = 1$ and $N_{\beta\beta} = M\bar{c}N_\beta/2$, whereas in the pure solvent $N_{\alpha\alpha} = \bar{c}N_\alpha/2$. The energy change is

$$\begin{aligned}
\Delta E_{\mathrm{m}} &= \bar{c}w_{\alpha\alpha}\phi_\alpha\frac{N_\alpha}{2} + w_{\beta\beta}M\bar{c}\phi_\beta\frac{N_\beta}{2} + w_{\alpha\beta}M\bar{c}\phi_\alpha N_\beta \\
&\quad - w_{\alpha\alpha}\bar{c}\frac{N_\alpha}{2} - w_{\beta\beta}M\bar{c}\frac{N_\beta}{2} \\
&= -\bar{c}w_{\alpha\alpha}\frac{N_\alpha}{2}\phi_\beta - \bar{c}Mw_{\beta\beta}\frac{N_\beta}{2}\phi_\alpha + w_{\alpha\beta}M\bar{c}\phi_\alpha N_\beta \\
&= \left(w_{\alpha\beta} - \frac{w_{\alpha\alpha} + w_{\beta\beta}}{2}\right)\bar{c}N\phi_\alpha\phi_\beta \\
&= -\frac{w}{2}\bar{c}N\phi_\alpha\phi_\beta = Nk_{\mathrm{B}}T\chi\phi_\alpha\phi_\beta,
\end{aligned} \tag{2.39}$$

with the Flory interaction parameter

$$\chi = -\frac{w\bar{c}}{2k_{\mathrm{B}}T}. \tag{2.40}$$

For the change in free energy, we find (Figs. 2.6–2.8)

$$\frac{\Delta F_{\mathrm{m}}}{Nk_{\mathrm{B}}T} = \frac{\Delta E_{\mathrm{m}}}{Nk_{\mathrm{B}}T} - \frac{\Delta S_{\mathrm{m}}}{Nk_{\mathrm{B}}} = \phi_\alpha \ln\phi_\alpha + \frac{\phi_\beta}{M}\ln\phi_\beta + \chi\phi_\alpha\phi_\beta. \tag{2.41}$$

2.3 Phase Transitions

2.3.1 Stability Criterion

In equilibrium, the free energy (if volume is constant) has a minimum value. Hence, a homogeneous system becomes unstable and separates into two phases

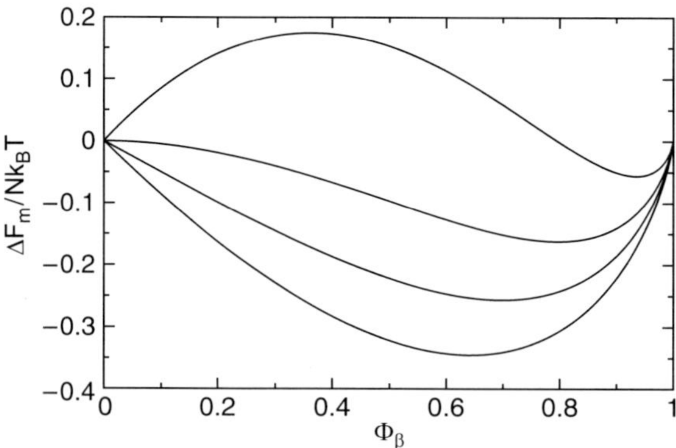

Fig. 2.6. Free energy as a function of solute concentration. $\Delta F_{\mathrm{m}}/Nk_{\mathrm{B}}T$ is shown for $M = 1,000$ and $\chi = 0.1, 0.5, 1.0, 2.0$

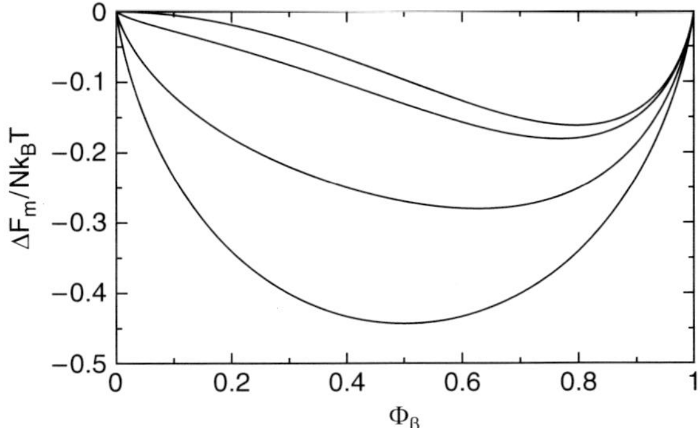

Fig. 2.7. Free energy as a function of solute concentration. $\Delta F_{\mathrm{m}}/Nk_{\mathrm{B}}T$ is shown for $\chi = 1.0$ and $M = 1,\ 2,\ 10,\ 1,000$

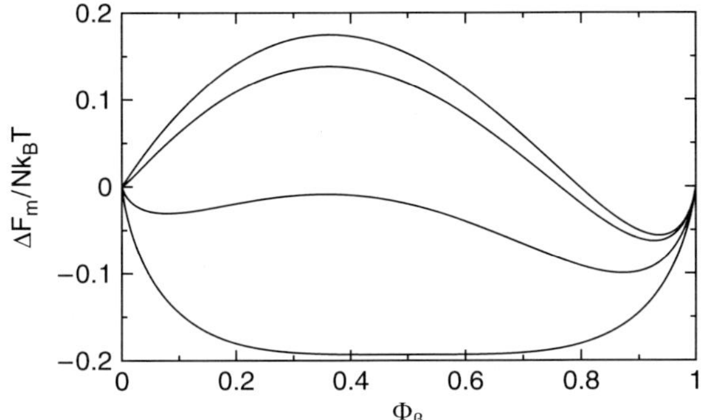

Fig. 2.8. Free energy as a function of solute concentration. $\Delta F_{\mathrm{m}}/Nk_{\mathrm{B}}T$ is shown for $\chi = 2.0$ and $M = 1,\ 2,\ 10,\ 1,000$

if the free energy of the two-phase system is lower, that is, the following equation can be fulfilled:

$$\Delta F_{\mathrm{m}}(\phi_\beta, N) > \Delta F_{\mathrm{m}}(\phi'_\beta, N') + \Delta F_{\mathrm{m}}(\phi''_\beta, N - N'). \tag{2.42}$$

But since $\Delta F_{\mathrm{m}} = N \Delta f_{\mathrm{m}}(\phi_\beta)$ is proportional to N, this condition becomes

$$N \Delta f_{\mathrm{m}}(\phi_\beta) > N' \Delta f_{\mathrm{m}}(\phi'_\beta) + (N - N')\Delta f_{\mathrm{m}}(\phi''_\beta). \tag{2.43}$$

Since the total numbers N_α and N_β are conserved, we have

$$N\phi_\beta = N'\phi'_\beta + (N - N')\phi''_\beta \tag{2.44}$$

or

$$N' = N\frac{\phi_\beta - \phi''_\beta}{\phi'_\beta - \phi''_\beta}, \quad (N - N') = N\frac{\phi'_\beta - \phi_\beta}{\phi'_\beta - \phi''_\beta}. \tag{2.45}$$

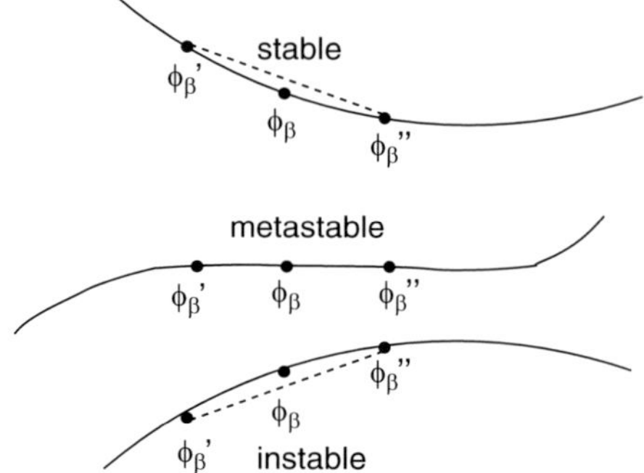

Fig. 2.9. Stability criterion for the free energy

But since N as well as $(N - N')$ should be positive numbers, there are two possible cases:

$$\phi_\beta' - \phi_\beta'' > 0, \quad \phi_\beta - \phi_\beta'' > 0, \quad \phi_\beta' - \phi_\beta > 0, \tag{2.46}$$

$$\phi_\beta' - \phi_\beta'' < 0, \quad \phi_\beta - \phi_\beta'' < 0, \quad \phi_\beta' - \phi_\beta < 0, \tag{2.47}$$

which means that either $\phi_\beta' < \phi_b < \phi_\beta''$ or $\phi_\beta' > \phi_b > \phi_\beta''$. By renaming, we always can choose the order

$$\phi_\beta' < \phi_\beta < \phi_\beta''. \tag{2.48}$$

The stability criterion becomes (Fig. 2.9)

$$\Delta f_{\mathrm{m}}(\phi_\beta) > \frac{\phi_\beta'' - \phi_\beta}{\phi_\beta'' - \phi_\beta'}\Delta f_{\mathrm{m}}(\phi_\beta') + \frac{\phi_\beta - \phi_\beta'}{\phi_\beta'' - \phi_\beta'}\Delta f_{\mathrm{m}}(\phi_\beta''), \tag{2.49}$$

which can be written with the abbreviation

$$h' = \phi_\beta - \phi_\beta', \quad h'' = \phi_\beta'' - \phi_\beta \tag{2.50}$$

as

$$\frac{\Delta f(\phi_\beta - h') - \Delta f(\phi_\beta)}{h'} + \frac{\Delta f(\phi_\beta + h'') - \Delta f(\phi_\beta)}{h''} < 0 \tag{2.51}$$

but that means the curvature has to be negative or locally

$$\frac{\partial^2 \Delta f(\phi_\beta)}{\partial \phi_\beta^2} < 0. \tag{2.52}$$

2.3.2 Critical Coupling

Instabilities appear above a certain value of $\chi = \chi_{\mathrm{c}}(M)$. To find this critical value, we have to look for the metastable case with

$$0 = \frac{\partial^2}{\partial^2 \phi_{\rm b}} \Delta f(\phi_{\rm b}) = \frac{1}{1 - \phi_{\rm b}} + \frac{1}{M \phi_{\rm b}} - 2\chi. \qquad (2.53)$$

In principle, the critical χ value can be found from solving this quadratic equation for $\phi_{\rm b}$ and looking for real roots in the range $0 \leq \phi_\beta \leq 1$. Here, however, a simpler strategy can be applied. At the boundaries of the interval $[0, 1]$ of possible $\phi_{\rm b}$-values, the second derivative is positive

$$\frac{\partial^2}{\partial^2 \phi_{\rm b}} \Delta f(\phi_{\rm b}) \rightarrow \frac{1}{M \phi_{\rm b}} > 0 \quad \text{for} \quad \phi_{\rm b} \rightarrow 0, \qquad (2.54)$$

$$\frac{\partial^2}{\partial^2 \phi_{\rm b}} \Delta f(\phi_{\rm b}) \rightarrow \frac{1}{1 - \phi_{\rm b}} > 0 \quad \text{for} \quad \phi_{\rm b} \rightarrow 1. \qquad (2.55)$$

Hence, we look for a minimum of the second derivative, that is, we solve

$$0 = \frac{\partial^3}{\partial^3 \phi_{\rm b}} \Delta f(\phi_{\rm b}) = \frac{1}{(1 - \phi_{\rm b})^2} - \frac{1}{M \phi_{\rm b}^2}. \qquad (2.56)$$

This gives immediately

$$\phi_{\rm bc} = \frac{1}{1 + \sqrt{M}}. \qquad (2.57)$$

Above the critical point, the minimum of the second derivative is negative. Hence we are looking for a solution of

$$\frac{\partial^2}{\partial^2 \phi_{\rm b}} \Delta f(\phi_{\rm b}) = \frac{\partial^3}{\partial^3 \phi_{\rm b}} \Delta f(\phi_{\rm b}) = 0. \qquad (2.58)$$

Inserting $\phi_{\rm bc}$ into the second derivative gives

$$0 = 1 + \frac{1}{M} + \frac{2}{\sqrt{M}} - 2\chi, \qquad (2.59)$$

which determines the critical value of χ

$$\chi_{\rm c} = \frac{(1 + \sqrt{M})^2}{2M} = \frac{1}{2} + \frac{1}{\sqrt{M}} + \frac{1}{2M}. \qquad (2.60)$$

From $\mathrm{d}F = -S\mathrm{d}T + \mu_\alpha \mathrm{d}N_\alpha + \mu_\beta \mathrm{d}N_\beta$, we obtain the change of the chemical potential as

$$\Delta \mu_\alpha = \mu_\alpha - \mu_\alpha^0 = \left. \frac{\partial \Delta F}{\partial N_\alpha} \right|_{N_\beta, T} = \left(\frac{\partial N}{\partial N_\alpha} \frac{\partial}{\partial N} + \frac{\partial \phi_\beta}{\partial N_\alpha} \frac{\partial}{\partial \phi_\beta} \right) N \Delta f(\phi_\beta)$$

$$= \Delta f(\phi_\beta) - \phi_\beta \Delta f'(\phi_\beta)$$

$$= k_{\rm B} T \left(\ln(1 - \phi_\beta) + \left(1 - \frac{1}{M} \right) \phi_\beta + \chi \phi_\beta^2 \right). \qquad (2.61)$$

Now the derivatives of $\Delta\mu_\alpha$ are

$$\frac{\partial}{\partial\phi_\beta}\Delta\mu_\alpha = -\phi_\beta\Delta f''(\phi_\beta), \tag{2.62}$$

$$\frac{\partial^2}{\partial\phi_\beta^2}\Delta\mu_\alpha = -\Delta f''(\phi_\beta) - \phi_\beta\Delta f'''(\phi_\beta). \tag{2.63}$$

Hence, the critical point can also be found by solving

$$\frac{\partial^2}{\partial\phi_\beta^2}\Delta\mu_\alpha = \frac{\partial}{\partial\phi_\beta}\Delta\mu_\alpha = 0. \tag{2.64}$$

Employing the ideal gas approximation

$$\mu = k_{\mathrm{B}}T\,\ln(p) + \text{const}, \tag{2.65}$$

this gives for the vapor pressure (Figs. 2.10 and 2.11)

$$\frac{p_\alpha}{p_\alpha^0} = \mathrm{e}^{(\mu_\alpha-\mu_\alpha^0)/k_{\mathrm{B}}T} = (1-\phi_\beta)\mathrm{e}^{\chi\phi_\beta^2+(1-1/M)\phi_\beta} \tag{2.66}$$

and since the exponential is a monotonous function, another condition for the critical point is

$$\frac{\partial}{\partial\phi_\beta}\frac{p_\alpha}{p_\alpha^0} = \frac{\partial^2}{\partial^2\phi_\beta}\frac{p_\alpha}{p_\alpha^0} = 0. \tag{2.67}$$

2.3.3 Phase Diagram

In the simple Flory–Huggins theory, the interaction parameter is proportional to $1/T$. Hence we can write it as

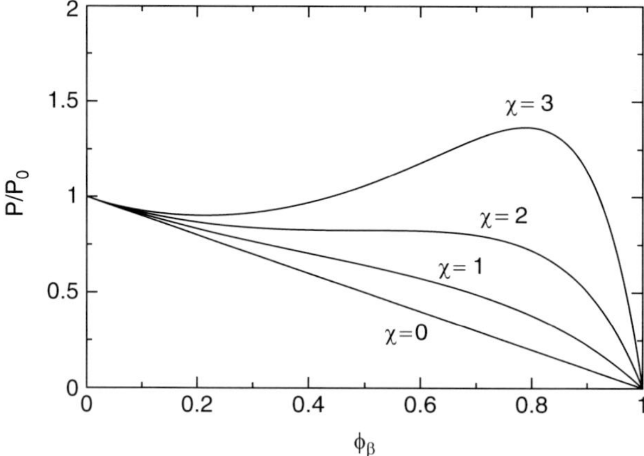

Fig. 2.10. Vapor pressure of a binary mixture with interaction ($M = 1$)

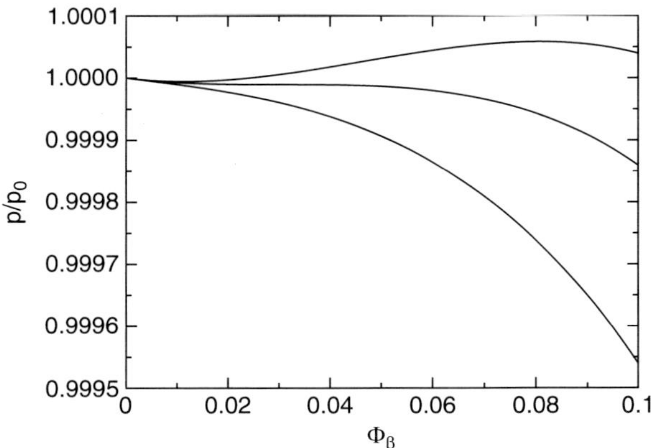

Fig. 2.11. Vapor pressure of a polymer solution with interaction for $M = 1,000, \chi = 0.5, 0.532, 0.55$

$$\chi = \frac{T_0 \chi_0}{T} \tag{2.68}$$

and discuss the free energy change as a function of ϕ_β and T:

$$\Delta F = N k_{\mathrm{B}} T \left((1 - \phi_\beta) \ln(1 - \phi_\beta) + \frac{\phi_\beta}{M} \ln \phi_\beta \right) + N k_{\mathrm{B}} T_0 \chi_0 \phi_\beta (1 - \phi_\beta). \tag{2.69}$$

The turning points follow from

$$0 = \frac{\partial^2}{\partial \phi_\beta^2} \Delta F = N k_{\mathrm{B}} \left(\frac{T}{1 - \phi_\beta} + \frac{T}{M \phi_\beta} - 2 T_0 \chi_0 \right) \tag{2.70}$$

as

$$\phi_{\beta,\mathrm{TP}} = \frac{1}{2} + \frac{T(1 - M) \pm \sqrt{T^2(M - 1)^2 + 4 T_0 \chi_0 M (T_0 \chi_0 M - T - 4MT)}}{4 T_0 \chi_0 M}. \tag{2.71}$$

This defines the spinodal curve which separates the instable from the metastable region (Fig. 2.12).

Which is the minimum free energy of a two-phase system? The free energy takes the form

$$\Delta F = \Delta F^1 + \Delta F^2 = N^1 k_{\mathrm{B}} T \Delta f(\phi_\beta^1) + N^2 k_{\mathrm{B}} T \Delta f(\phi_\beta^2) \tag{2.72}$$

with

$$N^j = N_\alpha^j + M N_\beta^j, \quad \phi_\beta^j = \frac{M N_\beta^j}{N^j}. \tag{2.73}$$

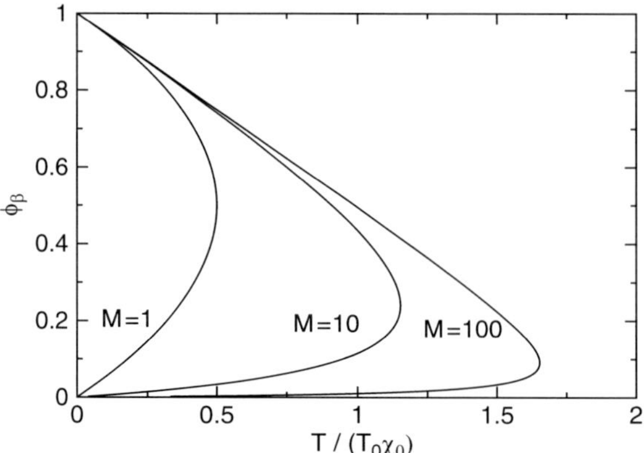

Fig. 2.12. Spinodal

The minimum free energy can be found from the condition that exchange of
solvent or solute molecules between the two phases does not change the free
energy, that is, the chemical potentials in the two phases are the same:

$$0 = \mathrm{d}\Delta F = \left(\frac{\partial \Delta F^1}{\partial N_\alpha^1} - \frac{\partial \Delta F^2}{\partial N_\alpha^2} \right) \mathrm{d}N_\alpha + \left(\frac{\partial \Delta F^1}{\partial N_\beta^1} - \frac{\partial \Delta F^2}{\partial N_\beta^2} \right) \mathrm{d}N_\beta, \quad (2.74)$$

$$0 = \mu_\alpha^1 - \mu_\alpha^2 = \mu_\beta^1 - \mu_\beta^2 \quad (2.75)$$

or

$$0 = \left(\Delta f^1 - \phi_\beta^1 \frac{\partial}{\partial \phi_\beta^1} \Delta f^1 \right) - \left(\Delta f^2 - \phi_\beta^2 \frac{\partial}{\partial \phi_\beta^2} \Delta f^2 \right), \quad (2.76)$$

$$0 = \left(\Delta f^1 + (1 - \phi_\beta^1) \frac{\partial}{\partial \phi_\beta^1} \right) \Delta f^1 - \left(\Delta f^2 + (1 - \phi_\beta^2) \frac{\partial}{\partial \phi_\beta^2} \Delta f^2 \right). \quad (2.77)$$

From the difference of these two equations, we find

$$\frac{\partial \Delta f^1}{\partial \phi_\beta^1} = \frac{\partial \Delta f^2}{\partial \phi_\beta^2} \quad (2.78)$$

and hence the slope of the free energy has to be the same for both phases.
    Inserting into the first equation then gives

$$\Delta f^1 - \Delta f^2 = (\phi_\beta^1 - \phi_\beta^2) \Delta f', \quad (2.79)$$

which shows that the concentrations of the two phases can be found by the
well-known "common tangent" construction (Fig. 2.13).
    These so-called binodal points give the border to the stable one-phase
region. Between spinodal and binodal the system is metastable. It is stable in
relation to small fluctuations since the curvature of the free energy is positive.
It is, however, instable against larger-scale fluctuations (Fig. 2.14).

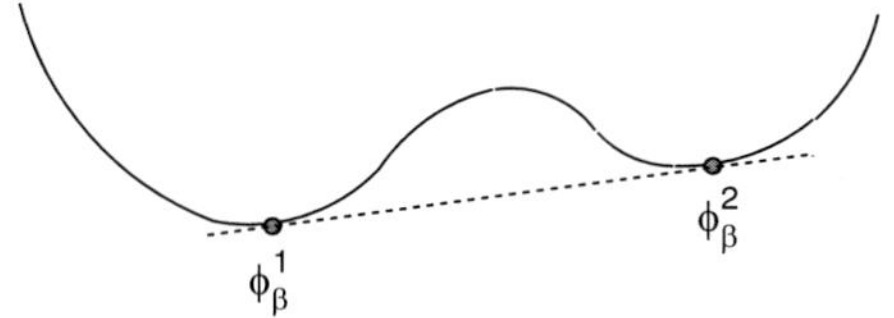

Fig. 2.13. Common tangent construction

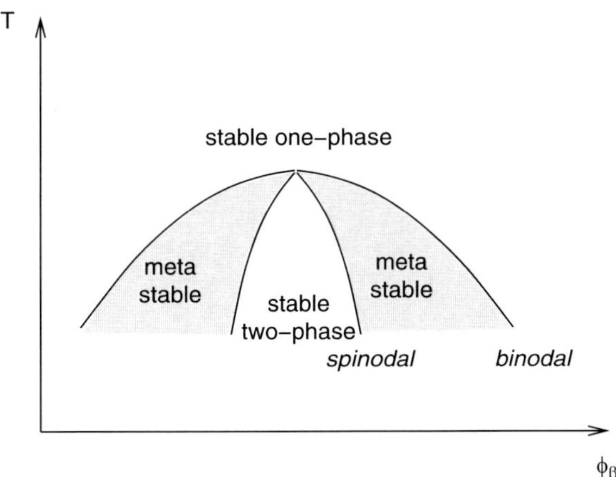

Fig. 2.14. Phase diagram

Problems

2.1. Osmotic Pressure of a Polymer Solution

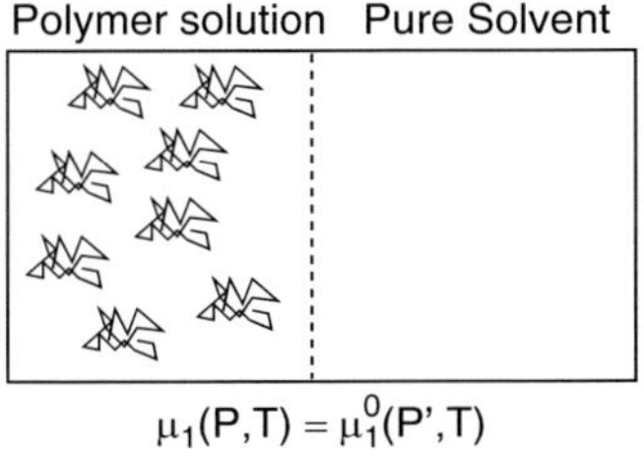

Calculate the osmotic pressure $\Pi = P - P'$ for the Flory–Huggins model of a polymer solution. The difference of the chemical potential of the solvent

$$\mu_\alpha(P,T) - \mu_\alpha^0(P,T) = \left.\frac{\partial \Delta F_{\mathrm{m}}}{\partial N_\alpha}\right|_{N_\beta,T}$$

can be obtained from the free energy change

$$\Delta F_{\mathrm{m}} = N k_{\mathrm{B}} T \left(\phi_\alpha \ln \phi_\alpha + \frac{\phi_\beta}{M} \ln \phi_\beta + \chi \phi_\alpha \phi_\beta \right)$$

and the osmotic pressure is given by

$$\mu_\alpha^0(P',T) - \mu_\alpha^0(P,T) = -\Pi \frac{\partial \mu_\alpha^0}{\partial P}.$$

Taylor series expansion gives the virial expansion of the osmotic pressure as a series of powers of ϕ_β. Truncate after the quadratic term and discuss qualitatively the dependence on the interaction parameter χ (and hence also on temperature).

2.2. Polymer Mixture

Consider a mixture of two types of polymers with chain lengths M_1 and M_2 and calculate the mixing free energy change ΔF_m. Determine the critical values ϕ_c and χ_c. Discuss the phase diagram for the symmetrical case $M_1 = M_2$.

Protein Electrostatics and Solvation

3

Implicit Continuum Solvent Models

Since an explicit treatment of all solvent atoms and ions is not possible in many cases, the effect of the solvent on the protein has to be approximated by implicit models which replace the large number of dynamic solvent modes by a continuous medium [8–10].

3.1 Potential of Mean Force

In solution, a protein occupies a conformation γ with the probability given by the Boltzmann factor

$$P(X,Y) = \frac{\mathrm{e}^{-U(X,Y)}}{\int \mathrm{d}X\mathrm{d}Y\mathrm{e}^{-U(X,Y)}}, \tag{3.1}$$

where X stands for the coordinates of the protein (including the protonation state) and Y stands for the coordinates of the solvent. The potential energy can be formally split into three terms:

$$U(X,Y) = U_{\mathrm{prot}}(X) + U_{\mathrm{solv}}(Y) + U_{\mathrm{prot,solv}}(X,Y). \tag{3.2}$$

The mean value of a physical quantity which depends only on the protein coordinates $Q(X)$ is

$$\overline{Q} = \int \mathrm{d}X\mathrm{d}Y, \quad Q(X)P(X,Y) = \int \mathrm{d}X Q(X)\widetilde{P}(X), \tag{3.3}$$

where we define a reduced probability distribution for the protein

$$\widetilde{P}(X) = \int \mathrm{d}Y P(X,Y), \tag{3.4}$$

which is represented by introducing the potential of mean force $W(x)$ as

$$\widetilde{P}(X) = \frac{e^{-W(X)/k_{\mathrm{B}}T}}{\int \mathrm{d}X e^{-W(X)/k_{\mathrm{B}}T}} \tag{3.5}$$

with

$$e^{-W(X)/k_{\mathrm{B}}T} = e^{-U_{\mathrm{prot}}(X)/k_{\mathrm{B}}T} \int \mathrm{d}Y e^{-(U_{\mathrm{solv}}(Y)+U_{\mathrm{prot,solv}}(X,Y))/k_{\mathrm{B}}T}$$

$$= e^{-(U_{\mathrm{prot}}(X)+\Delta W(X))/k_{\mathrm{B}}T}. \tag{3.6}$$

ΔW accounts implicitly but exactly for the solvents effect on the protein.

3.2 Dielectric Continuum Model

In the following, we discuss implicit solvent models, which treat the solvent as a dielectric continuum. In response to the partial charges of the protein q_i, polarization of the medium produces an electrostatic reaction potential ϕ^{R} (Fig. 3.1).

If the medium behaves linearly (no dielectric saturation), the reaction potential is proportional to the charges:

$$\phi_i^{\mathrm{R}} = \sum_j f_{ij} q_j. \tag{3.7}$$

Let us now switch on the charges adiabatically by introducing a factor

$$q_i \to q_i \lambda, \quad 0 < \lambda < 1. \tag{3.8}$$

The change of the free energy is

$$\mathrm{d}F = \sum_i \phi_i q_i \mathrm{d}\lambda = \sum_{i \neq j} \frac{q_i q_j \lambda}{4\pi \varepsilon r_{ij}} \mathrm{d}\lambda + \sum_{ij} f_{ij} q_j q_i \lambda \mathrm{d}\lambda \tag{3.9}$$

and thermodynamic integration gives the change of free energy due to Coulombic interactions:

$$\Delta F_{\mathrm{elec}} = \int_0^1 \lambda \mathrm{d}\lambda \left(\sum_{i \neq j} \frac{q_i q_j}{4\pi \varepsilon r_{ij}} + \sum_{ij} f_{ij} q_j q_i \right)$$

$$= \frac{1}{2} \sum_{i \neq j} \frac{q_i q_j}{4\pi \varepsilon r_{ij}} + \frac{1}{2} \sum_{ij} f_{ij} q_j q_i. \tag{3.10}$$

The first part is a property of the protein and hence included in U_{prot}. The second part is the mean force potential:

$$\Delta W_{\mathrm{elec}} = \frac{1}{2} \sum_{ij} f_{ij} q_i q_j. \tag{3.11}$$

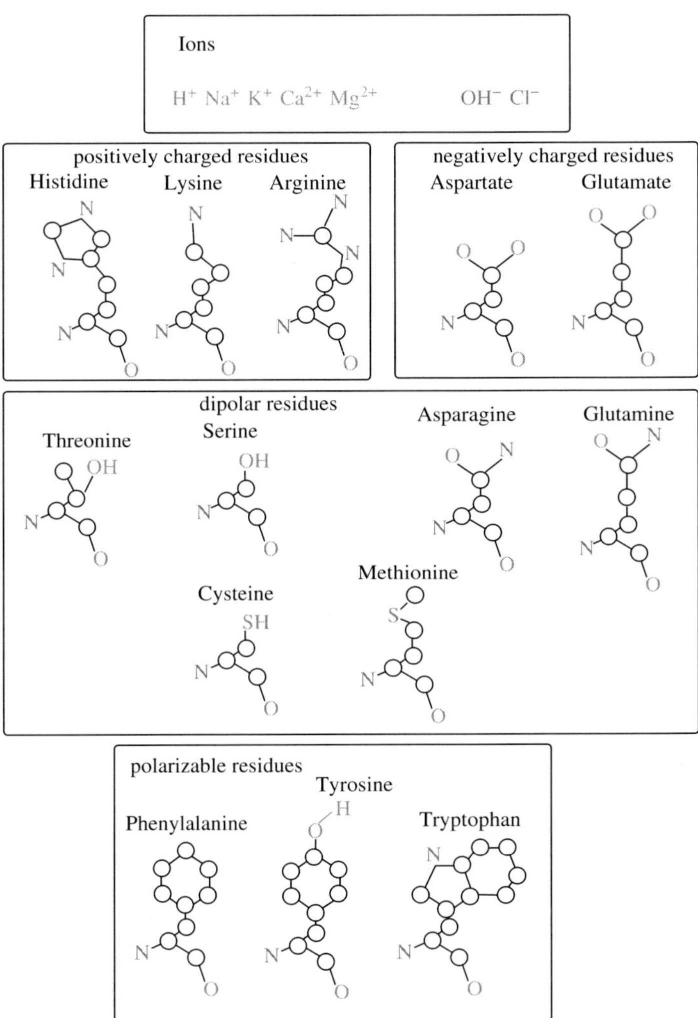

Fig. 3.1. Charged, polar, and polarizable groups in proteins. Biological macromolecules contain chemical compounds with certain electrostatic properties. These are often modeled using localized electric multipoles (partial charges, dipoles, etc.) and polarizabilities

3.3 Born Model

The Born model [11] is a simple continuum model to calculate the solvation free energy of an ion. Consider a point charge q in the center of a cavity with radius a (the so-called Born radius of the ion). The dielectric constant is ε_1 inside the sphere and ε_2 outside (Fig. 3.2).

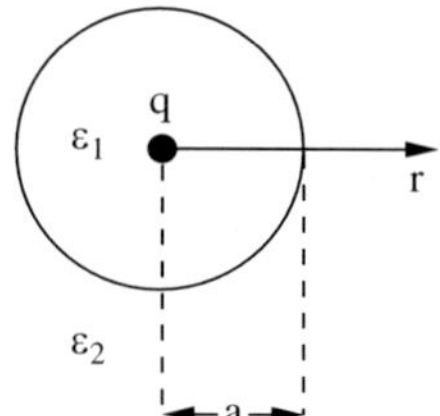

Fig. 3.2. Born model for ion solvation

The electrostatic potential is given outside the sphere by

$$\Phi = \frac{q}{4\pi\varepsilon_2 r} \tag{3.12}$$

and inside by

$$\Phi = \Phi_0 + \frac{q}{4\pi\varepsilon_1 r}. \tag{3.13}$$

From the continuity of the potential at the cavity surface, we find the reaction potential

$$\Phi_0 = \frac{q}{4\pi a}\left(\frac{1}{\varepsilon_2} - \frac{1}{\varepsilon_1}\right). \tag{3.14}$$

Since there is only one charge, we have

$$f_{1,1} = \frac{1}{4\pi a}\left(\frac{1}{\varepsilon_2} - \frac{1}{\varepsilon_1}\right) \tag{3.15}$$

and the solvation energy is given by the famous Born formula:

$$\Delta W_{\mathrm{elec}} = \frac{q^2}{8\pi a}\left(\frac{1}{\varepsilon_2} - \frac{1}{\varepsilon_1}\right). \tag{3.16}$$

3.4 Charges in a Protein

The continuum model can be applied to a more general charge distribution in a protein. An important example is an ion pair within a protein ($\varepsilon_{\mathrm{r}} = 2$) surrounded by water ($\varepsilon_{\mathrm{r}} = 80$). We study an idealized model where the protein is represented by a sphere (Fig. 3.3).

We will first treat a single charge within the sphere. A system of charges can then be treated by superposition of the individual contributions.

Using polar coordinates (r, θ, φ), the potential of a system with axial symmetry (no dependence on φ)can be written with the help of Legendre polynomials as

$$\phi = \sum_{n=0}^{\infty}(A_n r^n + B_n r^{-(n+1)})P_n(\cos\theta). \tag{3.17}$$

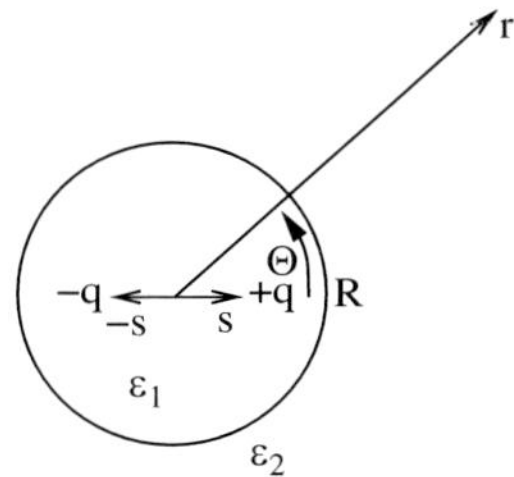

Fig. 3.3. Ion pair in a protein

The general solution can be written as the sum of a special solution and a harmonic function. The special solution is given by the multipole expansion:

$$\frac{q}{4\pi\varepsilon_1|\mathbf{r}-\mathbf{s}_+|} = \frac{q}{4\pi\varepsilon_1}\frac{1}{r}\sum_{n=0}^{\infty}\left(\frac{s}{r}\right)^n P_n(\cos\theta). \tag{3.18}$$

Since the potential has to be finite at large distances, outside it has the form

$$\phi_2 = \sum_{n=0}^{\infty} B_n r^{-(n+1)} P_n(\cos\theta) \tag{3.19}$$

and inside the potential is given by

$$\phi_1 = \sum_{n=0}^{\infty}\left(A_n r^n + \frac{qs^n}{4\pi\varepsilon_1}r^{-(n+1)}\right)P_n(\cos\theta). \tag{3.20}$$

At the boundary, we have two conditions

$$\phi_1(R) = \phi_2(R) \rightarrow B_n R^{-(n+1)} = A_n R^n + \frac{qs^n}{4\pi\varepsilon_1}R^{-(n+1)}, \tag{3.21}$$

$$\varepsilon_1\frac{\partial}{\partial r}\phi_1(R) = \varepsilon_2\frac{\partial}{\partial r}\phi_2(R) = 0, \tag{3.22}$$

$$\rightarrow -\frac{\varepsilon_2}{\varepsilon_1}(n+1)B_n R^{-(n+2)} = nA_n R^{n-1} - (n+1)\frac{qs^n}{4\pi\varepsilon_1}R^{-(n+2)}, \tag{3.23}$$

from which the coefficients can be easily determined:

$$A_n = \frac{qs^n}{4\pi\varepsilon_1}R^{-1-2n}\frac{(\varepsilon_1-\varepsilon_2)(n+1)}{n\varepsilon_1+(n+1)\varepsilon_2},$$

$$B_n = \frac{qs^n}{4\pi\varepsilon_1}\frac{(2n+1)\varepsilon_1}{n\varepsilon_1+(n+1)\varepsilon_2}. \tag{3.24}$$

The potential inside the sphere is

$$\phi_1 = \frac{q}{4\pi\varepsilon_1|\mathbf{r}-\mathbf{s}_+|} + \phi^{\mathrm{R}} \tag{3.25}$$

with the reaction potential

$$\phi^{\mathrm{R}} = \sum_{n=0}^{\infty} \frac{qs^n}{4\pi\varepsilon_1} R^{-1-2n} \frac{(\varepsilon_1 - \varepsilon_2)(n+1)}{n\varepsilon_1 + (n+1)\varepsilon_2} r^n P_n(\cos\theta) \qquad (3.26)$$

and the electrostatic energy is given (without the infinite self-energy) by

$$\frac{1}{2} q\phi(s, \cos\theta = 1) = \frac{q}{2} \sum_{n=1}^{\infty} \frac{qs^n}{4\pi\varepsilon_1} R^{-1-2n} \frac{(\varepsilon_1 - \varepsilon_2)(n+1)}{n\varepsilon_1 + (n+1)\varepsilon_2} s^n, \qquad (3.27)$$

which for $\varepsilon_2 \gg \varepsilon_1$ is approximately

$$\frac{q^2}{2} \frac{1}{4\pi R} \left(\frac{1}{\varepsilon_2} - \frac{1}{\varepsilon_1} \right) \sum \left(\frac{s}{R} \right)^{2n} = -\frac{q^2}{2} \frac{1}{4\pi R} \left(\frac{1}{\varepsilon_1} - \frac{1}{\varepsilon_2} \right) \frac{1}{1 - s^2/R^2}. \qquad (3.28)$$

Consider now two charges $\pm q$ at symmetric positions $\pm s$. The reaction potentials of the two charges add up to

$$\phi^{\mathrm{R}} = \phi_+^{\mathrm{R}} + \phi_-^{\mathrm{R}} \qquad (3.29)$$

and the electrostatic free energy is given by

$$\frac{-q^2}{4\pi\varepsilon_1(2s)} + \frac{1}{2} q\phi_+^{\mathrm{R}}(s) + \frac{1}{2} q\phi_-^{\mathrm{R}}(s) + \frac{1}{2}(-q)\phi_+^{\mathrm{R}}(-s) + \frac{1}{2}(-q)\phi_-^{\mathrm{R}}(-s). \qquad (3.30)$$

By comparison we find

$$\frac{q^2}{2} f_{++} = \frac{1}{2} q\phi_+^{\mathrm{R}}(s) = -\frac{q^2}{2} \frac{1}{4\pi R} \left(\frac{1}{\varepsilon_1} - \frac{1}{\varepsilon_2} \right) \frac{1}{1 - s^2/R^2}, \qquad (3.31)$$

$$\frac{(-q)^2}{2} f_{--} = \frac{1}{2}(-q)\phi_-^{\mathrm{R}}(-s) = -\frac{q^2}{2} \frac{1}{4\pi R} \left(\frac{1}{\varepsilon_1} - \frac{1}{\varepsilon_2} \right) \frac{1}{1 - s^2/R^2}, \qquad (3.32)$$

$$\frac{(-q)q}{2} f_{-+} = \frac{1}{2}(-q)\phi_+^{\mathrm{R}}(-s)$$

$$= \frac{(-q)}{2} \sum_{n=1}^{\infty} \frac{qs^n}{4\pi\varepsilon_1} R^{-1-2n} \frac{(\varepsilon_1 - \varepsilon_2)(n+1)}{n\varepsilon_1 + (n+1)\varepsilon_2} (-s)^n$$

$$\approx \frac{-q^2}{2} \frac{1}{4\pi R} \left(\frac{1}{\varepsilon_2} - \frac{1}{\varepsilon_1} \right) \sum_{n=0}^{\infty} (-)^n \left(\frac{s}{R} \right)^{2n}$$

$$= -\frac{(-q^2)}{2} \frac{1}{4\pi R} \left(\frac{1}{\varepsilon_1} - \frac{1}{\varepsilon_2} \right) \frac{1}{s^2/R^2 + 1}, \qquad (3.33)$$

$$\frac{q(-q)}{2} f_{+-} = \frac{1}{2} q\phi_-^{\mathrm{R}}(s) = \frac{(-q)q}{2} f_{-+}, \qquad (3.34)$$

and finally the solvation energy is

$$W_{\mathrm{elec}} = \sum_{n=1}^{\infty} \frac{q^2 s^n}{4\pi\varepsilon_1} R^{-1-2n} \frac{(\varepsilon_1 - \varepsilon_2)(n+1)}{n\varepsilon_1 + (n+1)\varepsilon_2}(s^n - (-s)^n), \qquad (3.35)$$

$$W_{\mathrm{elec}} \approx -q^2 \frac{1}{4\pi R}\left(\frac{1}{\varepsilon_1} - \frac{1}{\varepsilon_2}\right)\frac{1}{1 - s^2/R^2} - (-q^2)\frac{1}{4\pi R}\left(\frac{1}{\varepsilon_1} - \frac{1}{\varepsilon_2}\right)\frac{1}{s^2/R^2 + 1}$$

$$= -\frac{(2qs)^2}{8\pi R}\left(\frac{1}{\varepsilon_1} - \frac{1}{\varepsilon_2}\right)\frac{1}{1 - s^4/R^4}. \qquad (3.36)$$

If the extension of the system of charges in the protein is small compared to the radius $s \ll R$, the multipole expansion of the reaction potential converges rapidly. Since the total charge of the ion pair is zero, the monopole contribution($n = 0$)

$$W_{\mathrm{elec}}^{(1)} = \frac{Q^2}{8\pi R}\left(\frac{1}{\varepsilon_2} - \frac{1}{\varepsilon_1}\right) \qquad (3.37)$$

(which is nothing but the Born energy (3.16)) vanishes and the leading term is the dipole contribution[1] ($n = 1$)

$$W_{\mathrm{elec}}^{(2)} = \frac{p^2}{4\pi\varepsilon_1 R^3}\frac{(\varepsilon_1 - \varepsilon_2)}{\varepsilon_1 + 2\varepsilon_2}. \qquad (3.38)$$

3.5 Generalized Born Models

The generalized Born model approximates the protein as a sphere with a dielectric constant different from that of the solvent [12, 13].

Still and coworkers [13] proposed an approximate expression for the solvation free energy of an arbitrary distribution of N charges:

$$W_{\mathrm{elec}} = \frac{1}{8\pi}\left(\frac{1}{\varepsilon_2} - \frac{1}{\varepsilon_1}\right)\sum_{i,j=1}^{N}\frac{q_i q_j}{f_{\mathrm{GB}}(r_{ij}, a_{ij})} \qquad (3.39)$$

with the smooth function

$$f_{\mathrm{GB}} = \sqrt{r_{ij}^2 + a_i a_j \exp\left\{-\frac{r_{ij}^2}{2a_i a_j}\right\}}, \qquad (3.40)$$

where the effective Born radius a_i accounts for the effect of neighboring solute atoms. Several methods have been developed to calculate appropriate values of the a_i [13–16].

---

[1] This has been associated with several names (Bell, Onsager, and Kirkwood).

Expression (3.39) interpolates between the Born energy of a total charge Nq at short distances

$$W_{\text{elec}} \rightarrow \frac{1}{8\pi} \left(\frac{1}{\varepsilon_2} - \frac{1}{\varepsilon_1} \right) \frac{N^2 q^2}{a} \quad \text{for} \quad r_{ij} \rightarrow 0 \tag{3.41}$$

and the sum of the individual Born energies plus the change in Coulombic energy

$$W_{\text{elec}} \rightarrow \frac{1}{8\pi} \left(\frac{1}{\varepsilon_2} - \frac{1}{\varepsilon_1} \right) \left(\sum_i \frac{q_i^2}{a_i} + \sum_{i \neq j} \frac{q_i q_j}{r_{ij}} \right) \quad \text{for} \quad r_{ij} \gg \sqrt{a_i a_j} \tag{3.42}$$

for a set of well-separated charges. It gives a reasonable description in many cases without the computational needs of a full electrostatics calculation.

4

Debye–Hückel Theory

Mobile charges like Na^+ or Cl^- ions are very important for the functioning of Biomolecules. In a stationary state, their concentration depends on the local electrostatic field which is produced by the charge distribution of the Biomolecule, which in turn depends for instance on the protonation state and on the conformation. In this chapter, we present continuum models to describe the interaction between a Biomolecule and surrounding mobile charges [17–20].

4.1 Electrostatic Shielding by Mobile Charges

We consider a fully dissociated (strong) electrolyte containing N_i mobile ions of the sort $i = 1 \ldots$ with charges $Z_i e$ per unit volume. The charge density of the mobile charges is given by the average numbers of ions per volume:

$$\varrho_{\mathrm{mob}}(\mathbf{r}) = \sum_i Z_i e \overline{N}_i(\mathbf{r}). \tag{4.1}$$

The electrostatic potential is given by the Poisson equation:

$$\varepsilon \Delta \phi(\mathbf{r}) = -\varrho(\mathbf{r}) = -\varrho_{\mathrm{mob}}(\mathbf{r}) - \varrho_{\mathrm{fix}}(\mathbf{r}). \tag{4.2}$$

Debye and Hückel [21] used Boltzmann's theorem to determine the mobile charge density. Without the presence of fixed charges, the system is neutral:

$$0 = \varrho_{\mathrm{mob}}^0 = \sum_i Z_i e N_i^0 \tag{4.3}$$

and the constant value of the potential can be chosen as zero:

$$\phi^0 = 0. \tag{4.4}$$

The fixed charges produce a change of the potential. The electrostatic energy of an ion of sort i is

$$W_i = Z_i e \phi(\mathbf{r}) \tag{4.5}$$

and the density of such ions is given by a Boltzmann distribution:

$$\frac{N_i(\mathbf{r})}{N_i^0} = \frac{\mathrm{e}^{-Z_i e \phi(\mathbf{r})/k_\mathrm{B}T}}{\mathrm{e}^{-Z_i e \phi^0/k_\mathrm{B}T}} \tag{4.6}$$

or

$$N_i(\mathbf{r}) = N_i^0 \mathrm{e}^{-Z_i e \phi(\mathbf{r})/k_\mathrm{B}T}. \tag{4.7}$$

The total mobile charge density is

$$\varrho_\mathrm{mob}(\mathbf{r}) = \sum_i Z_i e N_i^0 \mathrm{e}^{-Z_i e \phi(\mathbf{r})/k_\mathrm{B}T} \tag{4.8}$$

and we obtain the Poisson–Boltzmann equation:

$$\varepsilon \Delta \phi(\mathbf{r}) = -\sum_i Z_i e N_i^0 \mathrm{e}^{-Z_i e \phi(\mathbf{r})/k_\mathrm{B}T} - \varrho_\mathrm{fix}(\mathbf{r}). \tag{4.9}$$

If the solution is very dilute, we can expect that the ion–ion interaction is much smaller than thermal energy:

$$Z_i e \phi \ll k_\mathrm{B}T \tag{4.10}$$

and linearize the Poisson–Boltzmann equation:

$$\varepsilon \Delta \phi(\mathbf{r}) = -\varrho_\mathrm{fix}(\mathbf{r}) - \sum_i Z_i e N_i^0 \left(1 - \frac{Z_i e}{k_\mathrm{B}T} \phi(\mathbf{r}) + \cdots \right). \tag{4.11}$$

The first summand vanishes due to electroneutrality and we find finally

$$\Delta \phi(\mathbf{r}) - \kappa^2 \phi(\mathbf{r}) = -\frac{1}{\varepsilon} \varrho_\mathrm{fix}(\mathbf{r}) \tag{4.12}$$

with the inverse Debye length

$$\lambda_\mathrm{Debye}^{-1} = \kappa = \sqrt{\frac{e^2}{\varepsilon k_\mathrm{B}T} \sum_i N_i^0 Z_i^2}. \tag{4.13}$$

4.2 1–1 Electrolytes

If there are only two types of ions with charges $Z_{1,2} = \pm 1$ (also in semiconductor physics), the Poisson–Boltzmann equation can be written as

$$\frac{e}{k_\mathrm{B}T} \Delta \phi(\mathbf{r}) + \frac{e^2}{\varepsilon k_\mathrm{B}T} N^0 (\mathrm{e}^{-e\phi(\mathbf{r})/k_\mathrm{B}T} - \mathrm{e}^{e\phi(\mathbf{r})/k_\mathrm{B}T}) = -\frac{e}{\varepsilon k_\mathrm{B}T} \varrho_\mathrm{fix}(\mathbf{r}) \tag{4.14}$$

which, after substitution

$$\widetilde{\phi}(\mathbf{r}) = \frac{e}{k_{\mathrm{B}}T}\phi(\mathbf{r}), \qquad (4.15)$$

takes the form

$$\Delta\widetilde{\phi}(\mathbf{r}) - \kappa^2 \sinh(\widetilde{\phi}(\mathbf{r})) = -\frac{e}{\varepsilon k_{\mathrm{B}}T}\varrho_{\mathrm{fix}}(\mathbf{r}) \qquad (4.16)$$

and with the scaled radius vector $\mathbf{r}' = \kappa\mathbf{r} \rightarrow \widetilde{\phi}(\mathbf{r}) = f(\mathbf{r}') = f(\kappa\mathbf{r})$

$$\Delta f(\mathbf{r}') - \sinh(f(\mathbf{r}')) = -\frac{e}{\kappa^2 \varepsilon k_{\mathrm{B}}T}\varrho_{\mathrm{fix}}(\kappa^{-1}\mathbf{r}'). \qquad (4.17)$$

4.3 Charged Sphere

We consider a spherical protein (radius R) with a charged sphere (radius a) in its center (Fig. 4.1).

For a spherical problem, we only have to consider the radial part of the Laplacian and the linearized Poisson–Boltzmann equation becomes outside the protein:

$$\frac{1}{r}\frac{\mathrm{d}^2}{\mathrm{d}r^2}\left(r\phi(r)\right) - \kappa^2\phi(r) = 0, \qquad (4.18)$$

which has the solution

$$\phi_2(r) = \frac{c_1\mathrm{e}^{-\kappa r} + c_2\mathrm{e}^{\kappa r}}{r}. \qquad (4.19)$$

Since the potential should vanish at large distances, we have $c_2 = 0$. Inside the protein ($a < r < R$), solution of the Poisson equation gives

$$\phi_1(r) = c_3 + \frac{Q}{4\pi\varepsilon_1 r}. \qquad (4.20)$$

At the boundary, we have the conditions

$$\phi_1(R) = \phi_2(R) \rightarrow c_3 = \frac{c_1\mathrm{e}^{-\kappa R}}{R} - \frac{Q}{4\pi\varepsilon_1 R}, \qquad (4.21)$$

$$\varepsilon_1\frac{\partial}{\partial r}\phi_1(R) = \varepsilon_2\frac{\partial}{\partial r}\phi_2(R) \rightarrow -\frac{Q}{4\pi R^2} = c_1\varepsilon_2\left(-\frac{\mathrm{e}^{-\kappa R}}{R^2} - \kappa\frac{\mathrm{e}^{-\kappa R}}{R}\right), \qquad (4.22)$$

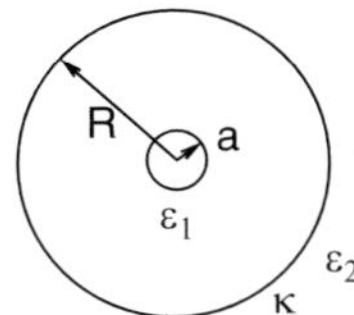

Fig. 4.1. Simple model of a charged protein

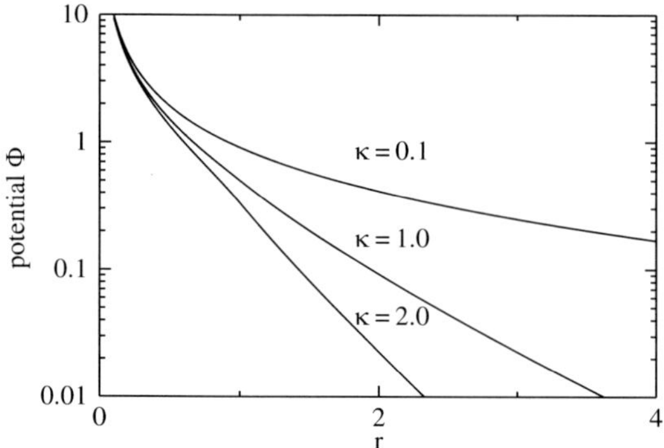

Fig. 4.2. Charged sphere in an electrolyte. The potential $\phi(r)$ is shown for $a = 0.1$, $R = 1.0$

which determine the constants:

$$c_1 = \frac{Q e^{\kappa R}}{4\pi\varepsilon_2(1 + \kappa R)} \tag{4.23}$$

and

$$c_3 = -\frac{Q}{4\pi\varepsilon_1 R} + \frac{Q}{4\pi\varepsilon_2 R(1 + \kappa R)}. \tag{4.24}$$

Together, we find the potential inside the sphere

$$\phi_1(r) = \frac{Q}{4\pi\varepsilon_1}\left(\frac{1}{r} - \frac{1}{R}\right) + \frac{Q}{4\pi\varepsilon_2 R(1 + \kappa R)} \tag{4.25}$$

and outside (Fig. 4.2)

$$\phi_2(r) = \frac{Q}{4\pi\varepsilon_2(1 + \kappa R)}\frac{e^{-\kappa(r-R)}}{r}. \tag{4.26}$$

The ion charge density is given by

$$\varrho_{\mathrm{mob}}(r) = \varepsilon_2\Delta\phi_2 = \varepsilon_2\kappa^2\phi_2; \tag{4.27}$$

hence, the ion charge at distances between r and $r + \mathrm{d}r$ is given by

$$\kappa\frac{Q}{(1 + \kappa R)}r e^{-\kappa(r-R)}\mathrm{d}r. \tag{4.28}$$

This function has a maximum at $r_{\mathrm{max}} = 1/\kappa$ and decays exponentially at larger distances (Fig. 4.3).

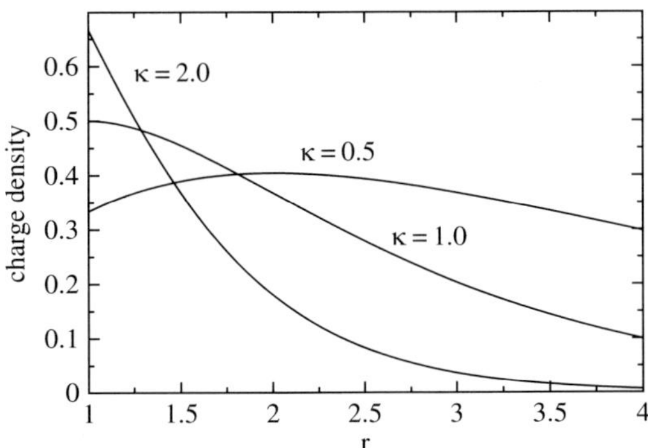

Fig. 4.3. Charge density around the charged sphere for $a = 0.1$, $R = 1.0$

Let the charge be concentrated on the surface of the inner sphere. Then we have

$$\phi_1(a) = \frac{Q}{4\pi\varepsilon_1}\left(\frac{1}{a} - \frac{1}{R}\right) + \frac{Q}{4\pi\varepsilon_2 R(1 + \kappa R)}. \tag{4.29}$$

Without the medium ($\varepsilon_2 = \varepsilon_1$, $\kappa = 0$), the potential would be

$$\phi_1^0(a) = \frac{Q}{4\pi\varepsilon_1 a}. \tag{4.30}$$

Hence, the solvation energy is

$$W = \frac{1}{2}Q\phi^R = \frac{1}{2}Q(\phi_1(a) - \phi_1^0(a)) = \frac{Q^2}{8\pi R}\left(\frac{1}{\varepsilon_2(1 + \kappa R)} - \frac{1}{\varepsilon_1}\right), \tag{4.31}$$

which for $\kappa = 0$ is given by the well-known Born formula:

$$W = -\frac{Q^2}{8\pi R}\left(\frac{1}{\varepsilon_1} - \frac{1}{\varepsilon_2}\right). \tag{4.32}$$

For $a \to R$, $\varepsilon_1 = \varepsilon_2$ this becomes the solvation energy of an ion in solution:

$$\Delta G_{\text{sol}} = W = -\frac{Q^2}{8\pi\varepsilon}\frac{\kappa}{(1 + \kappa R)}. \tag{4.33}$$

4.4 Charged Cylinder

Next, we discuss a cylinder of radius a and length $l \ll a$ carrying the net charge Ne uniformly distributed on its surface:

$$\sigma = \frac{Ne}{2\pi a l}. \tag{4.34}$$

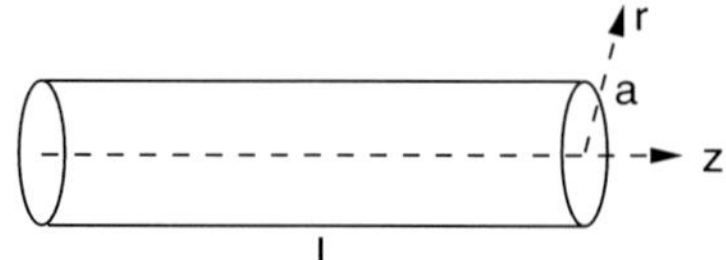

Fig. 4.4. Charged cylinder model

Outside the cylinder, this charge distribution is equivalent to a linear distribution of charges along the axis of the cylinder with a one-dimensional density (Fig. 4.4):

$$\frac{Ne}{l} = 2\pi a\sigma = \frac{e}{b}. \tag{4.35}$$

For the general case of a charged cylinder in an ionic solution, we have to restrict the discussion to the linearized Poisson–Boltzmann equation which becomes outside the cylinder

$$\frac{1}{r}\frac{\mathrm{d}}{\mathrm{d}r}\left(r\frac{\mathrm{d}}{\mathrm{d}r}\right)\phi(r) = \kappa^2\phi(r). \tag{4.36}$$

Substitution $r \to x = \kappa r$ gives the equation

$$\frac{\mathrm{d}^2}{\mathrm{d}x^2}\phi(x) + \frac{1}{x}\frac{\mathrm{d}}{\mathrm{d}x}\phi(x) - \phi(x) = 0. \tag{4.37}$$

Solutions of this equation are the modified Bessel functions of order zero denoted as $I_0(x)$ and $K_0(x)$. For large values of x,

$$\lim_{x\to\infty} I_0(x) = \infty, \quad \lim_{x\to\infty} K_0(x) = 0, \tag{4.38}$$

and hence the potential in the outer region has the form

$$\phi(r) = C_1 K_0(\kappa r). \tag{4.39}$$

Inside the cylinder surface, the electric field is given by Gauss' theorem:

$$2\pi rl\varepsilon_1 E(r) = Ne, \tag{4.40}$$

$$E(r) = -\frac{\mathrm{d}\phi(r)}{\mathrm{d}r} = \frac{Ne}{2\pi\varepsilon_1 rl}, \tag{4.41}$$

and hence the potential inside is

$$\phi(r) = C_2 - \frac{Ne}{2\pi\varepsilon_1 l}\ln r. \tag{4.42}$$

The boundary conditions are

$$\phi(a) = C_1 K_0(\kappa a) = C_2 - \frac{Ne}{2\pi\varepsilon_1 l}\ln a, \tag{4.43}$$

$$\varepsilon_1 \frac{d\phi(a)}{dr} = -\frac{Ne}{2\pi a l} = \varepsilon_2\frac{d\phi(a)}{dr} = \varepsilon_2 C_1(-\kappa K_1(\kappa a)), \tag{4.44}$$

from which we find

$$C_1 = \frac{Ne}{2\pi a l \varepsilon_2 \kappa K_1(\kappa a)} \tag{4.45}$$

and

$$C_2 = \frac{Ne}{2\pi a l \varepsilon_2 \kappa}\frac{K_0(\kappa a)}{K_1(\kappa a)} + \frac{Ne}{2\pi\varepsilon_1 l}\ln a. \tag{4.46}$$

The potential is then outside

$$\phi(r) = \frac{Ne}{2\pi a l \varepsilon_2 \kappa}\frac{K_0(\kappa r)}{K_1(\kappa a)} \tag{4.47}$$

and inside

$$\phi(r) = \frac{Ne}{2\pi a l \varepsilon_2 \kappa}\frac{K_0(\kappa a)}{K_1(\kappa a)} + \frac{Ne}{2\pi\varepsilon_1 l}\ln a - \frac{Ne}{2\pi\varepsilon_1 l}\ln r. \tag{4.48}$$

For small $\kappa a \to 0$, we use the asymptotic behavior of the Bessel functions

$$K_0(x) \to \ln\frac{2}{x} - \gamma + \cdots, \quad \gamma = 0.577\cdots,$$

$$K_1(x) \to \frac{1}{x} + \cdots \tag{4.49}$$

to have approximately

$$C_1 \approx \frac{Ne}{2\pi a l \varepsilon_2 \kappa}\kappa a = \frac{Ne}{2\pi l \varepsilon_2},$$

$$C_2 \approx \frac{Ne}{2\pi l \varepsilon_2}\left(\ln\frac{2}{\kappa a} - \gamma\right) + \frac{Ne}{2\pi\varepsilon_1 l}\ln a. \tag{4.50}$$

The potential outside is

$$\phi(r) = -\frac{Ne}{2\pi l \varepsilon_2}\left(\gamma + \ln\frac{\kappa}{2} + \ln r\right) \tag{4.51}$$

and inside is

$$\begin{aligned}\phi(r) &= \frac{Ne}{2\pi l \varepsilon_2}\left(\ln\frac{2}{\kappa a} - \gamma\right) + \frac{Ne}{2\pi\varepsilon_1 l}\ln a - \frac{Ne}{2\pi\varepsilon_1 l}\ln r \\ &= \frac{Ne}{2\pi l}\left(-\frac{\gamma}{\varepsilon_2} - \frac{1}{\varepsilon_2}\ln\frac{\kappa}{2} + \left(\frac{1}{\varepsilon_1} - \frac{1}{\varepsilon_2}\right)\ln a - \frac{1}{\varepsilon_1}\ln r\right). \end{aligned} \tag{4.52}$$

Outside the potential consists of the potential of the charged line ($\ln r$) and an additional contribution from the screening of the ions (Fig. 4.5).

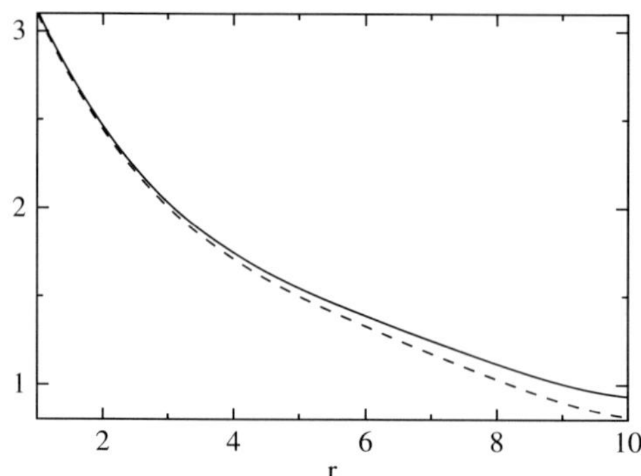

Fig. 4.5. Potential of a charged cylinder with unit radius $a = 1$ for $\kappa = 0.05$. *Solid curve*: $K_0(\kappa r)/(\kappa K_1(\kappa))$. *Dashed curve*: approximation by $-\ln(\kappa/2) - \ln r - \gamma$

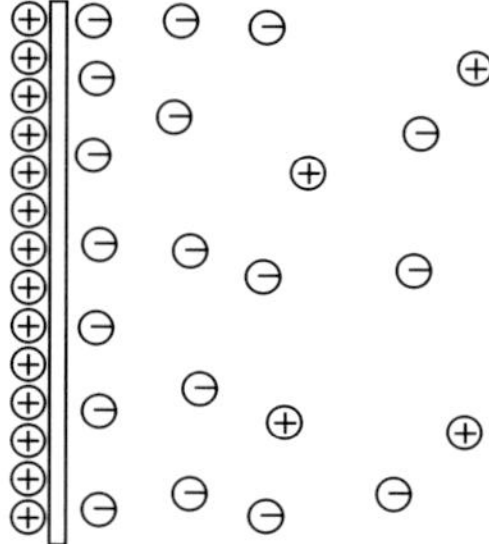

Fig. 4.6. Goüy–Chapman double layer

4.5 Charged Membrane (Goüy–Chapman Double Layer)

We approximate the surface charge of a membrane by a thin layer charged with a homogeneous charge distribution (Fig. 4.6).

Goüy [22, 23] and Chapman [24] derived the potential similar to Debye–Hückel theory. For a 1–1 electrolyte (e.g., NaCl), the one-dimensional Poisson–Boltzmann equation has the form (with transformed variables as above)

$$\frac{d^2}{dx^2} f(x) - \sinh(f(x)) = g(x), \tag{4.53}$$

where the source term $g(x) = -(e/\kappa^2 \varepsilon k_B T)\varrho(x/\kappa)$ has the character of a δ-function centered at $x = 0$. Consider an area A of the membrane and integrate along the x-axis:

$$\int dA \int_{-0}^{+0} \varrho(x)dx = \sigma_0 A, \tag{4.54}$$

$$\int \mathrm{d}A \int_{-0}^{+0} \mathrm{d}x g(x) = -\frac{e}{\kappa^2 \varepsilon k_{\mathrm B} T} \int \mathrm{d}A \int_{-0}^{+0} \kappa \mathrm{d}x' \varrho(x')$$
$$= -\frac{e}{\kappa \varepsilon k_{\mathrm B} T}\sigma_0 A. \tag{4.55}$$

Hence we identify

$$\varrho(x) = \sigma_0 \delta(x), \quad g(x) = -\frac{e\sigma_0}{\kappa \varepsilon k_{\mathrm B} T}\delta(x). \tag{4.56}$$

The Poisson–Boltzmann equation can be solved analytically in this simple case. But first, we study the linearized homogeneous equation:

$$\frac{\mathrm{d}^2}{\mathrm{d}x^2}f(x) - f(x) = 0 \tag{4.57}$$

with the solution

$$f(x) = f_0 e^{\pm x} \tag{4.58}$$

or going back to the potential

$$\phi(x) = \frac{k_{\mathrm B} T}{e}f_0 e^{\pm \kappa x} = \phi_0 e^{\pm \kappa x}. \tag{4.59}$$

The membrane potential is related to the surface charge density. Let us assume that on the left side ($x < 0$), the medium has a dielectric constant of ε_1 and on the right side ε. Since in one dimension the field in a dielectric medium does not decay, we introduce a shielding constant κ_1 on the left side and take the limit $\kappa_1 \to 0$ to remove contributions not related to the membrane charge. The potential then is given by

$$\phi(x) = \begin{cases} \phi_0 e^{-\kappa x} & x > 0, \\ \phi_0 e^{\kappa_1 x} & x < 0. \end{cases} \tag{4.60}$$

The membrane potential ϕ_0 is determined from the boundary condition:

$$\varepsilon \frac{\mathrm{d}\phi}{\mathrm{d}x}(+0) - \varepsilon_1 \frac{\mathrm{d}\phi}{\mathrm{d}x}(-0) = -\sigma_0, \tag{4.61}$$

which gives

$$-\varepsilon \kappa \phi_0 - \varepsilon_1 \kappa_1 \phi_0 = -\sigma_0. \tag{4.62}$$

In the limit $\kappa_1 \to 0$, we find

$$\phi_0 = \frac{\sigma_0}{\varepsilon \kappa}. \tag{4.63}$$

For $x < 0$ the potential is constant and for $x > 0$ the charge density is (Figs. 4.7 and 4.8)

$$\varrho(x) = -\varepsilon \frac{\mathrm{d}^2 \phi(x)}{\mathrm{d}x^2} = -\sigma_0 \kappa e^{-\kappa x}, \tag{4.64}$$

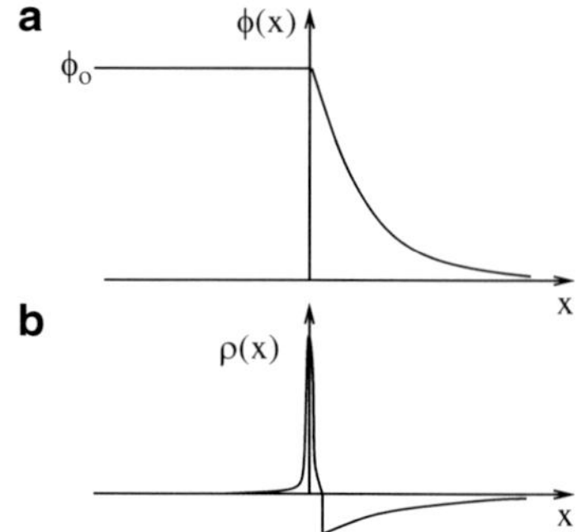

Fig. 4.7. Electrolytic double layer. (**a**) The potential is continuous and decreases exponentially in the electrolyte. (**b**) The charge density has a sharp peak on the membrane which is compensated by the exponentially decreasing charge in the electrolyte

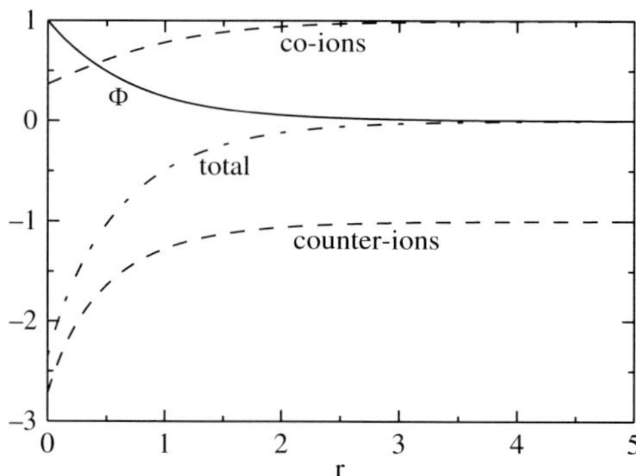

Fig. 4.8. Charge density of counter- and co-ions. For an exponentially decaying potential $\Phi(r)$ (*full curve*), the density of counter- and co-ions (*dashed curves*) and the total charge density (*dash-dotted curve*) are shown

which adds up to a total net charge per unit area of

$$\int_0^\infty \varrho(x)\mathrm{d}x = -\sigma_0. \tag{4.65}$$

Hence, the system is neutral and behaves like a capacity of

$$\frac{\sigma_0 A}{\phi_0} = \varepsilon \kappa A = \varepsilon \frac{A}{L_{\mathrm{Debye}}}. \tag{4.66}$$

The solution of the nonlinear homogeneous equation can be found multiplying the equation with $\mathrm{d}f(x)/\mathrm{d}x$:

$$\frac{\mathrm{d}f}{\mathrm{d}x}\frac{\mathrm{d}^2 f}{\mathrm{d}x^2} = \sinh(f)\frac{\mathrm{d}f}{\mathrm{d}x} \tag{4.67}$$

and rewriting this as

$$\frac{1}{2}\frac{\mathrm{d}}{\mathrm{d}x}\left(\frac{\mathrm{d}f}{\mathrm{d}x}\right)^2 = \frac{\mathrm{d}}{\mathrm{d}x}\cosh(f),\tag{4.68}$$

which can be integrated

$$\left(\frac{\mathrm{d}f}{\mathrm{d}x}\right)^2 = 2\left[\cosh(f) + C\right].\tag{4.69}$$

The constant C is determined by the asymptotic behavior[1]

$$\lim_{x\to\infty} f(x) = \lim_{x\to\infty}\frac{\mathrm{d}f}{\mathrm{d}x} = 0\tag{4.70}$$

and obviously has the value $C = -1$. Making use of the relation

$$\cosh(f) - 1 = 2\sinh\left(\frac{f}{2}\right)^2,\tag{4.71}$$

we find

$$\frac{\mathrm{d}}{\mathrm{d}x}f(x) = \left(\pm 2\sinh\left(\frac{f(x)}{2}\right)\right).\tag{4.72}$$

Separation of variables then gives

$$\frac{\mathrm{d}f}{2\sinh(f/2)} = \pm\mathrm{d}x\tag{4.73}$$

with the solution

$$f(x) = 2\ln\left(\pm\tanh\left(\frac{x}{2} + \frac{C}{2}\right)\right).\tag{4.74}$$

For $x > 0$, only the plus sign gives a physically meaningful expression. The constant C is generally complex-valued. It can be related to the potential at the membrane surface:

$$C = 2\,\mathrm{arctanh}(e^{f(0)/2}) = \ln\left(\frac{1 + e^{f(0)/2}}{1 - e^{f(0)/2}}\right).\tag{4.75}$$

For $f(0) > 0$, the argument of the logarithm becomes negative. Hence, we replace in (4.74) C by $C + i\pi$ to have (Fig. 4.9)

$$f(x) = 2\ln\left(\tanh\left(\frac{x}{2} + \frac{C}{2} + \frac{i\pi}{2}\right)\right) = -2\ln\left(\tanh\left(\frac{x}{2} + \frac{C}{2}\right)\right),\tag{4.76}$$

---

[1] We consider only solutions for the region $x > 0$ in the following.

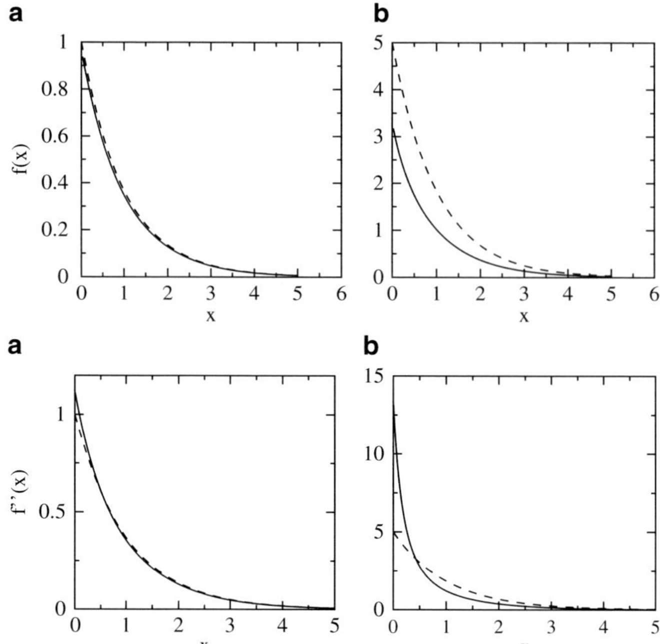

Fig. 4.9. One-dimensional Poisson–Boltzmann equation. The solutions of the full (*full curves*) and the linearized equation (*broken curves*) are compared for (**a**) $B = -1$ and (**b**) $B = -5$

where C is now given by

$$C = 2\,\mathrm{arctanh}(\mathrm{e}^{-f(0)/2}) = \ln\left(\frac{1 + \mathrm{e}^{-f(0)/2}}{1 - \mathrm{e}^{-f(0)/2}}\right). \tag{4.77}$$

The integration constant C is again connected to the surface charge density by

$$\frac{\mathrm{d}\phi}{\mathrm{d}x}(0) = -\frac{\sigma_o}{\varepsilon} \tag{4.78}$$

and from

$$\frac{\mathrm{d}}{\mathrm{d}x}\phi(x) = \frac{k_\mathrm{B}T}{e}\frac{\mathrm{d}}{\mathrm{d}x}f(\kappa x) = \frac{k_\mathrm{B}T}{e}\kappa f'(\kappa x), \tag{4.79}$$

we find

$$\frac{\sigma_0}{\varepsilon} = -\frac{k_\mathrm{B}T}{e}\kappa f'(0). \tag{4.80}$$

Now the derivative is in the case $f(0) > 0$ given by

$$f'(x) = \tanh\left(\frac{x}{2} + \frac{C}{2}\right) - \frac{1}{\tanh\left(\frac{x}{2} + \frac{C}{2}\right)} \tag{4.81}$$

and especially

$$f'(0) = \tanh\left(\frac{C}{2}\right) - \frac{1}{\tanh(C/2)} \tag{4.82}$$

and we have to solve the equation

$$t - \frac{1}{t} = -\frac{e\sigma_0}{k_{\mathrm{B}}T\kappa\varepsilon} = B, \tag{4.83}$$

which yields[2]

$$t = \frac{B}{2} + \frac{\sqrt{B^2 + 4}}{2}, \tag{4.84}$$

$$C = 2\mathrm{arctanh}\left(\frac{B}{2} + \frac{\sqrt{B^2 + 4}}{2}\right). \tag{4.85}$$

4.6 Stern Modification of the Double Layer

Real ions have a finite radius R. Therefore, they cannot approach the membrane closer than R and the ion density has a maximum possible value, which is reached when the membrane is occupied by an ion layer. To account for this, Stern [25] extended the Goüy–Chapman model by an additional ion layer between the membrane and the diffusive ion layer (Fig. 4.10).

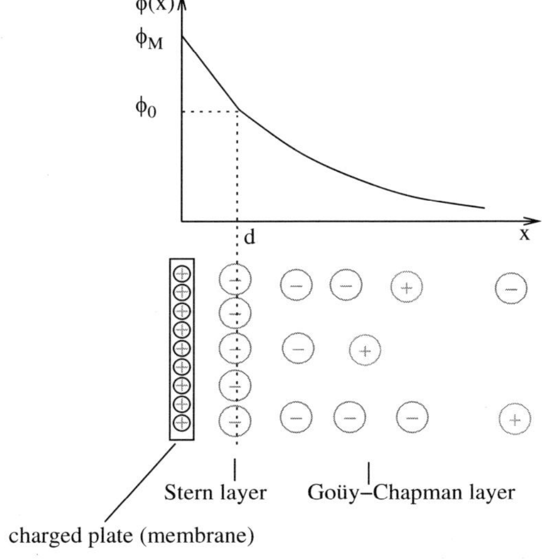

Fig. 4.10. Stern modification of the double layer

---

[2] The second root leads to imaginary values.

Within the Stern layer of thickness d, there are no charges and the potential drops linearly from the membrane potential ϕ_M to a value ϕ_0:

$$\phi(x) = \phi_\mathrm{M} - \frac{\phi_\mathrm{M} - \phi_0}{d}x, \quad 0 < x < d. \tag{4.86}$$

In the diffusive Goüy–Chapman layer, the potential decays exponentially

$$\phi(x) = \phi_0 \mathrm{e}^{-\kappa(x-d)}. \tag{4.87}$$

Assuming the same dielectric constant for both layers, we have to fulfill the boundary condition

$$\frac{\mathrm{d}\phi}{\mathrm{d}x}(d) = -\frac{\phi_\mathrm{M} - \phi_0}{d} = -\kappa\phi_0 \tag{4.88}$$

and hence

$$\phi_0 = \frac{\phi_\mathrm{M}}{1 + \kappa d}. \tag{4.89}$$

The total ion charge in the diffusive layer now is

$$q_\mathrm{diff} = \int_d^\infty \varrho(x)\mathrm{d}x = -A\varepsilon\kappa^2\phi_0 \int_d^\infty \mathrm{e}^{-\kappa(x-a)}\mathrm{d}x = -A\varepsilon\kappa\phi_0$$
$$= -A\frac{\varepsilon\kappa}{1 + \kappa d}\phi_\mathrm{M} \tag{4.90}$$

and the capacity is

$$C = \frac{-q_\mathrm{diff}}{\phi_\mathrm{M}} = \frac{A\varepsilon}{d + \kappa^{-1}}, \tag{4.91}$$

which are just the capacities of the two layers in series

$$\frac{1}{C} = \frac{Ad}{\varepsilon} + \frac{A\lambda_\mathrm{Debye}}{\varepsilon} = \frac{1}{C_\mathrm{Stern}} + \frac{1}{C_\mathrm{diff}}. \tag{4.92}$$

Problems

4.1. Membrane Potential

Consider a dielectric membrane in an electrolyte with an applied voltage V.

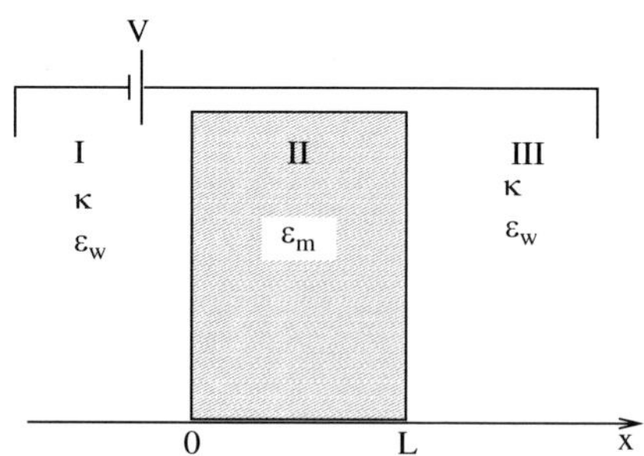

Solve the linearized Poisson–Boltzmann equation

$$\frac{\mathrm{d}^2}{\mathrm{d}x^2}\phi(x) = \kappa^2(\phi(x) - \phi^{(0)})$$

with boundary conditions

$$\phi^{(0)}(I) = \phi(-\infty) = 0,$$
$$\phi^{(0)}(III) = \phi(\infty) = V,$$

and determine the voltage difference

$$\Delta V = \phi(L) - \phi(0).$$

Calculate the charge density $\varrho_{\mathrm{mob}}(x)$ and the integrated charge on both sides of the membrane. What is the capacity of the membrane?

4.2. Ionic Activity

The chemical potential of an ion with charge Ze is given in terms of the activity a by $\mu = \mu^0 + k_\mathrm{B}T\ln a$. Assume that the deviation from the ideal behavior $\mu^{\mathrm{ideal}} = \mu^0 + k_\mathrm{B}T\ln c$ is due to electrostatic interactions only. Then for an ion with radius R, Debye–Hückel theory gives

$$\mu - \mu^{\mathrm{ideal}} = k_\mathrm{B}T\ln a - k_\mathrm{B}T\ln c = \Delta(\mu_\alpha - \mu_\alpha^0)G_{\mathrm{solv}} = -\frac{Z^2e^2}{8\pi\varepsilon}\frac{\kappa}{(1+\kappa R)}.$$

For a 1–1 electrolyte, calculate the mean activity coefficient

$$\gamma_\pm^{\mathrm{c}} = \sqrt{\gamma_+^{\mathrm{c}}\gamma_-^{\mathrm{c}}} = \sqrt{\frac{a_+\,a_-}{c_+\,c_-}}$$

and discuss the limit of extremely dilute solutions (Debye–Hückel limiting law).

5

Protonation Equilibria

Some of the amino acid residues building a protein can be in different protonation states and therefore differently charged states (Fig. 5.1).

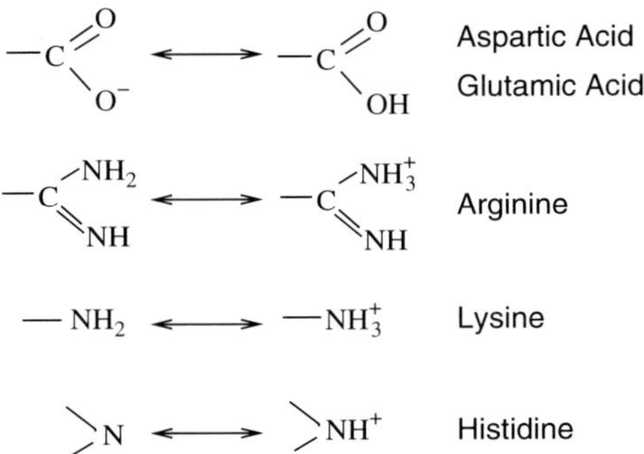

Fig. 5.1. Functional groups which can be in different protonation states

In this chapter, we discuss the dependence of the free energy of a protein on the electrostatic interactions of its charged residues. We investigate the chemical equilibrium between a large number of different protein conformations and the dependence on the pH value [26].

5.1 Protonation Equilibria in Solution

We consider a dilute aqueous solution containing N titrable molecules (Fig. 5.2) (i.e., which can be in two different protonation states $0 =$ deprotonated, $1 =$ protonated). For one molecule, we have for fixed protonation state

$$G_0 = F_0 + pV_0 = -k_\mathrm{B}T \ln z_0 + pV_0, \tag{5.1}$$

$$G_1 = F_1 + pV_1 = -k_\mathrm{B}T \ln z_1 + pV_1, \tag{5.2}$$

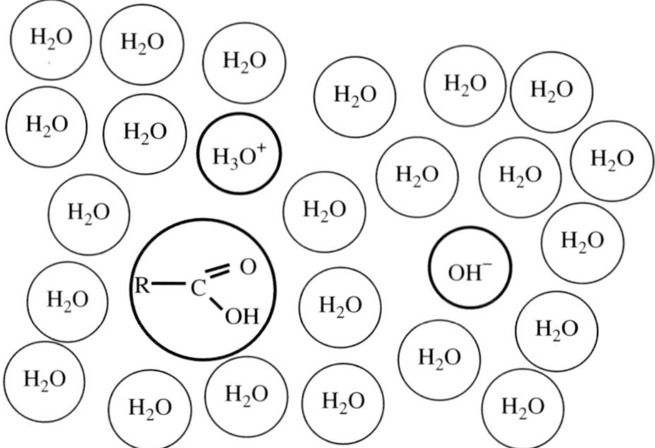

Fig. 5.2. Titration in solution

and the free enthalpy difference between protonated and deprotonated form
of the molecule is

$$G_1 - G_0 = -k_\mathrm{B}T \ln \frac{z_1}{z_0} + p\Delta V. \tag{5.3}$$

In the following, the volume change will be neglected.[1] If we now put such
N molecules into the solution, the number of protonated molecules can fluctuate by exchanging protons with the solvent. Removal of one proton from
the solvent costs a free enthalpy of

$$\Delta G = -\mu_{\mathrm{H}+}, \tag{5.4}$$

where $\mu_{\mathrm{H}+}$ is the chemical potential of the proton in solution.[2] Hence, we
come to the grand canonical partition function

$$\Xi = \sum Z_M e^{\mu M / k_\mathrm{B} T}, \tag{5.5}$$

where the partition function for fixed number of protonated molecules M is
given by

$$Z_M = \frac{N!}{M!(N-M)!} z_1^M z_0^{N-M} \tag{5.6}$$

if the N molecules cannot be distinguished. Hence we find

$$\Xi = \sum \frac{N!}{M!(N-M)!} (z_1 e^{\mu / k_\mathrm{B} T})^M z_0^{N-M} = (z_0 + z_1 e^{\mu / k_\mathrm{B} T})^N. \tag{5.7}$$

---

[1] At atmospheric pressure, the mechanic work term $p\Delta V$ is very small. We pre-
fer to discuss the free enthalpy G in the following since experimentally usually
temperature and pressure are constant.

[2] We omit the index H$^+$ in the following.

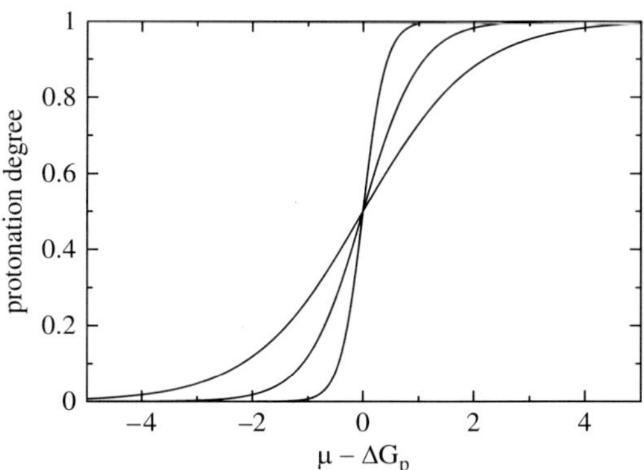

Fig. 5.3. Henderson–Hasselbalch titration curve. The protonation degree (5.9) is shown for $k_B\, T = 1, 2, 5$

The average number of protonated molecules can be found from

$$\overline{M} = \frac{\partial}{\partial(\mu/k_B T)}\ln \Xi = N\frac{z_1 e^{\mu/k_B T}}{z_0 + z_1 e^{\mu/k_B T}} \tag{5.8}$$

and the protonation degree, that is, the fraction of protonated molecules is (Fig. 5.3)

$$\frac{\overline{M}}{N} = \frac{1}{1 + (z_0/z_1)e^{-\mu/k_B T}} = \frac{1}{1 + e^{(G_1 - G_0 - \mu)/k_B T}}. \tag{5.9}$$

In physical chemistry [27, 28], the following quantities are usually introduced:

- The *activity* of a species a_i is defined by

$$\mu_i = \mu_i^0 + k_B T \ln a_i \tag{5.10}$$

in analogy to $\mu_i = \mu_i^0 + k_B T \ln p_i/p_i^0$ for the ideal gas. For very dilute solutions, it can be approximated by

$$\mu_i = k_B T \ln N_i - k_B T \ln z_i = \mu_i^0 + k_B T \ln \frac{c_i}{c_0}, \tag{5.11}$$

where (0) indicates the standard state (usually $p_0 = 1\,\mathrm{atm}$, $c_0 = 1\,\mathrm{mol/l}$).
- The concentration of protons is measured by the pH value[3]

$$\mathrm{pH} = -\log_{10} a_{H^+} = -\frac{\mu_{H^+}}{k_B T \ln(10)} \approx -\log_{10}\left(\frac{c(H^+)}{c_0}\right). \tag{5.12}$$

- The standard reaction enthalpy of an acid–base equilibrium

$$\mathrm{AH} \rightleftharpoons \mathrm{A}^- + \mathrm{H}^+, \tag{5.13}$$

---

[3] The standard enthalpy of formation for a proton is zero per definition.

where

$$0 = \sum \nu_i \mu_i = \Delta G_r^0 + k_B T \sum \nu_i \ln a_i, \tag{5.14}$$

$$K_a = e^{-\Delta G_r^0 / k_B T} = \frac{a(A^-)a(H^+)}{a(AH)}, \tag{5.15}$$

is measured by the pK_a value:

$$pK_a = -\log_{10}(K_a) = \frac{1}{\ln(10)} \frac{\Delta G_r^0}{k_B T} = -\log_{10}\left(\frac{a(A^-)a(H^+)}{a(AH)}\right)$$

$$\approx -\log_{10}\left(\frac{\frac{c(A^-)}{c_0}\frac{c(H^+)}{c_0}}{\frac{c(AH)}{c_0}}\right), \tag{5.16}$$

which is usually simply written as[4]

$$pK_a = -\log_{10}\left(\frac{c(A^-)c(H^+)}{c(AH)}\right), \tag{5.17}$$

which together with (5.12) gives the Henderson–Hasselbalch equation:

$$pH - pK_a = \log_{10}\left(\frac{c(A^-)}{c(AH)}\right). \tag{5.18}$$

The standard reaction enthalpy of the acid–base equilibrium is (with the approximation (5.11))

$$\Delta G_r^0 = \mu_{HA}^0 + \mu_{M^-}^0 + \mu_{H^+}^0$$

$$= -(k_B T \ln L - k_B T \ln z_1) + (k_B T \ln L - k_B T \ln z_0)$$

$$= -k_B T \ln \frac{z_0}{z_1} = -(G_1 - G_0). \tag{5.19}$$

In the language of physical chemistry, the protonation degree (5.9) is given by

$$\frac{c(AH)}{c(AH) + c(A^-)} = \frac{1}{1 + 10^{(pH - pK_a)}}. \tag{5.20}$$

5.2 Protonation Equilibria in Proteins

In the protein, there are additional steric and electrostatic interactions with other groups of the protein, which contribute to the energies of the titrable site (Fig. 5.4).

---

[4] But you have to be aware that all concentrations have to be taken in units of the standard concentration c_0. The argument of a logarithm should be dimensionless.

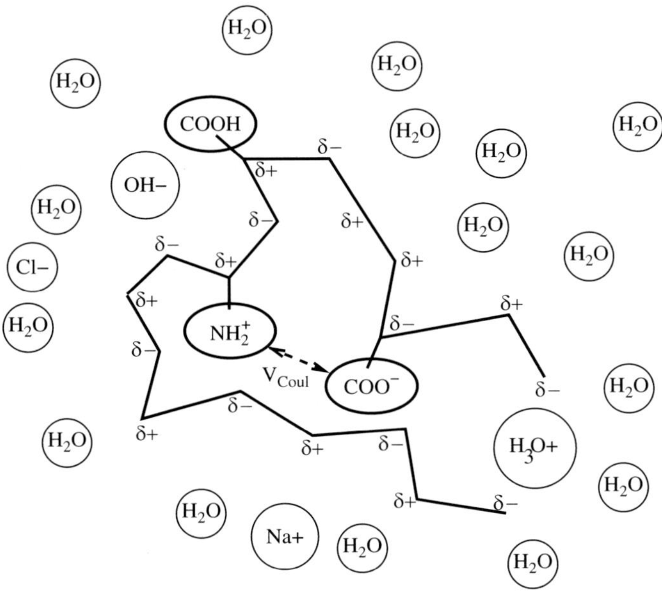

Fig. 5.4. Titration in a protein. Electrostatic interactions with fixed charges (charged residues and background charges) and mobile charges (ions) have to be taken into account

$$
\begin{array}{ccc}
\text{RMH} & \xrightarrow{\Delta G_{sol}} & \overset{-}{R}M + \overset{+}{H} \\
\Delta G_1 \downarrow & & \downarrow \Delta G_0 \\
\text{PMH} & \xrightarrow[\Delta G_{prot}]{} & PM^- + H^+
\end{array}
$$

Fig. 5.5. Thermodynamic cycle. The protonation enthalpy of a titrable group in the protein (PMH) differs from that of a model compound in solution (RMH) [29]

5.2.1 Apparent pK_a Values

The pK_a of a titrable group depends on the interaction with background charges, with all the other residues and with the solvent which contains dipolar molecules and free moving ions. As a consequence, the pK_a value is different from that of a model compound containing the titrable group in solution (Fig. 5.5).

The difference of protonation enthalpies

$$\Delta\Delta G = \Delta G_{\mathrm{prot}} - \Delta G_{\mathrm{solv}} = \Delta G_0 - \Delta G_1 \tag{5.21}$$

can be divided into three parts:

$$\Delta\Delta G = \Delta\Delta E + p\Delta\Delta V - T\Delta\Delta S. \tag{5.22}$$

In the following, the volume change will be neglected. The pK_a value of a group in the protein

$$\mathrm{p}K_{\mathrm{a}}^{\mathrm{prot}} = \frac{1}{k_{\mathrm{B}}T\ln(10)}\Delta G_{\mathrm{prot}} = \frac{1}{k_{\mathrm{B}}T\ln(10)}\left(\Delta G_{\mathrm{solv}} + \Delta\Delta G\right)$$

$$= \mathrm{p}K_{\mathrm{a}}^{\mathrm{model}} + \frac{\Delta\Delta G}{k_{\mathrm{B}}T\ln(10)} \tag{5.23}$$

is called the apparent pK_a of the group in the protein. It depends on the mutual interactions of the titrable groups and hence on the pH value. Therefore, titration of groups in a protein cannot be described by a simple Henderson–Hasselbalch equation.

5.2.2 Protonation Enthalpy

The amino acids forming a protein can be enumerated by their appearance in the primary structure and are therefore distinguishable. The protonation state of a protein with N titrable sites will be described by the protonation vector:

$$\mathbf{s} = (s_1, s_2, \ldots, s_S) \quad \text{with} \quad s_i = \begin{cases} 1 & \text{if group } i \text{ is protonated,} \\ 0 & \text{if group } i \text{ is deprotonated.} \end{cases} \tag{5.24}$$

The number of protonation states

$$N_{\mathrm{pstates}} = 2^N$$

can be very big for real proteins.[5] Proteins are very flexible and have a large number of configurations. These will be denoted by the symbol $\boldsymbol{\gamma}$ which summarizes all the orientational angles between neighboring residues of the protein. The apparent pK_a values will in general depend on this configuration vector, since for instance distances between the residues depend on the configuration.

The enthalpy change by protonating the ith residue is denoted as

$$G(s_1, \ldots, s_{i-1}, 1_i, s_{i+1}, \ldots, s_N, \boldsymbol{\gamma}) - G(s_1, \ldots, s_{i-1}, 0_i, s_{i+1}, \ldots, \boldsymbol{\gamma})$$

$$= \Delta G_{i,\mathrm{intr}} + \sum_{j\neq i}\left(E_{i,j}^{1,s_j} - E_{i,j}^{0,s_j}\right) \tag{5.25}$$

with the Coulombic interaction between two residues i, j in the protonation states s_i, s_j (Fig. 5.6):

$$E_{i,j}^{s_i,s_j}(\boldsymbol{\gamma}). \tag{5.26}$$

---

[5] We do not take into account that some residues can be in more than two protonation states.

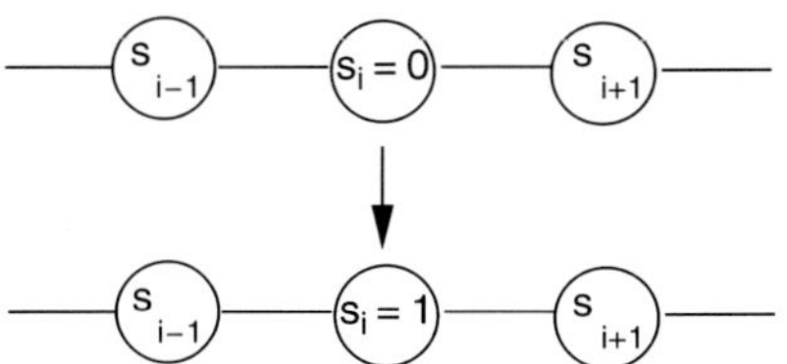

Fig. 5.6. Protonation of one residue. The protonation state of the ith residue changes together with its charge. Electrostatic interactions are divided into an intrinsic part and the interactions with all other residues

The so-called intrinsic protonation energy $\Delta G_{i,\mathrm{intr}}$ is the protonation energy of residue i if there are no Coulombic interactions with other residues, that is, if all other residues are in their neutral state. It can be estimated from the model energy (5.3) and (5.19), taking into account all remaining interactions with background charges[6] and the different solvation energies, which can be calculated using Born models (3.16) and (3.39) or by solving the Poisson–Boltzmann equation (4.9) and (4.12):

$$\Delta G_{i,\mathrm{intr}} \approx \Delta G_{i,\mathrm{solv}} + \Delta E_{i,\mathrm{bg}} + \Delta E_{i,\mathrm{Born}}. \tag{5.27}$$

Let us now calculate the protonation enthalpy of a protein with S of its N titrable residues protonated. The contributions from the intrinsic enthalpy changes can be written as

$$\sum_i s_i \Delta G_{i,\mathrm{intr}}. \tag{5.28}$$

The Coulombic interactions are divided into pairs of protonated residues

$$\sum_{i<j} s_i s_j E_{i,j}^{1,1}, \tag{5.29}$$

pairs of unprotonated residues

$$\sum_{i<j}(1 - s_i)(1 - s_j)E_{i,j}^{0,0} = \sum_{i<j} E_{i,j}^{0,0} + \sum_{i<j} s_i s_j E_{i,j}^{0,0} - \sum_{i \neq j} s_i E_{i,j}^{0,0}, \tag{5.30}$$

and interactions between one protonated and one unprotonated residue

$$\sum_{i<j}\left(s_i(1 - s_j)E_{i,j}^{1,0} + (1 - s_i)s_j E_{i,j}^{0,1}\right) = -\sum_{i<j} s_i s_j \left(E_{i,j}^{1,0} + E_{i,j}^{0,1}\right)$$
$$+ \sum_{i \neq j} s_i E_{i,j}^{1,0}. \tag{5.31}$$

---

[6] That is, the charge distribution of the protein backbone and the nontitrable residues.

$$AH \rightleftharpoons A^- + H^+ \qquad BH^+ \rightleftharpoons B + H^+$$

	1	0	1	0
s	1	0	1	0
s^0	1	1	0	0
$q = s - s^0$	0	-1	$+1$	0
$q^c = 1 - 2s^0$	-1	-1	1	1

Fig. 5.7. Protonation states. Protonation state s_i, protonation of the neutral state s_i^0, charge q_i, and charge of the non-neutral state q_i^c are correlated

Summing up these contributions and subtracting the Coulombic interactions of the fully deprotonated protein, we find the enthalpy change

$$\Delta G(\mathbf{s}) = \sum_i s_i \left(\Delta G_{i,\text{intr}} + \sum_{j \neq i} \left(E_{i,j}^{1,0} - E_{i,j}^{0,0} \right) \right) + \sum_{i<j} s_i s_j W_{i,j} \qquad (5.32)$$

with the interaction parameter

$$W_{ij} = E_{ij}(1,1) - E_{ij}(1,0) - E_{ij}(0,1) + E_{ij}(0,0). \qquad (5.33)$$

In fact for each pair i, j, only one of the summands is nonzero.

5.2.3 Protonation Enthalpy Relative to the Uncharged State

In the literature, the enthalpy is often taken relative to a reference state (s_i^0) where all titrable residues are in their uncharged state. As is illustrated in Fig. 5.7, the formal dimensionless charge of a residue is given by

$$q_i = s_i - s_i^0.$$

The enthalpy change relative to the uncharged reference state is

$$\Delta G(\mathbf{s}) - \Delta G(\mathbf{s}^0)$$

$$= \sum_{i=1}^{N} (s_i - s_i^0) \Delta G_{i,\text{intr}} + \sum_{i=1}^{N} (s_i - s_i^0) \sum_{j \neq i, j=1}^{N} (E_{i,j}^{1,0} - E_{i,j}^{0,0})$$

$$+ \frac{1}{2} \sum_i \sum_{j \neq i} W_{i,j} \left((s_i - s_i^0)(s_j - s_j^0) + s_i^0 s_j + s_i s_j^0 - 2 s_i^0 s_j^0 \right). \qquad (5.34)$$

Note that $W_{ij} = W_{ji}$ and therefore the last sum can be simplified

$$\frac{1}{2} \sum_i \sum_{j \neq i} W_{i,j} \left((s_i - s_i^0)(s_j - s_j^0) + s_i^0 s_j + s_i s_j^0 - 2 s_i^0 s_j^0 \right)$$

$$= \frac{1}{2} \sum_i \sum_{j \neq i} W_{i,j} (q_i q_j + 2(s_i - s_i^0) s_j^0). \qquad (5.35)$$

Consider now the expression

$$\Delta W = \sum_{i=1}^{N} q_i \sum_{j\neq i, j=1}^{N} (E_{i,j}^{1,0} - E_{i,j}^{0,0} + s_j^0 W_{i,j}).$$

(5.36)

For each residue $j \neq i$, there are only two alternatives[7]

$$s_j^0 = 0 \rightarrow E_{i,j}^{1,0} - E_{i,j}^{0,0} = 0,$$

(5.37)

$$s_j^0 = 1 \rightarrow E_{i,j}^{1,0} - E_{i,j}^{0,0} + W_{i,j} = E_{i,j}^{1,1} - E_{i,j}^{0,1} = 0,$$

(5.38)

and hence we have

$$\Delta W = 0,$$

(5.39)

$$\Delta G(\mathbf{s}) - \Delta G(\mathbf{s}^0) = \sum_{\substack{s_i \neq s_i^0}}^{N} q_i \Delta G_{i,\text{intr}} + \frac{1}{2} \sum_{i\neq j} W_{i,j} q_i q_j.$$

(5.40)

5.2.4 Statistical Mechanics of Protonation

The partition function for a specific total charge

$$Q = \sum q_i = \sum (s_i - s_i^0)$$

(5.41)

is given by

$$Z(Q) = \sum_{\{\mathbf{s}, \sum q_i = Q\}} \sum_{\gamma} e^{-\Delta G/k_B T}$$

$$= \sum_{\{\mathbf{s}, \sum q_i = Q\}} \sum_{\gamma} e^{-\left[\sum (s_i - s_i^0)\Delta G_{i,\text{intr}} + \frac{1}{2}\sum' W_{i,j} q_i q_j\right]/k_B T}.$$

(5.42)

We come to the grand canonical partition function by introducing the factor

$$e^{\mu Q/k_B T} = e^{\mu \sum (s_i - s_i^0)/k_B T}$$

(5.43)

and summing over all possible charge states of the protein

$$\Xi = \sum_{Q} Z(Q) e^{\mu Q/k_B T}$$

$$= \sum_{\gamma, \mathbf{s}} e^{-\left[\sum (s_i - s_i^0)(\Delta G_{i,\text{intr}}(\gamma) - \mu) + \frac{1}{2}\sum' W_{i,j}(\gamma) q_i q_j\right]/k_B T}.$$

(5.44)

---

[7] The Coulombic interaction vanishes if one of the residues is in the neutral state.

With the approximation (5.27) and (5.3), the partition function

$$Z(Q) \approx \sum_{\{\mathbf{s}, \sum q_i = Q\}} \left(\frac{z_1^{(i)}}{z_0^{(i)}} \right)^{q_i} Z_{\mathrm{conf}} \tag{5.45}$$

becomes the product of one factor which relates to the internal degrees of freedom which are usually assumed to be configuration-independent[8] and a second factor which depends only on the configurational degrees of freedom:

$$Z_{\mathrm{conf}}(\mathbf{s}) = \sum_{\gamma} e^{-\left[\sum (s_i - s_i^0)(\Delta E_{i,\mathrm{bg}} + \Delta E_{i,\mathrm{Born}}) + \frac{1}{2} \sum' W_{i,j} q_i q_j \right] / k_{\mathrm{B}} T}. \tag{5.46}$$

5.3 Abnormal Titration Curves of Coupled Residues

Let us consider a simple example of a model protein with only two titrable sites of the same type. The free enthalpies of the four possible states are (Fig. 5.8)

$$\Delta G(\mathrm{AH}, \mathrm{AH}) = \Delta G_{1,\mathrm{intr}} + \Delta G_{2,\mathrm{intr}} + E_{1,2}^{1,1} - E_{1,2}^{0,0},$$

$$\Delta G(\mathrm{A}^-, \mathrm{AH}) - \Delta G(\mathrm{AH}, \mathrm{AH}) = -\Delta G_{1,\mathrm{intr}},$$

$$\Delta G(\mathrm{AH}, \mathrm{A}^-) - \Delta G(\mathrm{AH}, \mathrm{AH}) = -\Delta G_{2,\mathrm{intr}},$$

$$\Delta G(\mathrm{A}^-, \mathrm{A}^-) - \Delta G(\mathrm{AH}, \mathrm{AH}) = -\Delta G_{1,\mathrm{intr}} - \Delta G_{2,\mathrm{intr}} + W_{12}$$

$$= -\Delta G_{2,\mathrm{intr}} - \Delta G_{1,\mathrm{intr}} + E_{1,2}^{0,0}. \tag{5.47}$$

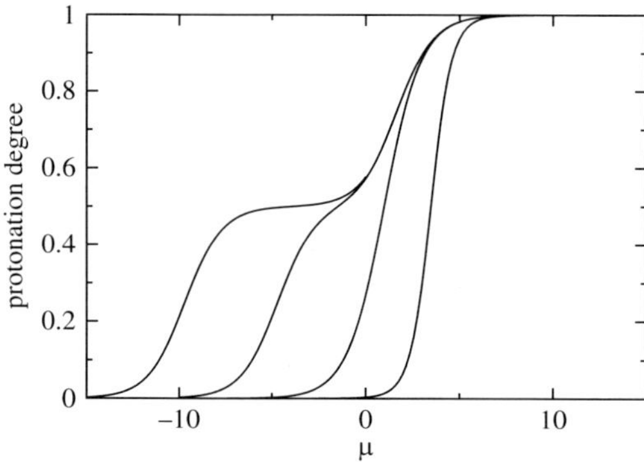

Fig. 5.8. Abnormal titration curves. Two interacting residues with neutral protonated states (AH), $\Delta G_{1,\mathrm{intr}} = \Delta G_{2,\mathrm{intr}} = 1.0$, $W = -5, 0, 5, 10$

---

[8] Protonation-dependent degrees of freedom can be important in certain cases [30].

The grand partition function is

$$\Xi = 1 + e^{-(-\Delta G_{1,\text{intr}}+\mu)/k_{\mathrm{B}}T} + e^{-(-\Delta G_{2,\text{intr}}+\mu)/k_{\mathrm{B}}T}$$
$$+ e^{-(-\Delta G_{2,\text{intr}}-\Delta G_{1,\text{intr}}+2\mu+W)/k_{\mathrm{B}}T}. \tag{5.48}$$

The average protonation values are

$$\overline{s_1} = \frac{1 + e^{-(-\Delta G_{2,\text{intr}}+\mu)/k_{\mathrm{B}}T}}{\Xi},$$
$$\overline{s_2} = \frac{1 + e^{-(-\Delta G_{1,\text{intr}}+\mu)/k_{\mathrm{B}}T}}{\Xi}. \tag{5.49}$$

Problems

5.1. Abnormal Titration Curves

Consider a simple example of a model protein with only two titrable sites of the same type. Determine the relative free enthalpies of the four possible states.

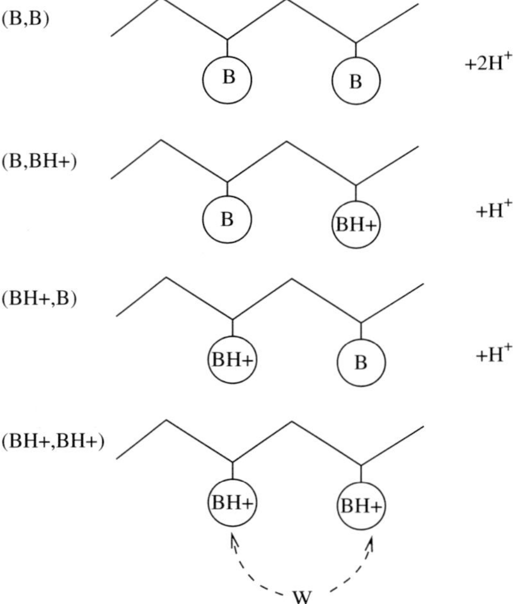

From the grand canonical partition function (the number of protons is not fixed), calculate the protonation degree for both sites and discuss them as a function of the interaction energy W.

Part III

Reaction Kinetics

6

Formal Kinetics

In this chapter, we discuss the phenomenological description of elementary chemical reactions and photophysical processes with the help of rate equations.

6.1 Elementary Chemical Reactions

The basic steps of chemical reactions can be divided into several classes of elementary reactions. They can be photoinduced or thermally activated, may involve the transfer of an electron or proton, and are accompanied by structural changes, like breaking and forming bonds or at least a reorganization of bond lengths and angles (Figs. 6.1 and 6.2).

All elementary reactions are reversible. There is a dynamical equilibrium between forward and back reaction, which are independent, for instance

$$H_2 + J_2 \rightleftharpoons 2HJ. \tag{6.1}$$

6.2 Reaction Variable and Reaction Rate

We consider a general stoichiometric equation for the reaction of several species[1]

$$\sum_i \nu_i A_i = 0 \tag{6.2}$$

and define a reaction variable x based on the concentration of the species A_i by

$$c_i = c_{i,0} + \nu_i x \tag{6.3}$$

---

[1] The stoichiometric coefficients ν_i are positive for products and negative for educts. This is the conventional definition. Products and educts can be exchanged at least in principle, since the back reaction is always possible.

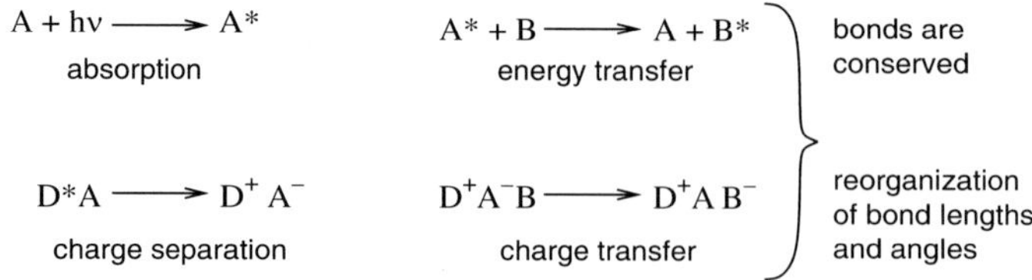

Fig. 6.1. Elementary reactions without bond reformation

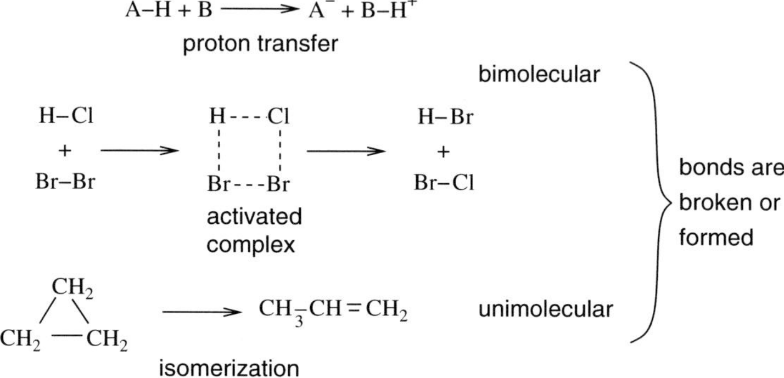

Fig. 6.2. Elementary reactions with bond reformation

as

$$x = \frac{c_i - c_{i,0}}{\nu_i} \tag{6.4}$$

and the reaction rate as its time derivative

$$r = \frac{\mathrm{d}x}{\mathrm{d}t} = \frac{1}{\nu_i}\frac{\mathrm{d}c_i}{\mathrm{d}t}. \tag{6.5}$$

6.3 Reaction Order

The progress of a chemical reaction can be frequently described by a simple
rate expression such as

$$r = k c_1^{n_1} c_2^{n_2} \cdots = k \prod_{i \in \text{educts}} c_i^{n_i} \tag{6.6}$$

with the rate constant k. For such a system, the exponent[2] of the ith term is
called the order of the reaction with respect to this substance and the sum of
all the exponents is called the overall reaction order.

---

[2] For more complicated reactions, the exponents need not be integers. For simple
reactions, they are given by the stoichiometric coefficients $n_i = |\nu_i|$ of the educts
(the products for the back reaction).

6.3.1 Zero-Order Reactions

Zero-order reactions proceed at the same rate regardless of concentration. The rate expression for a reaction of this type is

$$\frac{\mathrm{d}c}{\mathrm{d}t} = k_0 \tag{6.7}$$

which can be integrated

$$c = c_0 + k_0 t. \tag{6.8}$$

Zero-order reactions appear when the determining factor is an outside source of energy (light) or when the reaction occurs on the surface of a catalyst.

6.3.2 First-Order Reactions

First-order reactions describe the decay of an excited state, for instance a radioactive decay

$$A^* \rightarrow A. \tag{6.9}$$

The rate expression is

$$\frac{\mathrm{d}c_{A^*}}{\mathrm{d}t} = -\frac{\mathrm{d}c_A}{\mathrm{d}t} = -kc_{A^*}, \tag{6.10}$$

which gives an exponential decay

$$c_{A^*} = c_{A^*}(0)e^{-kt} \tag{6.11}$$

with a constant half-period

$$\tau_{1/2} = \frac{\ln(2)}{k}. \tag{6.12}$$

6.3.3 Second-Order Reactions

A second-order reaction between two different substances

$$A+B \rightarrow \cdots \tag{6.13}$$

obeys the equations

$$\frac{\mathrm{d}c_A}{\mathrm{d}t} = \frac{\mathrm{d}c_B}{\mathrm{d}t} = -k_2 c_A c_B, \tag{6.14}$$

which can be written down using the reaction variable x and the initial concentrations a, b as

$$c_A = a - x, \quad c_B = b - x, \tag{6.15}$$

$$\frac{\mathrm{d}x}{\mathrm{d}t} = k_2(a - x)(b - x). \tag{6.16}$$

This can be integrated to give

$$\frac{1}{a-b} \ln \frac{b(a-x)}{a(b-x)} = k_2 t. \tag{6.17}$$

If two molecules of the same type react with each other, we have instead

$$-\frac{\mathrm{d}c_A}{\mathrm{d}t} = -k_2 c_A^2 \tag{6.18}$$

which gives an algebraic decay

$$c_A(t) = \frac{1}{k_2 t + \frac{1}{a}}, \tag{6.19}$$

where the half-period now depends on the initial concentration

$$\tau_{1/2} = \frac{1}{k_2 a}. \tag{6.20}$$

An example is exciton–exciton annihilation in the light harvesting complex of photosynthesis:

$$A^* + A^* \rightarrow A + A. \tag{6.21}$$

6.4 Dynamical Equilibrium

We consider a first-order reaction together with the back reaction

$$A \; \underset{k_{-1}}{\overset{k_1}{\underset{\longleftarrow}{\longrightarrow}}} \; B. \tag{6.22}$$

The reaction variable of the back reaction will be denoted by y. The concentrations are

$$c_A(t) = a - x + y, \tag{6.23}$$
$$c_B(t) = b + x - y, \tag{6.24}$$

and the reaction rates are

$$\frac{\mathrm{d}x}{\mathrm{d}t} = k_1 c_A = k_1(a - x + y), \tag{6.25}$$

$$\frac{\mathrm{d}y}{\mathrm{d}t} = k_{-1} c_B = k_{-1}(b + x - y). \tag{6.26}$$

Introducing an overall reaction variable

$$z = x - y \tag{6.27}$$

and the equilibrium value

$$s = \frac{k_1 a - k_{-1} b}{k_1 + k_{-1}},$$ (6.28)

we have

$$\frac{\mathrm{d}z}{\mathrm{d}t} = k_1 a - k_{-1} b - (k_1 - k_{-1})z = (k_1 + k_{-1})(s - z)$$ (6.29)

which for $z(0) = 0$ has the solution

$$z = s(1 - \mathrm{e}^{-(k_1 + k_{-1})t}).$$ (6.30)

The reaction approaches the equilibrium with a rate constant $k_1 + k_{-1}$. In equilibrium, $z = s$ and $\mathrm{d}z/\mathrm{d}t = 0$. The equilibrium concentrations are

$$c_A = a - s = (a + b)\frac{k_{-1}}{k_1 + k_{-1}},$$ (6.31)

$$c_B = b + s = (a + b)\frac{k_1}{k_1 + k_{-1}},$$ (6.32)

and the equilibrium constant is

$$K = \frac{c_A}{c_B} = \frac{k_{-1}}{k_1}.$$ (6.33)

6.5 Competing Reactions

If one species decays via separate independent channels (fluorescence, electron transfer, radiationless transitions, etc.), the rates are additive:

$$\frac{\mathrm{d}c_A}{\mathrm{d}t} = -(k_1 + k_2 + \cdots)c_A.$$ (6.34)

6.6 Consecutive Reactions

We consider a chain consisting of two first-order reactions[3]

$$A \xrightarrow{k_1} B \xrightarrow{k_2} C.$$ (6.35)

The reaction variables are denoted by x, y and the initial concentrations by a, b, c. The concentrations are

$$c_A = a - x,$$ (6.36)

---

[3] With negligible back reactions

$$c_B = b + x - y, \tag{6.37}$$

$$c_C = c + y, \tag{6.38}$$

and their time derivatives are

$$\frac{dc_A}{dt} = -\frac{dx}{dt} = -k_1 c_A = -k_1(a - x), \tag{6.39}$$

$$\frac{dc_B}{dt} = \frac{dx}{dt} - \frac{dy}{dt} = k_1 c_A - k_2 c_B = k_1 a - k_2 b + (k_2 - k_1)x - k_2 y, \tag{6.40}$$

$$\frac{dc_C}{dt} = \frac{dy}{dt} = k_2 c_B = k_2(b + x - y). \tag{6.41}$$

The first equation gives an exponential decay:

$$c_A = a e^{-k_1 t}. \tag{6.42}$$

Integration of

$$\frac{dc_B}{dt} + k_2 c_B = k_1 a e^{-k_1 t} \tag{6.43}$$

gives the concentration of the intermediate state:

$$c_B = \frac{k_1 a}{k_2 - k_1} e^{-k_1 t} + \left(b - \frac{k_1 a}{k_2 - k_1} \right) e^{-k_2 t}. \tag{6.44}$$

If at time zero only the species A is present, the concentration of B has a maximum at

$$t_{\max} = \frac{1}{k_1 - k_2} \ln \frac{k_1}{k_2} \tag{6.45}$$

with the value

$$c_{B,\max} = a \left(\frac{k_1}{k_1 - k_2} \right) \left(\exp \left(\frac{k_2}{k_1 - k_2} \ln \frac{k_2}{k_1} \right) - \exp \left(\frac{k_1}{k_1 - k_2} \ln \frac{k_2}{k_1} \right) \right). \tag{6.46}$$

6.7 Enzymatic Catalysis

Enzymatic catalysis is very important for biochemical reactions. It can be described schematically by formation of an enzyme–substrate complex followed by decomposition into enzyme and product:

$$E + S \underset{k_{-1}}{\overset{k_1}{\rightleftharpoons}} ES \rightleftharpoons E + P. \tag{6.47}$$

We consider the limiting case of negligible $k_{-2} \ll k_2$ and large concentration of substrate $c_S \gg c_E$. Then we have to solve the equations

$$\frac{dc_S}{dt} \approx -k_1 c_E c_S^0 + k_{-1} c_{ES},$$

$$\frac{dc_E}{dt} \approx -k_1 c_E c_S^0 + (k_{-1} + k_2)c_{ES},$$

$$\frac{dc_{ES}}{dt} \approx k_1 c_E c_S^0 - (k_{-1} + k_2)c_{ES},$$

$$\frac{dc_P}{dt} = k_2 c_{ES}. \tag{6.48}$$

First, we solve the equations for dc_E/dt and dc_{ES}/dt:

$$\frac{d}{dt}\begin{pmatrix} c_{ES} \\ c_E \end{pmatrix} = \begin{pmatrix} -k_{-1} - k_2 & k_1 c_S^0 \\ k_{-1} + k_2 & -k_1 c_S^0 \end{pmatrix}\begin{pmatrix} c_{ES} \\ c_E \end{pmatrix}. \tag{6.49}$$

The matrix has one eigenvalue $\lambda = 0$ corresponding to a stationary solution:

$$\begin{pmatrix} -k_{-1} - k_2 & k_1 c_S^0 \\ k_{-1} + k_2 & -k_1 c_S^0 \end{pmatrix}\begin{pmatrix} \frac{k_1 c_S^0}{k_{-1} + k_2} \\ 1 \end{pmatrix} = \begin{pmatrix} 0 \\ 0 \end{pmatrix}. \tag{6.50}$$

The stationary concentration of the ES complex is

$$c_{ES}^{\text{stat}} = \frac{k_1}{k_{-1} + k_2} c_S c_E = \frac{c_S c_E}{K_M} \tag{6.51}$$

with the Michaelis constant

$$K_M = \frac{k_{-1} + k_2}{k_1}. \tag{6.52}$$

The second eigenvalue relates to the time constant for reaching the stationary state:

$$\begin{pmatrix} -k_{-1} - k_2 & k_1 c_S^0 \\ k_{-1} + k_2 & -k_1 c_S^0 \end{pmatrix}\begin{pmatrix} 1 \\ -1 \end{pmatrix} = -(k_1 c_S^0 + k_{-1} + k_2)\begin{pmatrix} 1 \\ -1 \end{pmatrix}. \tag{6.53}$$

For the initial conditions

$$c_{ES}(0) = c_P(0) = 0, \tag{6.54}$$

we find

$$c_{ES}(t) = \frac{c_E^0}{1 + \frac{K_M}{c_S^0}}(1 - e^{-(k_1 + k_{-1} + k_2)t}),$$

$$c_E(t) = \frac{c_E^0}{1 + \frac{c_S^0}{K_M}}\left(1 + \frac{c_S^0}{K_M}e^{-(k_1 + k_{-1} + k_2)t}\right). \tag{6.55}$$

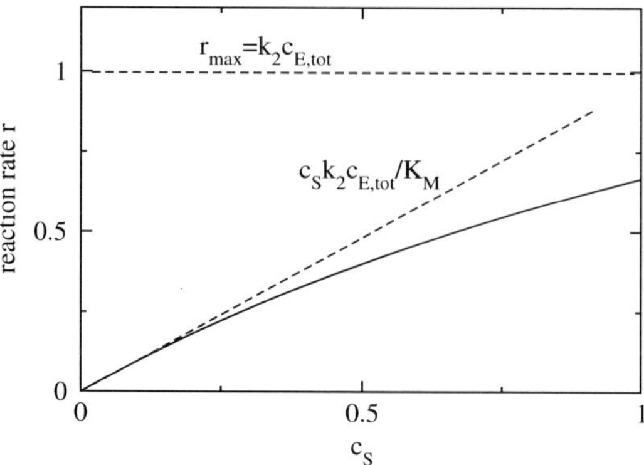

Fig. 6.3. Michaelis–Menten kinetics

The stationary state is stable, since any deviation will decrease exponentially. The overall rate of the enzyme-catalyzed reaction is given by the rate of product formation:

$$r = \frac{dc_P}{dt} = -\frac{dc_S}{dt} = k_2c_{ES} \tag{6.56}$$

and with the total concentration of enzyme

$$c_{E,tot} = c_E + c_{ES}, \tag{6.57}$$

we have

$$c_{ES} = \frac{c_E c_S}{K_M} = \frac{(c_{E,tot} - c_{ES})c_S}{K_M} \tag{6.58}$$

and hence

$$c_{ES} = \frac{c_{E,tot}c_S}{c_S + K_M}. \tag{6.59}$$

The overall reaction rate is given by the Michaelis–Menten equation (Fig. 6.3)

$$r = \frac{k_2c_{E,tot}c_S}{K_M + c_S}, \tag{6.60}$$

$$\frac{r}{r_{max}} = \frac{c_S}{c_S + K_M}, \quad r_{max} = k_2c_{E,tot}. \tag{6.61}$$

6.8 Reactions in Solutions

In solutions, the reacting molecules approach each other by diffusive motion, forming a reactive complex within a solvent cage which has a lifetime of typically $100\,\mathrm{ps}$ (Fig. 6.4).

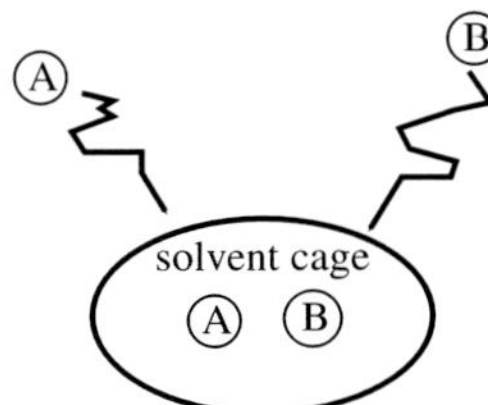

Fig. 6.4. Formation of a reactive complex

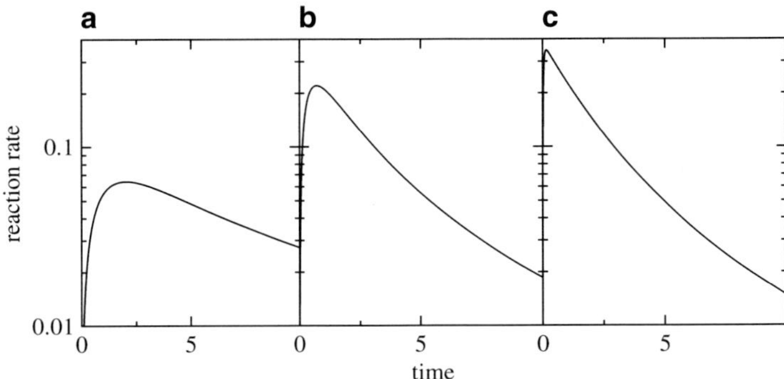

Fig. 6.5. Transition from the diffusion-controlled limit to the reaction-controlled limit. $r = k_2 c_{\{AB\}}$ is calculated numerically for $c_A(0) = c_B(0) = 1$, $k_2 = 1$: (**a**) $k_1 = k_{-1} = 0.1$, (**b**) $k_1 = k_{-1} = 1.0$, and (**c**) $k_1 = k_{-1} = 10$

Formally, this can be described by an equilibrium between the free reactants A and B and a reactive complex $\{AB\}$

$$\mathrm{A+B} \underset{k_{-1}}{\overset{k_1}{\rightleftharpoons}} \{\mathrm{A\,B}\} \overset{k_2}{\rightarrow} \mathrm{C}. \tag{6.62}$$

The concentrations obey the equations (Fig. 6.5)

$$\frac{dc_A}{dt} = \frac{dc_B}{dt} = -k_1 c_A c_B + k_{-1} c_{\{AB\}},$$

$$\frac{dc_C}{dt} = k_2 c_{\{AB\}},$$

$$\frac{dc_{\{AB\}}}{dt} = k_1 c_A c_B - (k_{-1} + k_2) c_{\{AB\}}. \tag{6.63}$$

Let us consider two limiting cases.

6.8.1 Diffusion-Controlled Limit

If the reaction rate k_2 is large compared to $k_{\pm 1}$, we find for the stationary solution approximately

$$k_2 c_{\{AB\}} \approx k_1 c_A c_B \tag{6.64}$$

and hence for the overall reaction rate

$$\frac{dc_C}{dt} = k_2 c_{\{AB\}} \approx k_1 c_A c_B. \tag{6.65}$$

The reaction rate is determined by the formation of the reactive complex.

6.8.2 Reaction-Controlled Limit

If, on the other hand $k_2 \ll k_{\pm 1}$, an equilibrium between reactants and reactive complex will be established

$$A + B \rightleftharpoons \{ A\,B \}, \quad \frac{c_{\{AB\}}}{c_A c_B} = K = \frac{k_1}{k_{-1}}. \tag{6.66}$$

Now, the overall reaction rate

$$\frac{dc_C}{dt} = k_2 c_{\{AB\}} = k_2 K c_A c_B \tag{6.67}$$

is determined by the reaction rate k_2 and the constant of the diffusion equilibrium.

Problems

6.1. pH Dependence of Enzyme Activity

Consider an enzymatic reaction where the substrate can be in two protonation states

$$S^- + H^+ \rightleftharpoons HS$$

and the enzyme reacts only with the deprotonated form

$$E + S^- \rightleftharpoons ES^- \rightarrow E + P.$$

Calculate the reaction rate as a function of the proton concentration c_{H^+}

6.2. Polymerization at the End of a Polymer

Consider the multiple equilibrium between the monomer M and the i-mer iM

$$c_{2M} = K c_M^2,$$

$$c_{3M} = K c_M c_{2M},$$

$$\vdots$$

$$c_{iM} = K c_M c_{(i-1)M},$$

where the equilibrium constant K is assumed to be independent of the degree of polymerization. Calculate the concentration of the i-mer c_{iM} and the mean degree of polymerization

$$\langle i \rangle = \frac{\sum_i i c_{iM}}{\sum_i c_{iM}}.$$

6.3. Primary Salt Effect

Consider the reaction of two ionic species A and B with charges $Z_{A,B}e$ which are in equilibrium with an activated complex X (with charge $(Z_A + Z_B)e$) which decays into the products C and D:

$$A+B \rightleftharpoons X \overset{k_1}{\longrightarrow} C + D.$$

The equilibrium constant is

$$K = \frac{a_X}{a_A a_B},$$

where the activities are given by the Debye–Hückel approximation

$$\mu_i = \mu_i^0 + k_B T \ln c_i - \frac{Z_i^2 e^2 \kappa}{8\pi\epsilon}, \qquad \kappa^2 = \frac{e^2}{\epsilon k_B T} \sum N_i Z_i^2.$$

Calculate the reaction rate

$$r = \frac{dc_C}{dt}.$$

Kinetic Theory: Fokker–Planck Equation

In this chapter, we consider a model system (protein) interacting with a surrounding medium which is only taken implicitly into account. We are interested in the dynamics on a time scale slower than the medium fluctuations. The interaction with the medium is described approximately as the sum of an average force and a stochastic force [31].

7.1 Stochastic Differential Equation for Brownian Motion

The simplest example describes one-dimensional Brownian motion of a big particle in a sea of small particles. The average interaction leads to damping of the motion which is described in terms of a velocity-dependent damping term:

$$\frac{\mathrm{d}v(t)}{\mathrm{d}t} = -\gamma v(t). \tag{7.1}$$

This equation alone leads to exponential relaxation $v = v(0)\mathrm{e}^{-\gamma t}$ which is not compatible with thermodynamics, since the average kinetic energy should be $(m/2)v^2 = k_B T/2$ in equilibrium. Therefore, we add a randomly fluctuating force which represents the collisions with many solvent molecules during a finite time interval τ. The result is the Langevin equation

$$\frac{\mathrm{d}v(t)}{\mathrm{d}t} = -\gamma v(t) + F(t) \tag{7.2}$$

with the formal solution

$$v(t) = v_0 \mathrm{e}^{-\gamma t} + \int_0^t \mathrm{e}^{\gamma(t'-t)} F(t')\mathrm{d}t'. \tag{7.3}$$

The average of the stochastic force has to be zero because the equation of motion for the average velocity should be

$$\frac{\mathrm{d}\langle v(t)\rangle}{\mathrm{d}t} = -\gamma\langle v(t)\rangle. \tag{7.4}$$

We assume that many collisions occur during τ and therefore forces at different times are not correlated:

$$\langle F(t)F(t')\rangle = C\delta(t-t'). \tag{7.5}$$

The velocity correlation function is

$$\langle v(t)v(t')\rangle = e^{-\gamma(t+t')}\left(v_0^2 + \int_0^t \mathrm{d}t_1 \int_0^{t'} \mathrm{d}t_2 e^{\gamma(t_1+t_2)}\langle F(t_1)F(t_2)\rangle\right). \tag{7.6}$$

Without losing generality, we assume $t' > t$ and substitute $t_2 = t_1 + s$ to find

$$\langle v(t)v(t')\rangle = v_0^2 e^{-\gamma(t+t')}$$
$$+ e^{-\gamma(t+t')}\int_0^t \mathrm{d}t_1 \int_{-t_1}^{t'-t_1} \mathrm{d}s\, e^{\gamma(2t_1+s)}\langle F(t_1)F(t_1+s)\rangle$$
$$= v_0^2 e^{-\gamma(t+t')} + e^{-\gamma(t+t')}\int_0^t \mathrm{d}t_1 e^{2\gamma t_1} C$$
$$= v_0^2 e^{-\gamma(t+t')} + e^{-\gamma(t+t')}\frac{e^{2\gamma t}-1}{2\gamma}C. \tag{7.7}$$

The exponential terms vanish very quickly and we find

$$\langle v(t)v(t')\rangle \to e^{-\gamma|t'-t|}\frac{C}{2\gamma}. \tag{7.8}$$

Now, C can be determined from the average kinetic energy as

$$\frac{m\langle v^2\rangle}{2} = \frac{k_\mathrm{B}T}{2} = \frac{m}{2}\frac{C}{2\gamma} \to C = \frac{2\gamma k_\mathrm{B}T}{m}. \tag{7.9}$$

The mean square displacement of a particle starting at x_0 with velocity v_0 is

$$\langle (x(t)-x(0))^2\rangle = \left\langle\left(\int_0^t \mathrm{d}t_1 v(t_1)\right)^2\right\rangle = \int_0^t \int_0^t \langle v(t_1)v(t_2)\rangle \mathrm{d}t_1\mathrm{d}t_2$$
$$= \int_0^t \int_0^t \left(v_0^2 e^{-\gamma(t_1+t_2)} + \frac{k_\mathrm{B}T}{m}e^{-\gamma|t_1-t_2|}\right)\mathrm{d}t \tag{7.10}$$

and since

$$\int_0^t \int_0^t e^{-\gamma(t_1+t_2)}\mathrm{d}t_1\mathrm{d}t_2 = \left(\frac{1-e^{-\gamma t}}{\gamma}\right)^2 \tag{7.11}$$

and

$$\int_0^t \int_0^t e^{-\gamma|t_1-t_2|} dt_1 dt_2 = 2 \int_0^t dt_1 \int_0^{t_1} e^{-\gamma(t_1-t_2)} dt_2$$

$$= \frac{2}{\gamma}t - \frac{2}{\gamma^2}(1 - e^{-\gamma t}), \tag{7.12}$$

we obtain

$$\langle (x(t) - x(0))^2 \rangle = \left(v_0^2 - \frac{k_B T}{m} \right) \frac{(1 - e^{-\gamma t})^2}{\gamma^2} + \frac{2k_B T}{m\gamma}t - \frac{2k_B T}{m\gamma^2}(1 - e^{-\gamma t}). \tag{7.13}$$

If we had started with an initial velocity distribution for the stationary state

$$\langle v_0^2 \rangle = k_B T/m, \tag{7.14}$$

then the first term in (7.12) would vanish. For very large times, the leading term is[1]

$$\langle (x(t) - x(0))^2 \rangle = 2Dt \tag{7.15}$$

with the diffusion coefficient

$$D = \frac{k_B T}{m\gamma}. \tag{7.16}$$

7.2 Probability Distribution

Now, we discuss the probability distribution $W(v)$. The time evolution can be described as

$$W(v, t + \tau) = \int P(v, t + \tau|v', t)W(v', t)dv'. \tag{7.17}$$

To derive an expression for the differential $\partial W(v, t)/\partial t$, we need the transition probability $P(v, t + \tau|v', t)$ for small τ. Introducing $\Delta = v - v'$, we expand the integrand in a Taylor series

$$P(v, t + \tau|v', t)W(v', t) = P(v, t + \tau|v - \Delta, t)W(v - \Delta, t)$$

$$= \sum_{n=0}^{\infty} \frac{(-1)^n}{n!} \Delta^n \left(\frac{\partial}{\partial v} \right)^n$$

$$\times P(v + \Delta, t + \tau|v, t)W(v, t). \tag{7.18}$$

Inserting this into the integral gives

$$W(v, t + \tau) = \sum_{n=0}^{\infty} \frac{(-1)^n}{n!} \left(\frac{\partial}{\partial v} \right)^n \left(\int \Delta^n P(v + \Delta, t + \tau|v, t)d\Delta \right) W(v, t) \tag{7.19}$$

---

[1] This is the well-known Einstein result for the diffusion constant D.

and assuming that the moments exist which are defined by

$$M_n(v', t, \tau) = \langle (v(t+\tau) - v(t))^n \rangle |_{v(t)=v'}$$

$$= \int (v - v')^n P(v, t+\tau | v', t)\,dv, \qquad (7.20)$$

we find

$$W(v, t+\tau) = \sum_{n=0}^{\infty} \frac{(-1)^n}{n!} \left(\frac{\partial}{\partial v} \right)^n M_n(v, t, \tau) W(v, t). \qquad (7.21)$$

Expanding the moments into a Taylor series

$$\frac{1}{n!} M_n(v, t, \tau) = \frac{1}{n!} M_n(v, t, 0) + D^{(n)}(v, t)\tau + \cdots, \qquad (7.22)$$

we have finally[2]

$$W(v, t+\tau) - W(v, t) = \sum_{1}^{\infty} \left(-\frac{\partial}{\partial v} \right)^n D^{(n)}(v, t) W(v, t)\tau + \cdots, \qquad (7.23)$$

which gives the equation of motion for the probability distribution[3]

$$\frac{\partial W(v, t)}{\partial t} = \sum_{1}^{\infty} \left(-\frac{\partial}{\partial v} \right)^n D^{(n)}(v, t) W(v, t). \qquad (7.24)$$

If this expansion stops after the second term,[4] the general form of the one-dimensional Fokker–Planck equation results

$$\frac{\partial W(v, t)}{\partial t} = \left(-\frac{\partial}{\partial v} D^{(1)}(v, t) + \frac{\partial^2}{\partial x^2} D^{(2)}(v, t) \right) W(v, t) \qquad (7.25)$$

7.3 Diffusion

Consider a particle performing a random walk in one dimension due to collisions. We use the stochastic differential equation[5]

$$\frac{dx}{dt} = v_0 + f(t), \qquad (7.26)$$

where the velocity has a drift component v_0 and a fluctuating part $f(t)$ with

$$\langle f(t) \rangle = 0, \quad \langle f(t) f(t') \rangle = q\delta(t - t'). \qquad (7.27)$$

---

[2] The zero-order moment does not depend on τ.
[3] This is known as the Kramers–Moyal expansion.
[4] It can be shown that this is the case for all Markov processes.
[5] This is a so-called Wiener process.

The formal solution is simply

$$x(t) - x(0) = v_0 t + \int_0^t f(t')\mathrm{d}t'. \tag{7.28}$$

The first moment

$$M_1(x_0, t, \tau) = \langle x(t+\tau) - x(t)\rangle|_{x(t)=x_0} = v_0\tau + \int_0^\tau \langle f(t')\rangle\mathrm{d}t' \tag{7.29}$$

gives

$$D^{(1)} = v_0. \tag{7.30}$$

The second moment is

$$M_2(x_0, t, \tau)$$
$$= v_0^2\tau^2 + v_0\tau\int_0^\tau \langle f(t')\rangle\mathrm{d}t' + \int_0^\tau\int_0^\tau \langle f(t_1)f(t_2)\rangle\mathrm{d}t_1\mathrm{d}t_2. \tag{7.31}$$

The second term vanishes and the only linear term in τ comes from the double integral

$$\int_0^\tau\int_0^\tau \langle f(t_1)f(t_2)\rangle\mathrm{d}t_1\mathrm{d}t_2 = \int_0^\tau \mathrm{d}t_1 \int_{-t_1}^{\tau-t_1} q\delta(t')\mathrm{d}t' = q\tau. \tag{7.32}$$

Hence

$$D^{(2)} = \frac{q}{2} \tag{7.33}$$

and the corresponding Fokker–Planck equation is the diffusion equation

$$\frac{\partial W(x,t)}{\partial t} = -v_0\frac{\partial W(x,t)}{\partial x} + D\frac{\partial^2 W(x,t)}{\partial x^2} \tag{7.34}$$

with the diffusion constant $D = D^{(2)}$.

7.3.1 Sharp Initial Distribution

We can easily find the solution for a sharp initial distribution $W(x,0) = \delta(x - x_0)$ by taking the Fourier transform:

$$\widetilde{W}(k,t) = \int_{-\infty}^{\infty} \mathrm{d}x\, W(x,t)\, \mathrm{e}^{-\mathrm{i}kx}. \tag{7.35}$$

We obtain the algebraic equation

$$\frac{\partial \widetilde{W}(k,t)}{\partial t} = (-Dk^2 + \mathrm{i}v_0 k)\widetilde{W}(k,t) \tag{7.36}$$

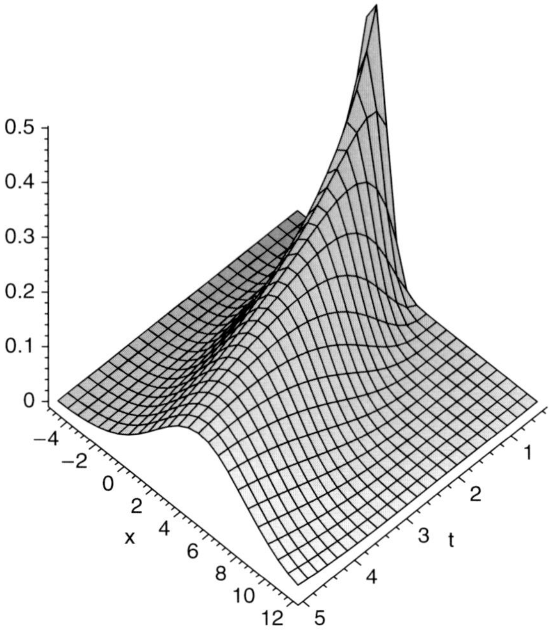

Fig. 7.1. Solution (7.38) of the diffusion equation for sharp initial conditions

which is solved by

$$\widetilde{W}(k,t) = \widetilde{W}_0 \exp\left\{(-Dk^2 + iv_0 k)t + ikx_0\right\}. \tag{7.37}$$

Inverse Fourier transformation then gives[6]

$$W(x,t) = \frac{1}{\sqrt{4\pi Dt}} \exp\left\{-\frac{(x - x_0 - v_0 t)^2}{4Dt}\right\}, \tag{7.38}$$

which is a Gaussian distribution centered at $x_c = x_0 + v_0 t$ with a variance of $\langle (x - x_c)^2 \rangle = 4Dt$ (Fig. 7.1).

7.3.2 Absorbing Boundary

Consider a particle from species A which can undergo a chemical reaction with a particle from species B at position $x_A = 0$

$$\text{A} + \text{B} \rightarrow \text{AB}. \tag{7.39}$$

If the reaction rate is very fast, then the concentration of A vanishes at $x = 0$ which gives an additional boundary condition

---

[6] with the proper normalization factor

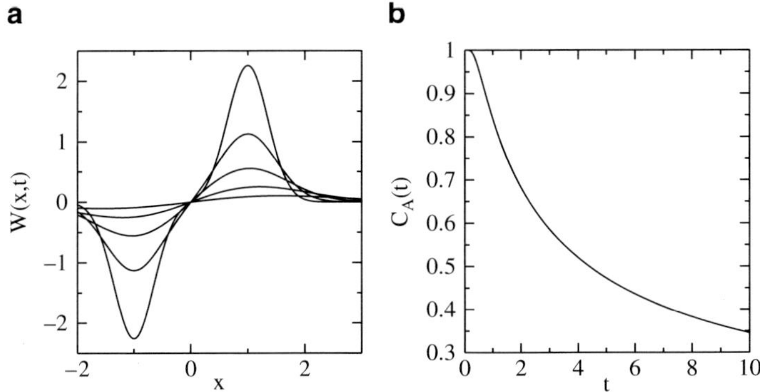

Fig. 7.2. Solution from the mirror principle. (**a**) The probability density distribution (7.42) at $t = 0.25, 0.5, 1, 2, 5$ and (**b**) the decay of the total concentration (7.43) are shown for $4D = 1$

$$W(x = 0, t) = 0. \tag{7.40}$$

Starting again with a localized particle at time zero with

$$W(x, 0) = \delta(x - x_0), \quad v_0 = 0, \tag{7.41}$$

the probability distribution

$$W(x, t) = \frac{1}{\sqrt{4\pi Dt}} \left(e^{-\frac{(x-x_0)^2}{4Dt}} - e^{-\frac{(x+x_0)^2}{4Dt}} \right) \tag{7.42}$$

is a solution which fulfills the boundary conditions. This solution is similar to the mirror principle known from electrostatics. The total concentration of species A in solution is then given by (Fig. 7.2)

$$C_A(t) = \int_0^\infty dx\, W(x, t) = \operatorname{erf}\left(\frac{x_0}{\sqrt{4Dt}} \right). \tag{7.43}$$

7.4 Fokker–Planck Equation for Brownian Motion

For Brownian motion, we have from the formal solution

$$v(\tau) = v_0(1 - \gamma\tau + \cdots) + \int_0^\tau (1 + \gamma(t_1 - \tau) + \cdots)F(t_1)dt_1. \tag{7.44}$$

The first moment[7]

$$M_1(v_0, t, \tau) = \langle v(\tau) - v(0) \rangle = -\gamma\tau v_0 + \cdots \tag{7.45}$$

---

[7] Here and in the following, we use $\langle F(t) \rangle = 0$.

gives

$$D^{(1)}(v,t) = -\gamma v. \tag{7.46}$$

The second moment follows from

$$\langle (v(\tau) - v_0)^2 \rangle$$
$$= (v_0 \gamma \tau)^2 + \int_0^\tau \int_0^\tau (1 + \gamma(t_1 + t_2 - 2\tau \cdots)) F(t_1) F(t_2) dt_1 dt_2. \tag{7.47}$$

The double integral gives

$$\int_0^\tau dt_1 \int_{-t_1}^{\tau - t_1} dt' (1 + \gamma(2t_1 + t' - 2\tau + \cdots)) \frac{2\gamma k_\mathrm{B} T}{m} \delta(t')$$
$$= \int_0^\tau dt_1 \frac{2\gamma k_\mathrm{B} T}{m} (1 + \gamma(2t_1 - 2\tau + \cdots))$$
$$= \frac{2\gamma k_\mathrm{B} T}{m} \tau + \cdots \tag{7.48}$$

and we have

$$D^{(2)} = \frac{\gamma k_\mathrm{B} T}{m}. \tag{7.49}$$

The higher moments have no contributions linear in τ and the resulting Fokker–Planck equation is

$$\frac{\partial W(v,t)}{\partial t} = \gamma \frac{\partial}{\partial v}(vW(v,t)) + \frac{\gamma k_\mathrm{B} T}{m} \frac{\partial^2}{\partial v^2} W(v,t). \tag{7.50}$$

7.5 Stationary Solution to the Focker–Planck Equation

The Fokker–Planck equation can be written in the form of a continuity equation:

$$\frac{\partial W(v,t)}{\partial t} = -\frac{\partial}{\partial v} S(v,t) \tag{7.51}$$

with the probability current

$$S(v,t) = -\frac{\gamma k T}{m} \left(\frac{mv}{k_\mathrm{B} T} W(v,t) + \frac{\partial}{\partial v} W(v,t) \right). \tag{7.52}$$

The probability current has to vanish for a stationary solution (with open boundaries $-\infty < v < \infty$)

$$\frac{\partial}{\partial v} W(v,t) = -\frac{mv}{k_\mathrm{B} T} W(v,t), \tag{7.53}$$

which has the Maxwell distribution as its solution

$$W_\mathrm{stat}(v,t) = \sqrt{\frac{m}{2\pi k_\mathrm{B} T}} e^{-mv^2/2k_\mathrm{B} T}. \tag{7.54}$$

Therefore, we conclude that the Fokker–Planck equation describes systems that reach thermal equilibrium, starting from a nonequilibrium distribution. In the following, we want to look at the relaxation process itself. We start with

$$\frac{\partial W(v,t)}{\partial t} = \gamma W(v,t) + \gamma v \frac{\partial W(v,t)}{\partial v} + D \frac{\partial^2 W(v,t)}{\partial v^2} \tag{7.55}$$

and introduce the new variables

$$\rho = v e^{\gamma t}, \quad y(\rho,t) = W(\rho e^{-\gamma t}, t) \tag{7.56}$$

which transform the differentials according to

$$\frac{\partial W}{\partial v} = \frac{\partial y}{\partial \rho}\frac{\partial \rho}{\partial v} = e^{\gamma t}\frac{\partial y}{\partial \rho},$$

$$\frac{\partial^2 W}{\partial v^2} = e^{2\gamma t}\frac{\partial^2 y}{\partial \rho^2},$$

$$\frac{\partial W}{\partial t} = \frac{\partial y}{\partial t} + \frac{\partial y}{\partial \rho}\frac{\partial \rho}{\partial t} = \frac{\partial y}{\partial t} + \gamma\rho\frac{\partial y}{\partial \rho}. \tag{7.57}$$

This leads to the new differential equation:

$$\frac{\partial y}{\partial t} = \gamma y + D e^{2\gamma t}\frac{\partial^2 y}{\partial \rho^2}. \tag{7.58}$$

To solve this equation, we introduce new variables again

$$y = \chi e^{\gamma t}, \tag{7.59}$$

which results in

$$\frac{\partial \chi}{\partial t} = D e^{2\gamma t}\frac{\partial^2 \chi}{\partial \rho^2}. \tag{7.60}$$

Now, we introduce a new time scale

$$\theta = \frac{1}{2\gamma}(e^{2\gamma t} - 1), \tag{7.61}$$

$$d\theta = e^{2\gamma t}dt, \tag{7.62}$$

satisfying the initial condition $\theta(t = 0) = 0$. Finally, we have to solve a diffusion equation

$$\frac{\partial \chi}{\partial \theta} = D\frac{\partial^2 \chi}{\partial \rho^2} \tag{7.63}$$

which gives (7.38)

$$\chi(\rho, \theta, \rho_0) = \frac{1}{\sqrt{4\pi D\theta}} \exp\left(-\frac{(\rho - \rho_0)^2}{4\pi D\theta}\right). \tag{7.64}$$

After back substitution of all variables, we find

$$W(v,t) = \sqrt{\frac{m}{2\pi k_{\mathrm{B}}T(1 - e^{-2\gamma t})}}\, \exp\left\{ -\frac{m}{2k_{\mathrm{B}}T}\frac{(v - v_0 e^{-\gamma t})^2}{(1 - e^{-2\gamma t})} \right\}. \tag{7.65}$$

This solution shows that the system behaves initially like

$$W(v,t) \approx \frac{1}{\sqrt{4\pi Dt}}\exp\left\{ -\frac{(v - v_0)^2}{4Dt} \right\} \tag{7.66}$$

and relaxes to the Maxwell distribution with a time constant $\Delta t = 1/2\gamma$.

7.6 Diffusion in an External Potential

We consider motion of a particle under the influence of an external (mean) force $K(x) = -(\mathrm{d}/\mathrm{d}x)U(x)$. The stochastic differential equations for position and velocity are

$$\dot{x} = v, \tag{7.67}$$

$$\dot{v} = -\gamma v + \frac{1}{m}K(x) + F(t). \tag{7.68}$$

We will calculate the moments for the Kramers–Moyal expansion. For small τ, we have

$$M_x = \langle x(\tau) - x(0)\rangle = \int_0^\tau v(t)\mathrm{d}t = v_0\tau + \cdots,$$

$$M_v = \langle v(\tau) - v(0)\rangle = \int_0^\tau \left(-\gamma v(t) + \frac{1}{m}K(x(t)) + \langle F(t)\rangle \right)\mathrm{d}t$$

$$= \left(-\gamma v_0 + \frac{1}{m}K(x_0) \right)\tau + \cdots,$$

$$M_{xx} = \langle (x(\tau) - x(0))^2\rangle = \int_0^\tau \int_0^\tau v(t_1)v(t_2)\mathrm{d}t_1\mathrm{d}t_2 = v_0^2\tau^2 + \cdots,$$

$$M_{vv} = \langle (v(\tau) - v(0))^2\rangle$$

$$= \left(-\gamma v_0 + \frac{1}{m}K(x_0) \right)^2\tau^2 + \int_0^\tau \int_0^\tau F(t_1)F(t_2)\mathrm{d}t_1\mathrm{d}t_2$$

$$= \frac{2\gamma k_{\mathrm{B}}T}{m}\tau + \cdots \tag{7.69}$$

The drift and diffusion coefficients are

$$D^{(x)} = v,$$

$$D^{(v)} = -\gamma v + \frac{1}{m}K(x),$$

$$D^{(xx)} = 0,$$

$$D^{(vv)} = \frac{\gamma k_{\mathrm{B}}T}{m}, \tag{7.70}$$

which leads to the Klein–Kramers equation

$$\frac{\partial W(x,v,t)}{\partial t} = \left[-\frac{\partial}{\partial x}D^{(x)} - \frac{\partial}{\partial v}D^{(v)} + \frac{\partial^2}{\partial v^2}D^{(vv)} \right] W(x,v,t),$$

$$\frac{\partial W(x,v,t)}{\partial t} = \left[-\frac{\partial}{\partial x}v + \frac{\partial}{\partial v}\left(\gamma v - \frac{K(x)}{m}\right) + \frac{\gamma k_B T}{m}\frac{\partial^2}{\partial v^2} \right] W(x,v,t).$$

$$(7.71)$$

This equation can be divided into a reversible and an irreversible part:

$$\frac{\partial W}{\partial t} = (\mathcal{L}_{\text{rev}} + \mathcal{L}_{\text{irrev}})W, \tag{7.72}$$

$$\mathcal{L}_{\text{rev}} = \left[-v\frac{\partial}{\partial x} + \frac{1}{m}\frac{\partial U}{\partial x}\frac{\partial}{\partial v} \right], \quad \mathcal{L}_{\text{irrev}} = \left[\frac{\partial}{\partial v}\gamma v + \frac{\gamma k_B T}{m}\frac{\partial^2}{\partial v^2} \right]. \tag{7.73}$$

The reversible part corresponds to the Liouville operator for a particle moving in the potential without friction

$$\mathcal{L} = \left[\frac{\partial \mathfrak{H}}{\partial x}\frac{\partial}{\partial p} - \frac{\partial \mathfrak{H}}{\partial p}\frac{\partial}{\partial x} \right], \quad \mathfrak{H} = \frac{p^2}{2m} + U(x). \tag{7.74}$$

Obviously

$$\mathcal{L}\mathfrak{H} = 0 \tag{7.75}$$

and

$$\mathcal{L}_{\text{irrev}} \exp\left\{ -\frac{\mathfrak{H}}{k_B T} \right\}$$

$$= \exp\left\{ -\frac{\mathfrak{H}}{k_B T} \right\} \left[\gamma - \gamma v\frac{mv}{k_B T} + \frac{\gamma k_B T}{m}\left(\left(\frac{mv}{k_B T}\right)^2 - \frac{m}{k_B T} \right) \right] = 0. \tag{7.76}$$

Therefore, the Klein–Kramers equation has the stationary solution

$$W_{\text{stat}}(x,v,t) = Z^{-1}e^{-(mv^2/2 + U(x))/k_B T} \tag{7.77}$$

with

$$Z = \int\int dv dx\, e^{-(mv^2/2 + U(x))/k_B T}. \tag{7.78}$$

The Klein–Kramers equation can be written in the form of a continuity equation:

$$\frac{\partial}{\partial t}W = -\frac{\partial}{\partial x}S_x - \frac{\partial}{\partial v}S_v \tag{7.79}$$

with the probability current

$$S_x = vW, \tag{7.80}$$

$$S_v = -\left[\gamma v + \frac{1}{m}\frac{\partial U}{\partial x} \right]W - \frac{\gamma k_B T}{m}\frac{\partial W}{\partial v}. \tag{7.81}$$

7.7 Large Friction Limit: Smoluchowski Equation

For large friction constant γ, we may neglect the second derivative with respect to time and obtain the stochastic differential equation

$$\dot{x} = \frac{1}{m\gamma} K(x) + \frac{1}{\gamma} F(t) \tag{7.82}$$

and the corresponding Fokker–Planck equation is the Smoluchowski equation:

$$\frac{\partial W(x,t)}{\partial t} = \left[-\frac{1}{m\gamma} \frac{\partial}{\partial x} K(x) + \frac{k_{\mathrm{B}}T}{m\gamma} \frac{\partial^2}{\partial x^2} \right] W(x,t), \tag{7.83}$$

which can be written with the mean force potential $U(x)$ as

$$\frac{\partial W(x,t)}{\partial t} = \frac{1}{m\gamma} \frac{\partial}{\partial x} \left[k_{\mathrm{B}}T \frac{\partial}{\partial x} + \frac{\partial U}{\partial x} \right] W(x,t). \tag{7.84}$$

7.8 Master Equation

The master equation is a very general linear equation for the probability density. If the variable x takes on only integer values, it has the form

$$\frac{\partial W_n}{\partial t} = \sum_m \left(w_{m \to n} W_m - w_{n \to m} W_n \right), \tag{7.85}$$

where W_n is the probability to find the integer value n and $w_{m \to n}$ is the transition probability. For continuous x, the summation has to be replaced by an integration

$$\frac{\partial W(x,t)}{\partial t} = \int \left(w_{x' \to x} W(x',t) - w_{x \to x'} W(x,t) \right) \mathrm{d}x'. \tag{7.86}$$

The Fokker–Planck equation is a special form of the master equation with

$$w_{x' \to x} = \left(-\frac{\partial}{\partial x} D^{(1)}(x) + \frac{\partial^2}{\partial x^2} D^{(2)}(x) \right) \delta(x - x'). \tag{7.87}$$

So far, we have discussed only Markov processes where the change of probability at time t only depends on the probability at time t. If memory effects are included, the generalized Master equation results.

Problems

7.1. Smoluchowski Equation

Consider a one-dimensional random walk. At times $t_n = n\Delta t$, a particle at position $x_j = j\Delta x$ jumps either to the left side $j - 1$ with probability w_j^- or to the right side $j+1$ with probability $w_j^{\shortmid} = 1 - w_j$. The probability to find a particle at site j at the time $n+1$ is then given by

$$P_{n+1,j} = w_{j-1}^{+} P_{n,j-1} + w_{j+1}^{-} P_{n,j+1}.$$

Show that in the limit of small $\Delta x, \Delta t$, the probability distribution $P(t,x)$ obeys a Smoluchowski equation

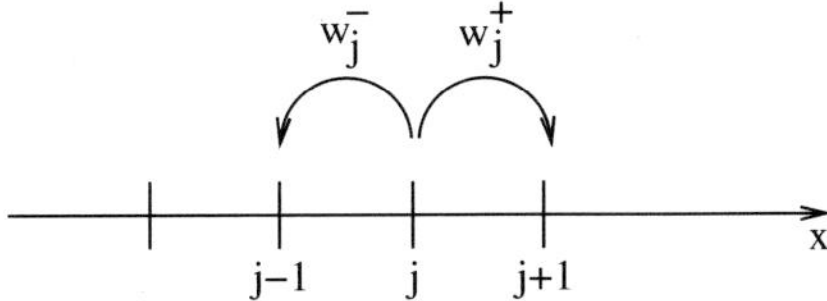

7.2. Eigenvalue Solution to the Smoluchowski Equation

Consider the one-dimensional Smoluchowski equation

$$\frac{\partial W(x,t)}{\partial t} = \frac{1}{m\gamma}\frac{\partial}{\partial x}\left[k_{\mathrm{B}}T\frac{\partial}{\partial x} + \frac{\partial U}{\partial x}\right]W(x,t) = -\frac{\partial}{\partial x}S(x)$$

for a harmonic potential

$$U(x) = \frac{m\omega^2}{2}x^2.$$

Show that the probability current can be written as

$$S(x,t) = -\frac{k_{\mathrm{B}}T}{m\gamma}\mathrm{e}^{-U(x)/kT}\left(\frac{\partial}{\partial x}\mathrm{e}^{U(x)/k_{\mathrm{B}}T}W(x,t)\right)$$

and that the Fokker–Planck operator can be written as

$$\mathfrak{L}_{\mathrm{FP}} = \frac{1}{m\gamma}\frac{\partial}{\partial x}\left[k_{\mathrm{B}}T\frac{\partial}{\partial x} + \frac{\partial U}{\partial x}\right] = \frac{k_{\mathrm{B}}T}{m\gamma}\frac{\partial}{\partial x}\mathrm{e}^{-U(x)/k_{\mathrm{B}}T}\frac{\partial}{\partial x}\mathrm{e}^{U(x)/k_{\mathrm{B}}T}$$

and can be transformed into a hermitian operator by

$$\mathfrak{L} = \mathrm{e}^{U(x)/2k_{\mathrm{B}}T}\mathfrak{L}_{\mathrm{FP}}\mathrm{e}^{-U(x)/2k_{\mathrm{B}}T}.$$

Solve the eigenvalue problem

$$\mathfrak{L}\psi_n(x) = \lambda_n\psi_n(x)$$

and use the function $\psi_0(x)$ to construct a special solution

$$W(x,t) = \mathrm{e}^{\lambda_0 t}\mathrm{e}^{-U(x)/2k_{\mathrm{B}}T}\psi_0(x).$$

7.3. Diffusion Through a Membrane

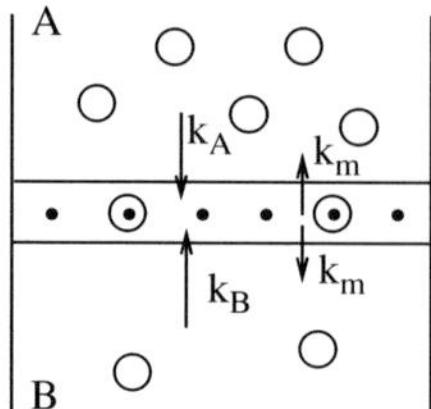

A membrane with M pore proteins separates two half-spaces A and B. An ion X may diffuse through M pore proteins in the membrane from A to B or vice versa. The rate constants for the formation of the ion-pore complex are k_A and k_B, respectively, while k_m is the constant for the decay of the ion–pore complex independent of the side to which the ion escapes. Let $P_N(t)$ denote the probability that there are N ion–pore complexes at time t. The master equation for the probability is

$$\frac{\mathrm{d}P_N(t)}{\mathrm{d}t} = -\left[(k_A + k_B)(M - N) + 2k_m N\right] P_N(t)$$
$$+ (k_A + k_B)(M - N + 1)P_{N-1}(t) + 2k_m(N + 1)P_{N+1}(t). \quad (7.88)$$

Calculate mean and variance of the number of complexes

$$\overline{N} = \sum_N N P_N,$$

$$\sigma^2 = \overline{N^2} - \overline{N}^2$$

and the mean diffusion current

$$J_{AB} = \frac{\mathrm{d}N_B}{\mathrm{d}t} - \frac{\mathrm{d}N_A}{\mathrm{d}t},$$

where $N_{A,B}$ is the number of ions in the upper of lower half-space.

Kramers' Theory

Kramers [32] used the concept of Brownian motion to describe motion of particles over a barrier as a model for chemical reactions in solution. The probability distribution of a particle moving in an external potential is described by the Klein–Kramers equation (7.71):

$$\frac{\partial W(x,v,t)}{\partial t} = \left[-\frac{\partial}{\partial x}v + \frac{\partial}{\partial v}\left(\gamma v - \frac{K(x)}{m}\right) + \frac{\gamma k_B T}{m}\frac{\partial^2}{\partial v^2} \right] W(x,v,t)$$

$$= -\frac{\partial}{\partial x}S_x - \frac{\partial}{\partial v}S_v \tag{8.1}$$

and the rate of the chemical reaction is related to the probability current S_x across the barrier.

8.1 Kramers' Model

A particle in a stable minimum A has to reach the transition state B by diffusive motion and then converts to the product C. The minimum well and the peak of the barrier are both approximated by parabolic functions (Fig. 8.1)

$$U_A = \frac{m}{2}\omega_a^2(x - x_0)^2, \tag{8.2}$$

$$U^* = E_a - \frac{m}{2}\omega^{*2}x^2. \tag{8.3}$$

Without the chemical reaction, the stationary solution is

$$W_{\text{stat}} = Z^{-1}\exp\left\{-\mathfrak{H}/k_B T\right\}. \tag{8.4}$$

We assume that the perturbation due to the reaction is small and make the following ansatz (Fig. 8.2)

$$W(x,v) = W_{\text{stat}}\, y(x,v), \tag{8.5}$$

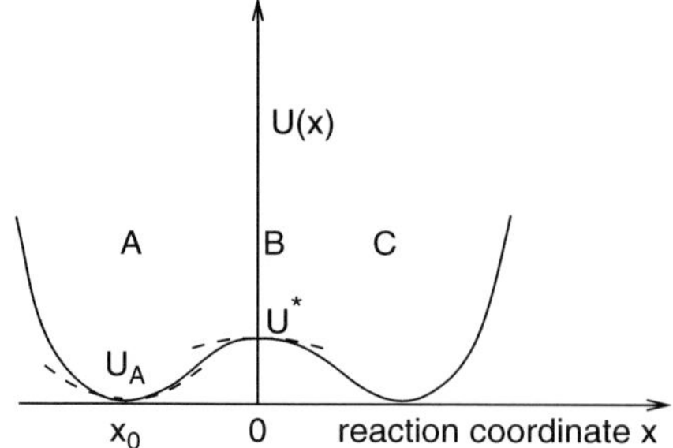

Fig. 8.1. Kramers' model

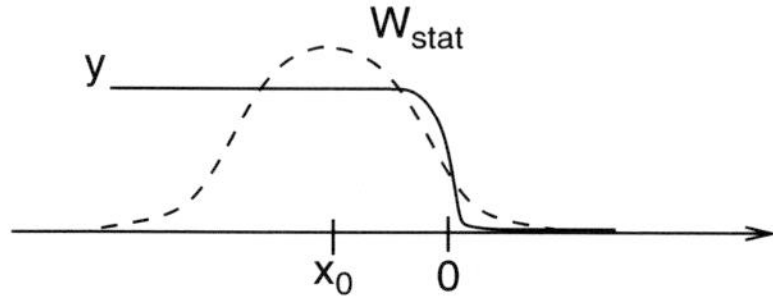

Fig. 8.2. Ansatz function. The stationary solution of a harmonic well is multiplied with the function $y(x,v)$ which switches from 1 to 0 at the saddle point

where the partition function is approximated by the harmonic oscillator expression

$$Z = \frac{2\pi k_{\mathrm{B}}T}{m\omega_{\mathrm{a}}}. \tag{8.6}$$

The probability distribution should fulfill two boundary conditions:

1. In the minimum of the A-well, the particles should be in thermal equilibrium and therefore

$$W(x,v) = Z^{-1}\exp\left\{-\frac{\mathfrak{H}}{k_{\mathrm{B}}T}\right\} \rightarrow y = 1 \quad \text{if} \quad x \approx x_0. \tag{8.7}$$

2. On the right side of the barrier $(x > 0)$, all particles A are converted into C and there will be no particles of the A species here

$$W(x,v) = 0 \rightarrow y = 0 \quad \text{if} \quad x > 0. \tag{8.8}$$

8.2 Kramers' Calculation of the Reaction Rate

Let us now insert the ansatz function into the Klein–Kramers equation. For the reversible part, we have

$$\mathcal{L}_{\mathrm{rev}}\exp\left\{-\frac{\mathfrak{H}}{k_{\mathrm{B}}T}\right\}y(x,v) = \exp\left\{-\frac{\mathfrak{H}}{k_{\mathrm{B}}T}\right\}\mathcal{L}_{\mathrm{rev}}y(x,v) \tag{8.9}$$

and for the irreversible part

$$\left[\frac{\partial}{\partial v}\gamma v + \frac{\gamma k_B T}{m}\frac{\partial^2}{\partial v^2}\right]\exp\left\{-\frac{\mathfrak{H}}{k_B T}\right\}y(x,v)$$
$$= \left\{\exp\left\{-\frac{\mathfrak{H}}{k_B T}\right\}\left[-\gamma v\frac{\partial}{\partial v} + \frac{\gamma k_B T}{m}\frac{\partial^2}{\partial v^2}\right]y\right\}. \tag{8.10}$$

Note the subtle difference. The operator $(\partial/\partial v)v$ is replaced by $-v(\partial/\partial v)$. Together, we have the following equation for y

$$0 = -v\frac{\partial y}{\partial x} + \frac{1}{m}\frac{\partial U}{\partial x}\frac{\partial y}{\partial v} - \gamma v\frac{\partial y}{\partial v} + \frac{\gamma k_B T}{m}\frac{\partial^2 y}{\partial v^2}, \tag{8.11}$$

which becomes in the vicinity of the top

$$0 = -v\frac{\partial y}{\partial x} - \omega^{*2}x\frac{\partial y}{\partial v} - \gamma v\frac{\partial y}{\partial v} + \frac{\gamma k_B T}{m}\frac{\partial^2 y}{\partial v^2}. \tag{8.12}$$

Now, we make a transformation of variables

$$(x,v) \rightarrow (\eta,\xi) = (x, v - \lambda x). \tag{8.13}$$

With

$$x = \eta, \quad v = \xi + \lambda\eta, \quad \frac{\partial}{\partial x} = \frac{\partial}{\partial \eta} - \lambda\frac{\partial}{\partial \xi}, \quad \frac{\partial}{\partial v} = \frac{\partial}{\partial \xi}, \tag{8.14}$$

(8.12) is transformed to

$$0 = (\xi + \lambda\eta)\frac{\partial}{\partial \eta}y + ((\lambda - \gamma)\xi + (\lambda^2 - \lambda\gamma - \omega^2)\eta)\frac{\partial}{\partial \xi}y + \frac{\gamma k_B T}{m}\frac{\partial^2}{\partial \xi^2}y. \tag{8.15}$$

Now choose

$$\lambda = \frac{\gamma}{2} + \sqrt{\frac{\gamma^2}{4} + \omega^{*2}} \tag{8.16}$$

to have the simplified equation

$$0 = (\xi + \lambda\eta)\frac{\partial}{\partial \eta}y + ((\lambda - \gamma)\xi)\frac{\partial}{\partial \xi}y + \frac{\gamma k_B T}{m}\frac{\partial^2}{\partial \xi^2}y \tag{8.17}$$

which obviously has solutions which depend only on ξ and obey

$$\xi\frac{\partial}{\partial \xi}\Phi(\xi) = -\frac{\gamma k_B T}{m(\lambda - \gamma)}\frac{\partial^2}{\partial \xi^2}\Phi(\xi). \tag{8.18}$$

The general solution of this equation is[1]

$$\Phi(\xi) = C_1 + C_2\mathrm{erf}\left(\xi\sqrt{\frac{m(\lambda - \gamma)}{2\gamma k_B T}}\right). \tag{8.19}$$

---

[1] Note that $\lambda - \gamma$ is always positive. This would not be true for the second root.

Now, we impose the boundary condition $\Phi \to 0$ for $x \to \infty$, which means $\xi \to -\infty$. From this, we find

$$C_1 = -C_2 \mathrm{erf}(-\infty) = C_2. \tag{8.20}$$

Together, we have for the probability density

$$W(x,v)$$
$$= C_2 \frac{m\omega_\mathrm{a}}{2\pi k_\mathrm{B}T} \exp\left\{ -\frac{\frac{m}{2}v^2 + U(x)}{k_\mathrm{B}T} \right\} \left\{ 1 + \mathrm{erf}\left(\sqrt{\frac{m(\lambda-\gamma)}{2\gamma k_\mathrm{B}T}}(v - \lambda x) \right) \right\} \tag{8.21}$$

and the flux over the barrier is approximately

$$S = \int \mathrm{d}v\, v W(0,v)$$
$$= C_2 \frac{m\omega_\mathrm{a}}{2\pi k_\mathrm{B}T} e^{-U(x^*)/k_\mathrm{B}T} \int \mathrm{d}v\, v e^{-mv^2/2k_\mathrm{B}T} \left\{ 1 + \mathrm{erf}\left(\sqrt{\frac{m(\lambda-\gamma)}{2\gamma k_\mathrm{B}T}}v \right) \right\}$$
$$= C_2 \frac{m\omega_\mathrm{a}}{2\pi k_\mathrm{B}T} e^{-U(0)/k_\mathrm{B}T} \frac{2k_\mathrm{B}T}{m} \sqrt{1 - \frac{\gamma}{\lambda}}. \tag{8.22}$$

In the A-well, we have approximately

$$W(x,v) \approx 2C_2 \frac{m\omega_\mathrm{a}}{2\pi k_\mathrm{B}T} \exp\left\{ -\frac{\frac{mv^2}{2} + \frac{m\omega_\mathrm{a}^2(x-x_\mathrm{a})^2}{2}}{k_\mathrm{B}T} \right\}. \tag{8.23}$$

The total concentration is approximately

$$[\mathrm{A}] = \int \mathrm{d}x \int \mathrm{d}v W_\mathrm{A}(x,v) = 2C_2. \tag{8.24}$$

Hence, we find

$$S = [\mathrm{A}]\frac{\omega_\mathrm{a}}{2\pi} e^{-U(0)/k_\mathrm{B}T} \sqrt{1 - \frac{\gamma}{\lambda}}. \tag{8.25}$$

The square root can be written as

$$\sqrt{1 - \frac{\gamma}{\frac{\gamma}{2} + \sqrt{\frac{\gamma^2}{4} + \omega^{*2}}}} = \sqrt{\frac{-\frac{\gamma}{2} + \sqrt{\frac{\gamma^2}{4} + \omega^{*2}}}{\frac{\gamma}{2} + \sqrt{\frac{\gamma^2}{4} + \omega^{*2}}}}$$
$$= \sqrt{\frac{\left(-\frac{\gamma}{2} + \sqrt{\frac{\gamma^2}{4} + \omega^{*2}}\right)^2}{-\frac{\gamma^2}{4} + \left(\frac{\gamma^2}{4} + \omega^{*2}\right)}}$$
$$= \frac{-\frac{\gamma}{2} + \sqrt{\frac{\gamma^2}{4} + \omega^{*2}}}{\omega^*} \tag{8.26}$$

and finally, we arrive at Kramers' famous result

$$k = \frac{S}{[\mathrm{A}]} = \frac{\omega_\mathrm{a}}{2\pi\omega^*}\mathrm{e}^{-U(0)/k_\mathrm{B}T}\left(-\frac{\gamma}{2} + \sqrt{\frac{\gamma^2}{4} + \omega^{*2}}\right). \tag{8.27}$$

The bracket is Kramers' correction to the escape rate. In the limit of high friction, series expansion in γ^{-1} gives

$$k = \gamma^{-1}\frac{\omega_\mathrm{a}\omega^*}{2\pi}\mathrm{e}^{-U(0)/k_\mathrm{B}T}. \tag{8.28}$$

In the limit of low friction, the result is

$$k = \frac{\omega_\mathrm{a}}{2\pi}\mathrm{e}^{-U(0)/k_\mathrm{B}T}. \tag{8.29}$$

Dispersive Kinetics

In this chapter, we consider the decay of an optically excited state of a donor molecule in a fluctuating medium. The fluctuations are modeled by time-dependent decay rates k (electron transfer), k_{-1} (back reaction), k_{da} (deactivation by fluorescence or radiationless transitions), and k_{cr} (charge recombination to the ground state). (Fig. 9.1)

The time evolution is described by the system of rate equations:

$$\frac{\mathrm{d}}{\mathrm{d}t} W(\mathrm{D}^*) = -k(t)W(\mathrm{D}^*) + k_{-1}(t)W(\mathrm{D}^+\mathrm{A}^-) - k_{\mathrm{da}}W(\mathrm{D}^*),$$

$$\frac{\mathrm{d}}{\mathrm{d}t} W(\mathrm{D}^+\mathrm{A}^-) = k(t)W(\mathrm{D}^*) - k_{-1}(t)W(\mathrm{D}^+\mathrm{A}^-) - k_{\mathrm{cr}}W(\mathrm{D}^+\mathrm{A}^-), \quad (9.1)$$

which has to be combined with suitable equations describing the dynamics of the environment. First, we will apply a very simple dichotomous model [33].

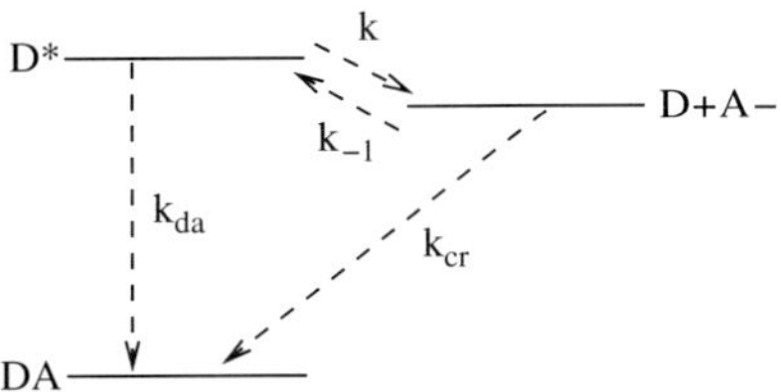

Fig. 9.1. Electron transfer in a fluctuating medium. The rates are time dependent

9.1 Dichotomous Model

The fluctuations of the rates are modeled by random jumps between two different configurations ($\pm$) of the environment which modulate the values of the rates. The probabilities of the two states are determined by the master equation:

$$\frac{\mathrm{d}}{\mathrm{d}t}\begin{pmatrix} W(+) \\ W(-) \end{pmatrix} = \begin{pmatrix} -\alpha & \beta \\ \alpha & -\beta \end{pmatrix}\begin{pmatrix} W(+) \\ W(-) \end{pmatrix},\tag{9.2}$$

which has the general solution:

$$\begin{aligned} W(+) &= C_1 + C_2 e^{-(\alpha+\beta)t}, \\ W(-) &= C_1\frac{\alpha}{\beta} - C_2 e^{-(\alpha+\beta)t}. \end{aligned}\tag{9.3}$$

Obviously, the equilibrium values are

$$W_{\mathrm{eq}}(+) = \frac{\beta}{\alpha+\beta}, \quad W_{\mathrm{eq}}(-) = \frac{\alpha}{\alpha+\beta}\tag{9.4}$$

and the correlation function is (with $Q_\pm = \pm 1$)

$$\begin{aligned} \langle Q(t)Q(0)\rangle &= W_{\mathrm{eq}}(+)(P(+,t|+,0) - P(-,t|+,0)) \\ &\quad + W_{\mathrm{eq}}(-)(P(-,t|-,0) - P(+,t|-,0)) \\ &= (W_{\mathrm{eq}}(+) - W_{\mathrm{eq}}(-))^2 + 4W_{\mathrm{eq}}(+)W_{\mathrm{eq}}(-)e^{-(\alpha+\beta)t} \\ &= \langle Q\rangle^2 + (\langle Q^2\rangle - \langle Q\rangle^2)e^{-(\alpha+\beta)t}. \end{aligned}\tag{9.5}$$

Combination of the two systems of equations (9.1) and (9.2) gives the equation of motion

$$\frac{\mathrm{d}}{\mathrm{d}t}\mathbf{W} = A\mathbf{W}\tag{9.6}$$

for the four-component state vector

$$\mathbf{W} = \begin{pmatrix} W(\mathrm{D}^*,+) \\ W(\mathrm{D}^*,-) \\ W(\mathrm{D}^+\mathrm{A}^-,+) \\ W(\mathrm{D}^+\mathrm{A}^-,-) \end{pmatrix}\tag{9.7}$$

with the rate matrix

$$A = \begin{pmatrix} -\alpha - k^+ - k_{\mathrm{da}} & \beta & k^+_{-1} & 0 \\ \alpha & -\beta - k^- - k_{\mathrm{da}} & 0 & k^-_{-1} \\ k^+ & 0 & -\alpha - k^+_{-1} - k_{\mathrm{cr}} & \beta \\ 0 & k^- & \alpha & -\beta - k^-_{-1} - k_{\mathrm{cr}} \end{pmatrix}.\tag{9.8}$$

Generally, the solution of this equation can be expressed by using the left and right eigenvectors and the eigenvalues λ of the rate matrix which obey

$$A\mathbf{R}_\nu = \lambda_\nu \mathbf{R}_\nu,\tag{9.9}$$

$$\mathbf{L}_\nu A = \lambda_\nu \mathbf{L}_\nu.\tag{9.10}$$

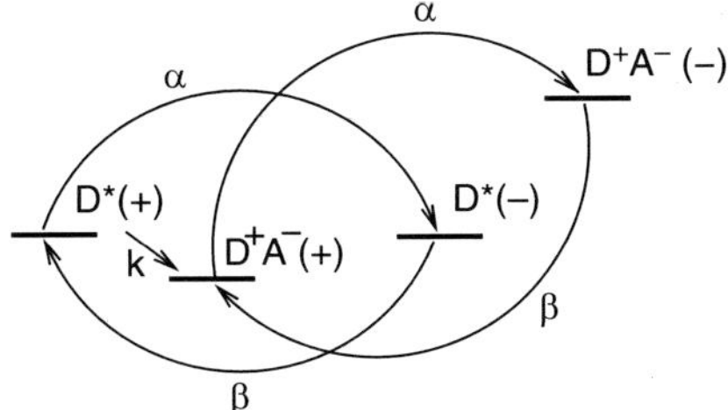

Fig. 9.2. Gated electron transfer

For initial values $\mathbf{W}(0)$, the solution is given by[1]

$$\mathbf{W}(t) = \sum_{\nu=1}^{4} \frac{(\mathbf{L}_\nu \cdot \mathbf{W}(0))}{(\mathbf{L}_\nu \cdot \mathbf{R}_\nu)} \mathbf{R}_\nu e^{\lambda_\nu t}. \tag{9.11}$$

In the following, we consider a simplified case of gated electron transfer with $k_{\mathrm{da}} = k_{\mathrm{cr}} = k_{-1}^{\pm} = k^- = 0$ (Fig. 9.2). Then the rate matrix becomes

$$\begin{pmatrix} -\alpha - k^+ & \beta & 0 & 0 \\ \alpha & -\beta & 0 & 0 \\ k^+ & 0 & -\alpha & \beta \\ 0 & 0 & \alpha & -\beta \end{pmatrix}. \tag{9.12}$$

As initial values, we chose

$$\mathbf{W_0} = \begin{pmatrix} W_{\mathrm{eq}}(+) \\ W_{\mathrm{eq}}(-) \\ 0 \\ 0 \end{pmatrix} = \begin{pmatrix} \frac{\beta}{\alpha+\beta} \\ \frac{\alpha}{\alpha+\beta} \\ 0 \\ 0 \end{pmatrix}. \tag{9.13}$$

There is one eigenvalue $\lambda_1 = 0$ corresponding to the eigenvectors:

$$\mathbf{R_1} = \begin{pmatrix} 0 \\ 0 \\ \beta \\ \alpha \end{pmatrix}, \quad \mathbf{L_1} = \begin{pmatrix} 1 & 1 & 1 & 1 \end{pmatrix}. \tag{9.14}$$

This reflects simply conservation of $\sum_{\nu=1}^{4} W_\nu$ in this special case. The contribution of the zero eigenvector is

$$\frac{\mathbf{L_1} \cdot \mathbf{W_0}}{\mathbf{L_1} \cdot \mathbf{R_1}} \mathbf{R_1} = \frac{1}{\alpha+\beta} \begin{pmatrix} 0 \\ 0 \\ \beta \\ \alpha \end{pmatrix} = \begin{pmatrix} 0 \\ 0 \\ W_{\mathrm{eq}}(+) \\ W_{\mathrm{eq}}(-) \end{pmatrix}. \tag{9.15}$$

---

[1] In the case of degenerate eigenvalues, linear combinations of the corresponding vectors can be found such that $\mathbf{L}_\nu \cdot \mathbf{L}_{\nu'} = 0$ for $\nu \neq \nu'$.

A second eigenvalue $\lambda_2 = -(\alpha + \beta)$ corresponds to the equilibrium in the final state D^+A^- where no further reactions take place

$$\mathbf{R}_2 = \begin{pmatrix} 0 \\ 0 \\ 1 \\ -1 \end{pmatrix}, \quad \mathbf{L}_2 = \begin{pmatrix} \alpha & -\beta & \alpha & -\beta \end{pmatrix}. \tag{9.16}$$

The contribution of this eigenvalue is

$$\frac{\mathbf{L}_2 \cdot \mathbf{W}_0}{\mathbf{L}_2 \cdot \mathbf{R}_2} \mathbf{R}_2 = 0 \tag{9.17}$$

since we assumed equilibrium in the initial state. The remaining two eigenvalues are

$$\lambda_{3,4} = -\frac{\alpha + \beta + k}{2} \pm \frac{1}{2}\sqrt{(\alpha + \beta + k)^2 - 4\beta k} \tag{9.18}$$

and the resulting decay will be in general biexponential. We consider two limits.

9.1.1 Fast Solvent Fluctuations

In the limit of small k, we expand the square root to find

$$\lambda_{3,4} = -\frac{\alpha + \beta}{2} \pm \frac{\alpha + \beta}{2} - \frac{k}{2} \pm \frac{\alpha - \beta}{\alpha + \beta} \frac{k}{2} + \cdots . \tag{9.19}$$

One of the eigenvalues is

$$\lambda_3 = -(\alpha + \beta) - \frac{\alpha}{\alpha + \beta} k + \cdots \tag{9.20}$$

In the limit of $k \to 0$, the corresponding eigenvectors are

$$\mathbf{R}_3 = \begin{pmatrix} 1 \\ -1 \\ -1 \\ 1 \end{pmatrix}, \quad \mathbf{L}_3 = \begin{pmatrix} \alpha & -\beta & 0 & 0 \end{pmatrix} \tag{9.21}$$

and will not contribute significantly. The second eigenvalue

$$\lambda_4 = -\frac{\beta}{\alpha + \beta} k + \cdots = -W_{\mathrm{eq}}(+)k \tag{9.22}$$

is given by the average rate. The eigenvectors are

$$\mathbf{R}_4 = \begin{pmatrix} \beta \\ \alpha \\ -\beta \\ -\alpha \end{pmatrix}, \quad \mathbf{L}_4 = \begin{pmatrix} 1 & 1 & 0 & 0 \end{pmatrix} \tag{9.23}$$

and the contribution to the dynamics is

$$\frac{(\mathbf{L}_4 \cdot \mathbf{W}_0)}{(\mathbf{L}_4 \cdot \mathbf{R}_4)} \mathbf{R}_4 e^{\lambda_4 t} = \frac{1}{\alpha + \beta} \begin{pmatrix} \beta \\ \alpha \\ -\beta \\ -\alpha \end{pmatrix} e^{-\lambda_4 t}. \tag{9.24}$$

The total time dependence is approximately given by

$$\mathbf{W}(t) = \begin{pmatrix} W_{\mathrm{eq}}(+)e^{-\lambda_4 t} \\ W_{\mathrm{eq}}(-)e^{-\lambda_4 t} \\ W_{\mathrm{eq}}(+)(1 - e^{-\lambda_4 t}) \\ W_{\mathrm{eq}}(-)(1 - e^{-\lambda_4 t}) \end{pmatrix}. \tag{9.25}$$

9.1.2 Slow Solvent Fluctuations

In the opposite limit, we expand the square root for small k^{-1} to find

$$\lambda_{3,4} = -\frac{\alpha + \beta}{2} - \frac{k}{2} \pm \frac{1}{2}\left(k + (\alpha - \beta) + 2\alpha\beta k^{-1} + \cdots\right), \tag{9.26}$$

$$\lambda_3 = -\beta + \frac{\alpha\beta}{k} + \cdots, \tag{9.27}$$

$$\mathbf{R}_3 = \begin{pmatrix} 0 \\ 1 \\ 0 \\ -1 \end{pmatrix}, \quad \mathbf{L}_3 = \begin{pmatrix} \alpha & k & 0 & 0 \end{pmatrix}, \tag{9.28}$$

$$\lambda_4 = -k - \alpha + \cdots, \tag{9.29}$$

$$\mathbf{R}_4 = \begin{pmatrix} k \\ -\alpha \\ -k \\ \alpha \end{pmatrix}, \quad \mathbf{L}_4 = \begin{pmatrix} 1 & 0 & 0 & 0 \end{pmatrix} \tag{9.30}$$

and the time evolution is approximately

$$\mathbf{W}(t) = \begin{pmatrix} \frac{\beta}{\alpha+\beta} e^{-kt} \\ \frac{\alpha}{\alpha+\beta}\left(e^{-\beta t} - \frac{\beta}{k}e^{-kt}\right) \\ \frac{\beta}{\alpha+\beta}(1 - e^{-kt}) \\ \frac{\alpha}{\alpha+\beta}\left(1 - e^{-\beta t} + \frac{\beta}{k}e^{-kt}\right) \end{pmatrix}. \tag{9.31}$$

This corresponds to an inhomogeneous situation. One part of the ensemble is in a favorable environment and decays with the fast rate k. The rest has to wait for a suitable fluctuation which appears with a rate of β.

9.1.3 Numerical Example (Fig. 9.3)

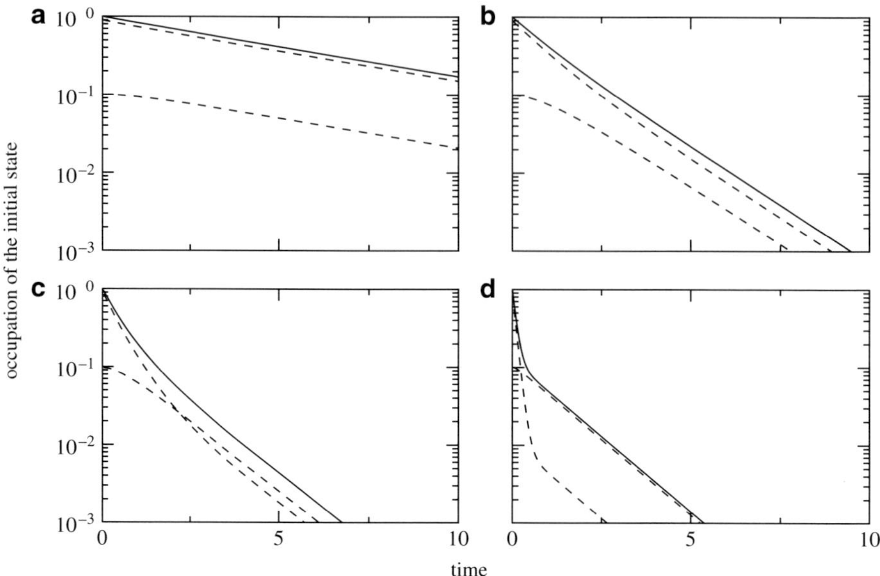

Fig. 9.3. Nonexponential decay. Numerical solutions of (9.12) are shown for $\alpha = 0.1$, $\beta = 0.9$, (**a**) $k = 0.2$, (**b**) $k = 2$, (**c**) $k = 5$, and (**d**) $k = 10$. *Dotted curves* show the two components of the initial state and *solid curves* show the total occupation of the initial state

9.2 Continuous Time Random Walk Processes

Diffusive motion can be modeled by random walk processes along a one-dimensional coordinate.

9.2.1 Formulation of the Model

The fluctuations of the coordinate $X(t)$ are described as random jumps [34, 35]. The time intervals between the jumps (waiting time) and the coordinate changes are random variables with independent distribution functions:

$$\psi(t_{n+1} - t_n) \quad \text{and} \quad f(X_{n+1}, X_n). \tag{9.32}$$

The probability that no jump happened in the interval $0 \ldots t$ is given by the survival function:

$$\Psi_0(t) = 1 - \int_0^t \mathrm{d}t' \psi(t') = \int_t^\infty \mathrm{d}t' \psi(t') \tag{9.33}$$

and the probability of finding the walker at position X at time t is given by (Fig. 9.4)

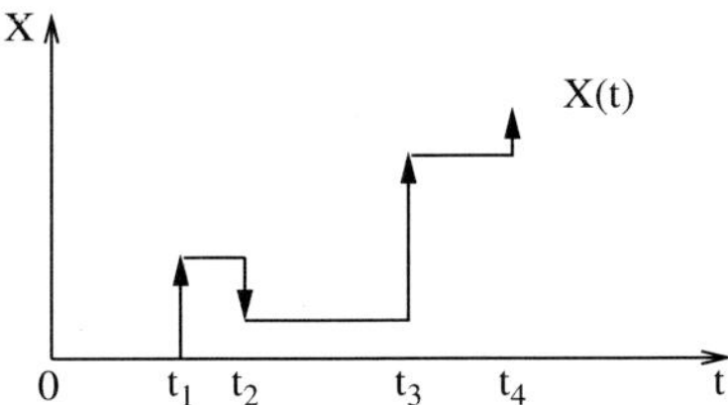

Fig. 9.4. Continuous time random walk

$$P(X,t) = P(X,0) \int_t^\infty \psi(t')\mathrm{d}t'$$
$$+ \int_0^t \mathrm{d}t' \int_{-\infty}^\infty \mathrm{d}X' \psi(t-t')f(X,X')P(X',t'). \qquad (9.34)$$

Two limiting cases are well known from the theory of collisions. The correlated process with

$$f(X,X') = f(X-X') \qquad (9.35)$$

corresponds to weak collisions. It includes normal diffusion processes as a special case. For instance if we chose

$$\psi(t_{n+1} - t_n) = \delta(t_{n+1} - t_n - \Delta t) \qquad (9.36)$$

and

$$f(X-X') = p\delta(X - X' - \Delta X) + (1-p)\delta(X - X' + \Delta X), \qquad (9.37)$$

we have

$$P(X,t+\Delta t) = pP(X-\Delta X,t) + qP(X+\Delta X,t), \quad p+q = 1 \qquad (9.38)$$

and in the limit $\Delta t \to 0$ and $\Delta X \to 0$, Taylor expansion gives

$$P(X,t) + \frac{\partial P}{\partial t}\Delta t + \cdots = P(X,t) + \frac{\partial P}{\partial X}(q-p)\Delta X + \frac{\partial^2 P}{\partial X^2}\Delta X^2 + \cdots \qquad (9.39)$$

The leading terms constitute a diffusion equation

$$\frac{\partial P}{\partial t}P = (q-p)\frac{\Delta X}{\Delta t}\frac{\partial P}{\partial X} + \frac{\Delta X^2}{\Delta t}\frac{\partial^2 P}{\partial X^2} \qquad (9.40)$$

with drift velocity $(q-p)(\Delta X/\Delta t)$ and diffusion constant $\Delta X^2/\Delta t$.

The uncorrelated process, on the other hand, with

$$f(X,X') = f(X) \qquad (9.41)$$

corresponds to strong collisions. This kind of process can be analyzed analytically and will be applied in the following.

The (normalized) stationary distribution Φ_{eq} of the uncorrelated process obeys

$$\Phi_{\mathrm{eq}}(X) = \Phi_{\mathrm{eq}}(X) \int_t^\infty \psi(t')\mathrm{d}t' + f(X) \int_0^t \mathrm{d}t'\psi(t - t') \int_{-\infty}^\infty \mathrm{d}X'\phi_{\mathrm{eq}}(X')$$

$$= \Phi_{\mathrm{eq}}(X) \int_t^\infty \psi(t')\mathrm{d}t' + f(X) \int_0^t \mathrm{d}t'\psi(t'), \tag{9.42}$$

which shows that

$$f(X) = \Phi_{\mathrm{eq}}(X). \tag{9.43}$$

9.2.2 Exponential Waiting Time Distribution

Consider an exponential distribution of waiting times

$$\psi(t) = \tau^{-1}\mathrm{e}^{-t/\tau}, \quad \Psi_0(t) = \int_t^\infty \tau^{-1}\mathrm{e}^{-t'/\tau}\mathrm{d}t' = \mathrm{e}^{-t/\tau}. \tag{9.44}$$

It can be obtained from a Poisson process which corresponds to the master equation:

$$\frac{\mathrm{d}P_n}{\mathrm{d}t} = -\tau^{-1}P_n + \tau^{-1}P_{n-1}, \quad n = 0, 1, 2, \ldots \tag{9.45}$$

with the solution

$$P_n(0) = \delta_{n,0}, \quad P_n(t) = \frac{(t/\tau)^n}{n!}\mathrm{e}^{-t/\tau} \tag{9.46}$$

if we identify the survival function with the probability to be in the initial state P_0

$$\Psi(t) = P_0(t) = \mathrm{e}^{-t/\tau}. \tag{9.47}$$

The general uncorrelated process (9.34) becomes for an exponential distribution

$$P(X, t) = P(X, 0)\mathrm{e}^{-t/\tau} + \int_0^t \mathrm{d}t'\tau^{-1}\mathrm{e}^{-(t-t')/\tau} \int \mathrm{d}X'f(X, X')P(X', t'). \tag{9.48}$$

Laplace transformation gives

$$\tilde{P}(X, s) = P(X, 0)\frac{1}{s + \tau^{-1}} + \frac{\tau^{-1}}{s + \tau^{-1}} \int \mathrm{d}X'f(X, X')\tilde{P}(X', s), \tag{9.49}$$

which can be simplified

$$(s + \tau^{-1})\tilde{P}(X,s) = P(X,0) + \tau^{-1} \int \mathrm{d}X' f(X,X')\tilde{P}(X',s). \tag{9.50}$$

Back transformation gives

$$\left(\frac{\mathrm{d}}{\mathrm{d}t} + \tau^{-1}\right) P(X,t) = \tau^{-1} \int \mathrm{d}X' f(X,X') P(X',t) \tag{9.51}$$

and finally

$$\frac{\partial}{\partial t} P(X,t) = -\frac{1}{\tau} P(X,t) + \frac{1}{\tau} \int \mathrm{d}X' f(X,X') P(X',t), \tag{9.52}$$

which is obviously a Markovian process, since it involves only the time t. For the special case of an uncorrelated process with exponential waiting time distribution, the motion can be described by

$$\frac{\partial}{\partial t} P(X,t) = \mathcal{L}P(X,t), \tag{9.53}$$

$$\mathcal{L}P(X,t) = -\frac{1}{\tau}\left(P(X,t) - \phi_{\mathrm{eq}}(X)\langle P(t)\rangle\right). \tag{9.54}$$

9.2.3 Coupled Equations

Coupling of motion along the coordinate X with the reactions gives the following system of equations [36, 37]:

$$\frac{\partial}{\partial t} P(X,t) = \left(-k(X) + \mathcal{L}_1 - \tau_1^{-1}\right) P(X,t) + k_{-1}(X) C(X,t),$$

$$\frac{\partial}{\partial t} C(X,t) = \left(-k_{-1}(X) + \mathcal{L}_2 - \tau_2^{-1}\right) C(X,t) + k(X) P(X,t), \tag{9.55}$$

where $P(X,t)\Delta X$ and $C(X,t)\Delta X$ are the probabilities of finding the system in the electronic state D^* and $\mathrm{D}^+\mathrm{A}^-$, respectively; $\mathcal{L}_{1,2}$ are operators describing the motion in the two states and the rates $\tau_{1,2}^{-1}$ account for depopulation via additional channels. For the uncorrelated Markovian process (9.54), the rate equations take the form

$$\frac{\partial}{\partial t} \begin{pmatrix} P(X,t) \\ C(X,t) \end{pmatrix}$$

$$= -\begin{pmatrix} k(X) + \tau_1^{-1} + \tau^{-1} & -k_{-1}(X) \\ -k(X) & k_{-1}(X) + \tau_2^{-1} + \tau^{-1} \end{pmatrix} \begin{pmatrix} P(X,t) \\ C(X,t) \end{pmatrix}$$

$$+\tau^{-1} \begin{pmatrix} \phi_1(X) & \\ & \phi_2(X) \end{pmatrix} \begin{pmatrix} \langle P(t)\rangle \\ \langle C(t)\rangle \end{pmatrix}, \tag{9.56}$$

which can be written in matrix notation as

$$\frac{\partial}{\partial t}\mathbf{R}(X,t) = -A(X)\mathbf{R}(X,t) + \tau^{-1}B(X)\langle\mathbf{R}(t)\rangle. \tag{9.57}$$

Substitution

$$\mathbf{R}(\tilde{X},t) = \exp\left\{-A(X)\mathbf{U}(X,t)\right\} \tag{9.58}$$

gives

$$-A(X)\;\mathbf{R}(X,t) + \exp\left\{-A(X)\frac{\partial}{\partial t}\mathbf{U}(X,t)\right\}$$
$$= -A(X)\mathbf{R}(X,t) + \tau^{-1}B(X)\langle\mathbf{R}(t)\rangle, \tag{9.59}$$

$$\frac{\partial}{\partial t}\mathbf{U}(X,t) = \tau^{-1}\exp\left\{A(X)t\right\}B(X)\langle\mathbf{R}(t)\rangle. \tag{9.60}$$

Integration gives

$$\mathbf{U}(X,t) = \mathbf{U}(X,0) + \tau^{-1}\int_0^t \exp\left\{A(X)t'\right\}B(X)\langle\mathbf{R}(t')\rangle\mathrm{d}t', \tag{9.61}$$

$$\mathbf{R}(X,t) = \exp(-A(X)t)\mathbf{R}(X,0)$$
$$+ \tau^{-1}\int_0^t \exp(A(X)(t'-t))B(X)\langle\mathbf{R}(t')\rangle\mathrm{d}t' \tag{9.62}$$

and the total populations obey the integral equation:

$$\langle\mathbf{R}(t)\rangle = \langle\exp(-At)\mathbf{R}(0)\rangle$$
$$+ \tau^{-1}\int_0^t \langle\exp(A(t'-t))B\rangle\langle\mathbf{R}(t')\rangle\mathrm{d}t', \tag{9.63}$$

which can be solved with the help of a Laplace transformation:

$$\tilde{\mathbf{R}}(s) = \int_0^\infty \mathrm{e}^{-st}\langle\mathbf{R}(t)\rangle\mathrm{d}t, \tag{9.64}$$

$$\int_0^\infty \mathrm{e}^{-st}\exp(-At)\mathrm{d}t = (s+A)^{-1}, \tag{9.65}$$

$$\int_0^\infty \mathrm{e}^{-st}\mathrm{d}t\int_0^t \langle\exp(A(t'-t))B\rangle\langle\mathbf{R}(t')\rangle\mathrm{d}t'$$
$$= \langle(s+A)^{-1}B\rangle\tilde{\mathbf{R}}(s). \tag{9.66}$$

The Laplace transform of the integral equation

$$\tilde{\mathbf{R}}(s) = \langle(s+A)^{-1}\mathbf{R}(0)\rangle + \tau^{-1}\langle(s+A)^{-1}B\rangle\tilde{\mathbf{R}}(s) \tag{9.67}$$

is solved by

$$\tilde{\mathbf{R}}(s) = \left[1 - \tau^{-1}\langle (s + A)^{-1}B\rangle\right]^{-1}\langle (s + A)^{-1}\mathbf{R}(0)\rangle. \tag{9.68}$$

We assume that initially, the system is in the initial state D* and the motion is equilibrated:

$$\mathbf{R}(X,0) = \begin{pmatrix} \phi_1(X) \\ 0 \end{pmatrix}. \tag{9.69}$$

For simplicity, we treat here only the case of $\tau_{12} \to \infty$. Then we have

$$A = \begin{pmatrix} k + \tau^{-1} & -k_{-1} \\ -k & k_{-1} + \tau^{-1} \end{pmatrix}, \tag{9.70}$$

$$(s + A)^{-1} = \frac{1}{(s + \tau^{-1})(s + \tau^{-1} + k + k_{-1})}\begin{pmatrix} s + \tau^{-1} + k_{-1} & k_{-1} \\ k & s + \tau^{-1} + k \end{pmatrix} \tag{9.71}$$

and with the abbreviations

$$\alpha = \left(1 + \frac{1}{s + \tau^{-1}}(k + k_{-1})\right)^{-1} \tag{9.72}$$

and

$$\langle f(X)\rangle_{1,2} = \int \phi_{1,2}(X)f(X)\mathrm{d}X, \tag{9.73}$$

we find

$$\langle (s + A)^{-1}\mathbf{R}(0)\rangle$$
$$= \left\langle \frac{\phi_1\alpha}{(s + \tau^{-1})^2}\begin{pmatrix} \alpha^{-1}(s + \tau^{-1}) - k \\ k \end{pmatrix}\right\rangle = \begin{pmatrix} \frac{1}{s+\tau^{-1}} - \frac{1}{(s+\tau^{-1})^2}\langle \alpha k\rangle_1 \\ \frac{1}{(s+\tau^{-1})^2}\langle \alpha k\rangle_1 \end{pmatrix} \tag{9.74}$$

as well as

$$\langle (s + A)^{-1}B\rangle$$
$$= \begin{pmatrix} \frac{1}{s+\tau^{-1}} - \frac{1}{(s+\tau^{-1})^2}\langle \alpha k\rangle_1 & \frac{1}{(s+\tau^{-1})^2}\langle \alpha k_{-1}\rangle_2 \\ \frac{1}{(s+\tau^{-1})^2}\langle \alpha k\rangle_1 & \frac{1}{s+\tau^{-1}} - \frac{1}{(s+\tau^{-1})^2}\langle \alpha k_{-1}\rangle_2 \end{pmatrix} \tag{9.75}$$

and the final result becomes

$$\tilde{\mathbf{R}}(s) = \frac{1}{s(s^2 + \tau^{-1}(s + \langle \alpha k\rangle_1 + \langle \alpha k_{-1}\rangle_2))}$$
$$\times \begin{pmatrix} s(s + \tau^{-1}) - s\langle \alpha k\rangle_1 + \tau^{-1}\langle \alpha k_{-1}\rangle_2 \\ (s + \tau^{-1})\langle \alpha k\rangle_1 \end{pmatrix}. \tag{9.76}$$

Let us discuss the special case of thermally activated electron transfer. Here

$$\langle \alpha k \rangle_1, \langle \alpha k_{-1} \rangle_2 \ll 1 \tag{9.77}$$

and the decay of the initial state is approximately given by

$$P(s) = \frac{(s + \tau^{-1}) + \tau^{-1} s^{-1} \langle \alpha k_{-1} \rangle_2}{s^2 + \tau^{-1} s^{-1} (1 + \langle \alpha k \rangle_1 + \langle \alpha k_{-1} \rangle_2)} = \frac{s + K_2}{s^2 + s(K_1 + K_2)} \tag{9.78}$$

with

$$K_2 = \tau^{-1} \left(1 + s^{-1} \int dX \phi_2(X) \frac{k_{-1}(X)}{1 + \frac{1}{s + \tau^{-1}}(k(X) + k_{-1}(X))} \right) \tag{9.79}$$

$$\approx \int dX \phi_2(X) \frac{k_{-1}(X)}{1 + \tau(k(X) + k_{-1}(X))}, \tag{9.80}$$

$$K_1 = \int dX \phi_1(X) \frac{k(X)}{1 + \tau(k(X) + k_{-1}(X))}. \tag{9.81}$$

This can be visualized as the result of a simplified kinetic scheme

$$\frac{d}{dt} \langle P \rangle = -K_1 \langle P \rangle + K_2 \langle C \rangle, \tag{9.82}$$

$$\frac{d}{dt} \langle C \rangle = K_1 \langle P \rangle - K_2 \langle C \rangle \tag{9.83}$$

with the Laplace transform

$$s\tilde{P} - P(0) = -K_1 \tilde{P} + K_2 \tilde{C}, \tag{9.84}$$

$$s\tilde{C} - C(0) = K_1 \tilde{P} + K_2 \tilde{C}, \tag{9.85}$$

which has the solution

$$P = \frac{sP_0 + K_2(P_0 + C_0)}{s(s + K_1 + K_2)}, \quad C = \frac{sC_0 + K_1(C_0 + P_0)}{s(s + K_1 + K_2)}. \tag{9.86}$$

In the time domain, we find

$$P(t) = \frac{K_2 + K_1 e^{-(K_1 + K_2)t}}{K_1 + K_2}, \quad C(t) = \frac{K_1}{K_1 + K_2}(1 - e^{-(K_1 + K_2)t}). \tag{9.87}$$

Let us now consider the special case that the back reaction is negligible and $k(X) = k\Theta(X)$. Here, we have

$$\tilde{P}(s) = \frac{s(s + \tau^{-1}) - s\langle \alpha k \rangle_1}{s(s^2 + \tau^{-1}(s + \langle \alpha k \rangle_1))}, \tag{9.88}$$

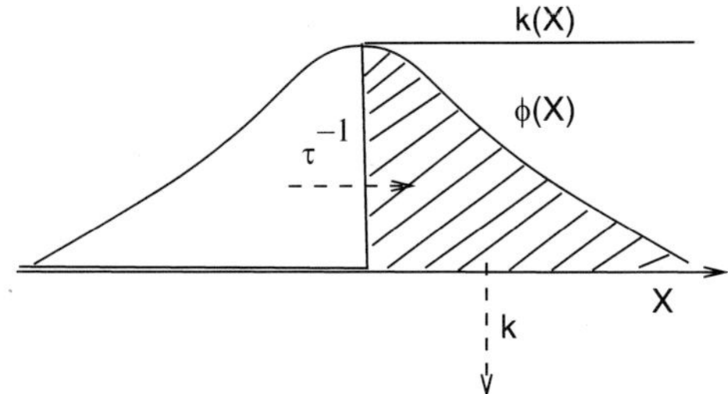

Fig. 9.5. Slow solvent limit

$$\langle \alpha k \rangle_1 = \int \mathrm{d}X \phi_1(X) \frac{k(X)}{1 + \frac{k(X)}{s + \tau^{-1}}} = \int_0^\infty \phi_1(X) \frac{k}{1 + \frac{k}{s + \tau^{-1}}} \mathrm{d}X$$

$$= bk \frac{s + \tau^{-1}}{k + s + \tau^{-1}}, \quad b = \int_0^\infty \phi_1(X) \mathrm{d}X, \quad a = 1 - b = \int_{-\infty}^0 \phi_1(X) \mathrm{d}X, \quad (9.89)$$

$$\tilde{P}(s) = \frac{(s + \tau^{-1} - bk\frac{s+\tau^{-1}}{k+s+\tau^{-1}})}{(s^2 + \tau^{-1}(s + bk\frac{s+\tau^{-1}}{k+s+\tau^{-1}}))} = \frac{s + \tau^{-1} + k(1-b)}{s^2 + s(\tau^{-1} + k) + bk\tau^{-1}}. \quad (9.90)$$

Inverse Laplace transformation gives a biexponential behavior

$$P(t) = \frac{(\mu_+ + k(1-2b))\mathrm{e}^{-t(k+\mu_-)/2} - (\mu_- + k(1-2b))\mathrm{e}^{-t(k+\mu_+)/2}}{\mu_+ - \mu_-} \quad (9.91)$$

with

$$\mu_\pm = \tau^{-1} \pm \sqrt{k^2 + \tau^{-2} + 2k\tau^{-1}(1-2b)}. \quad (9.92)$$

If the fluctuations are slow $\tau^{-1} \ll k$, then

$$\sqrt{k^2 + \tau^{-2} + 2k\tau^{-1}(1-2b)} = k + (1-2b)\tau^{-1} + \cdots, \quad (9.93)$$

$$\mu_+ = k + 2(1-b)\tau^{-1} + \cdots, \quad \mu_- = -k + 2b\tau^{-1} + \cdots, \quad (9.94)$$

and the two time constants are approximately (Fig. 9.5)

$$\frac{k + \mu_+}{2} = k + \cdots, \quad \frac{k + \mu_-}{2} = b\tau^{-1} + \cdots \quad (9.95)$$

9.3 Power Time Law Kinetics

The last example can be generalized to describe the power time law as observed for CO rebinding in myoglobin at low temperatures. The protein motion is now modeled by a more general uncorrelated process.[2]

---

[2] A much more detailed discussion is given in [37].

We assume that the rate k is negligible for $X < 0$ and very large for $X > 0$. Consequently, only jumps $X < 0 \to X > 0$ are considered. Then the probability obeys the equation

$$P(X,t)_{|X<0}$$

$$= P(X,0) \int_t^\infty \psi(t')\mathrm{d}t' + \int_{-\infty}^0 \mathrm{d}X' \int_0^t \mathrm{d}t'\psi(t-t')f(X)P(X',t')$$

$$= \phi_{\mathrm{eq}}(X)\Psi_0(t) + \phi_{\mathrm{eq}}(X) \int_0^t \mathrm{d}t'\psi(t-t') \int_{-\infty}^0 \mathrm{d}X'P(X',t') \tag{9.96}$$

$$\Psi_0(t) = \int_t^\infty \psi(t')\mathrm{d}t', \quad \tilde{\Psi}_0(s) = \frac{1-\tilde{\psi}(s)}{s} \tag{9.97}$$

and the total occupation of inactive configurations is

$$P_<(t) = \int_{-\infty}^0 \mathrm{d}X\,\phi_{\mathrm{eq}}(X) \left(\Psi_0(t) + \int_0^t \mathrm{d}t'\psi(t-t')P_<(t') \right)$$

$$= a \left(\Psi_0(t) + \int_0^t \mathrm{d}t'\psi(t-t')P_<(t') \right). \tag{9.98}$$

Laplace transformation gives

$$\tilde{P}_<(s) = a \left(\tilde{\Psi}_0(s) + \tilde{\psi}(s)\tilde{P}_<(s) \right) \tag{9.99}$$

with

$$a = \int_{-\infty 0}^0 \mathrm{d}X\,\phi_{\mathrm{eq}}(X) \tag{9.100}$$

and the decay of the initial state is given by

$$\tilde{P}_<(s) = \frac{a\tilde{\Psi}_0(s)}{1 - a\tilde{\psi}(s)} = \frac{1}{s + \frac{1-a}{a\tilde{\Psi}_0(s)}}. \tag{9.101}$$

For a simple Poisson process (9.44) with

$$\tilde{\Psi}_0 = \frac{1}{s + \tau^{-1}}, \tag{9.102}$$

this gives

$$\tilde{P}_<(s) = \frac{a}{s + (1-a)\tau^{-1}}, \tag{9.103}$$

which reproduces the exponential decay found earlier in the slow solvent limit (9.95)

$$P_<(t) = a e^{-t(1-a)/\tau}. \tag{9.104}$$

The long-time behavior is given by the asymptotic behavior for $s \to 0$. As $P_<(t) \to 0$ for $t \to \infty$, this is also the case for $\tilde{P}_<(s)$ in the limit $s \to 0$. Hence, the asymptotic behavior must be

$$\tilde{P}_<(s) \approx \frac{a\tilde{\Psi}_0(s)}{1-a} \to 0, \quad s \to 0,$$

(9.105)

$$P_<(t) \to \frac{a}{1-a}\Psi_0(t), \quad t \to \infty.$$

(9.106)

To describe a power time law at long times

$$P_<(t) \to t^{-\beta}, \quad t \to \infty,$$

(9.107)

$$\tilde{P}_<(s) \to s^{\beta-1}, \quad s \to 0,$$

(9.108)

the waiting time distribution has to be chosen as

$$\Psi_0(t) \sim \frac{1}{(zt)^\beta}, \quad t \to \infty,$$

which implies

$$\tilde{\Psi}_0(s) \sim z^{-\beta}s^{\beta-1},$$

(9.109)

where z^{-1} is the characteristic time for reaching the asymptotics. Finally, we find

$$\tilde{P}_<(s) \sim \frac{1}{s + \frac{1-a}{a}z^\beta s^{1-\beta}} = \frac{1}{s(1 + (\tilde{z}/s)^\beta)}.$$

(9.110)

In the time domain, this corresponds to the Mittag–Leffler function[3]

$$P_<(t) = \sum_{l=0}^{\infty} \frac{(-1)^l(\tilde{z}t)^{\beta l}}{\Gamma(\beta l + 1)} = E_\beta(-(\tilde{z}t)^\beta)$$

(9.111)

which can be approximated by the simpler function:

$$\frac{1}{1 + (t/\tau)^\beta}.$$

(9.112)

Problems

9.1. Dichotomous Model for Dispersive Kinetics

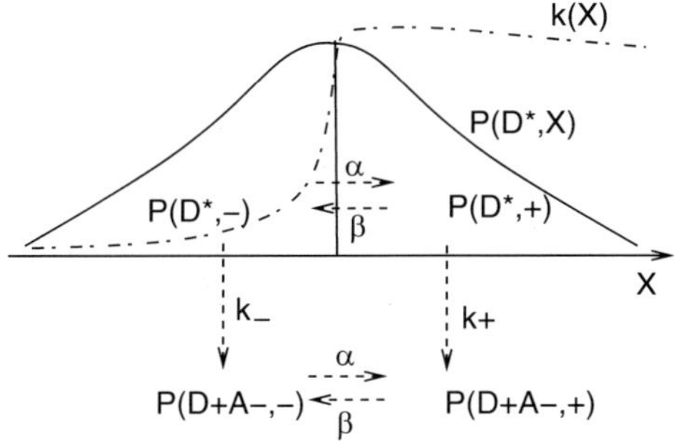

---

[3] which has also been discussed for nonexponential relaxation in inelastic solids and dipole relaxation processes corresponding to Cole–Cole spectra.

Consider the following system of rate equations

$$\frac{\mathrm{d}}{\mathrm{d}t}\begin{pmatrix} P(\mathrm{D}^*,+) \\ P(\mathrm{D}^*,-) \\ P(\mathrm{D}^+\mathrm{A}^-,+) \\ P(\mathrm{D}^+\mathrm{A}^-,-) \end{pmatrix} = \begin{pmatrix} -k_+ - \alpha & \beta & 0 & 0 \\ \alpha & -k_- - \beta & 0 & 0 \\ k_+ & 0 & -\alpha & \beta \\ 0 & k_- & \alpha & -\beta \end{pmatrix} \begin{pmatrix} P(\mathrm{D}^*,+) \\ P(\mathrm{D}^*,-) \\ P(\mathrm{D}^+\mathrm{A}^-,+) \\ P(\mathrm{D}^+\mathrm{A}^-,-). \end{pmatrix}$$

Determine the eigenvalues of the rate matrix M. Calculate the left and right eigenvectors approximately for the two limiting cases:

(a) Fast fluctuations $k_\pm \ll \alpha, \beta$. Show that the initial state decays with an average rate.

(b) Slow fluctuations $k_\pm \gg \alpha, \beta$. Show that the decay is nonexponential.

Part IV

Transport Processes

Nonequilibrium Thermodynamics

Biological systems are far from thermodynamic equilibrium. Concentration gradients and electrostatic potential differences are the driving forces for diffusive currents and chemical reactions (Fig. 10.1).

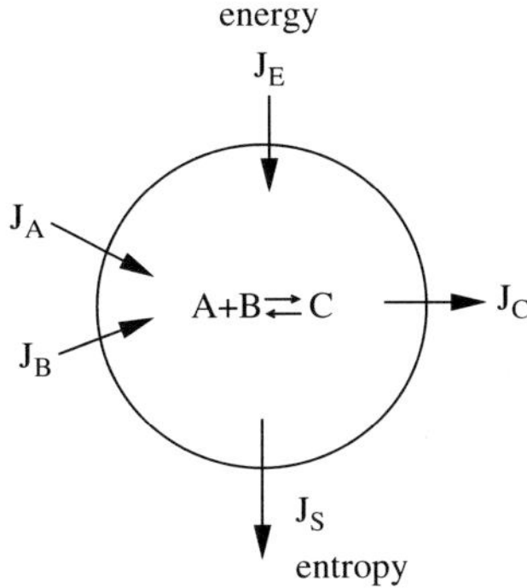

Fig. 10.1. Open nonequilibrium system. Energy and particles are exchanged with the environment. Chemical reactions change the concentrations and consume or release energy. Entropy is produced by irreversible processes like diffusion and heat transport

In this chapter, we consider a system composed of n species $k = 1, \ldots, n$ which can undergo r chemical reactions $j = 1, \ldots, r$. We assume that the system is locally in equilibrium and that all thermodynamic quantities are locally well defined.

10.1 Continuity Equation for the Mass Density

We introduce the partial mass densities

$$\varrho_k = \frac{m_k}{V} N_k = c_k m_k, \qquad (10.1)$$

the total density

$$\varrho = \sum_k \varrho_k = \frac{1}{V}\sum_k m_k N_k = \frac{M}{V}, \tag{10.2}$$

and the mass fraction

$$x_k = \frac{m_k}{M}N_k = \frac{m_k N_k}{\sum_k m_k N_k} = \frac{\varrho_k}{\varrho}. \tag{10.3}$$

From the conservation of mass, we have

$$\frac{\mathrm{d}}{\mathrm{d}t}M_k = m_k\frac{\mathrm{d}}{\mathrm{d}t}N_k = \int_V \frac{\partial}{\partial t}\varrho_k \mathrm{d}V$$

$$= -\int_{(V)} \varrho_k \mathbf{v}_k \mathrm{d}\mathbf{A} + \sum_{j=1}^r \int_V m_k \nu_{kj} r_j \mathrm{d}V \tag{10.4}$$

which can also be expressed in the form of a continuity equation

$$\frac{\partial}{\partial t}\varrho_k = -\mathrm{div}(\varrho_k \mathbf{v}_k) + \sum_{j=1}^r m_k \nu_{kj} r_j$$

$$= -\mathrm{div}(m_k \mathbf{J}_k) + \sum_{j=1}^r m_k \nu_{kj} r_j \tag{10.5}$$

with the diffusion fluxes $\mathbf{J}_k$. For the total mass density

$$\varrho = \sum_k \varrho_k, \tag{10.6}$$

we have

$$\frac{\partial}{\partial t}\varrho = -\mathrm{div}\left(\sum_k \varrho_k \mathbf{v}_k\right) + \sum_k \sum_{j=1}^r m_k \nu_{kj} r_j. \tag{10.7}$$

Due to conservation of mass for each reaction, the last term vanishes

$$\sum_k m_k \nu_{kj} = 0 \tag{10.8}$$

and with the center of mass velocity

$$\mathbf{v} = \frac{1}{\varrho}\sum_k \varrho_k \mathbf{v}_k, \tag{10.9}$$

we have

$$\frac{\partial}{\partial t}\varrho = -\mathrm{div}(\varrho \mathbf{v}). \tag{10.10}$$

10.2 Energy Conservation

We define the specific values of enthalpy, entropy, and volume h, s, ϱ^{-1}, respectively, by

$$H = h\,\varrho\,V,$$
$$S = s\,\varrho\,V,$$
$$V = \varrho^{-1}\,\varrho\,V. \tag{10.11}$$

The differential of the enthalpy is

$$dH = T dS + V dp + \sum_k \mu_k dN_k, \tag{10.12}$$

where μ_k is a generalized chemical potential, which also includes the potential energy of electrostatic or gravitational fields which are assumed to be time-independent. For the specific quantities, we find

$$V(h d\varrho + \varrho\,dh) = TV(s d\varrho + \varrho\,ds) + V dp$$
$$+ \sum_k \mu_k \frac{V}{m_k}(x_k d\varrho + \varrho\,dx_k) \tag{10.13}$$

and if we combine all terms with $d\varrho$,

$$\varrho\,dh - T\varrho\,ds - dp - \sum_k \frac{\mu_k}{m_k}\varrho\,dx_k = \left(Ts + \sum_k \mu_k \frac{x_k}{m_k} - h\right) d\varrho. \tag{10.14}$$

The right-hand side can be written as

$$\frac{1}{\varrho V}\left(TS + \sum_k \mu_k N_k - H\right) d\varrho, \tag{10.15}$$

which vanishes in equilibrium. Hence, we have for the specific quantities

$$dh = T ds + \varrho^{-1} dp + \sum_k \mu_k \frac{1}{m_k} dx_k. \tag{10.16}$$

In the following, we consider a resting solvent without convection. We neglect the center of mass motion and its contribution to the total energy. The diffusion currents are taken relative to the solvent.[1] For constant pressure, we have

$$\frac{\partial(\varrho h)}{\partial t} = T\frac{\partial(\varrho s)}{\partial t} + \sum_k \frac{\mu_k}{m_k}\frac{\partial \varrho_k}{\partial t} \tag{10.17}$$

and the enthalpy obeys the continuity equation

$$\frac{\partial(\varrho h)}{\partial t} = -\mathrm{div}(\mathbf{J}_h) \tag{10.18}$$

with the enthalpy flux $\mathbf{J}_h$.

---

[1] A more general discussion can be found in [38].

10.3 Entropy Production

The entropy change

$$dS = d\left(\int_V \varrho s\, dV\right) \tag{10.19}$$

has contributions from the local entropy production dS_i and from exchange with the surroundings dS_e:

$$dS = dS_i + dS_e. \tag{10.20}$$

The local entropy production can only be positive:

$$dS_i \geq 0, \tag{10.21}$$

whereas entropy exchange with the surroundings can be positive or negative (Fig. 10.2).

We denote by σ the entropy production per volume:

$$\frac{dS_i}{dt} = \int_V \sigma\, dV. \tag{10.22}$$

The entropy exchange is connected with the total entropy flux:

$$\frac{dS_e}{dt} = -\int_{(V)} \mathbf{J}_s d\mathbf{A}. \tag{10.23}$$

Together, we have

$$\frac{\partial(\varrho s)}{\partial t} = -\mathrm{div}(\mathbf{J}_s) + \sigma, \tag{10.24}$$

$$\sigma \geq 0. \tag{10.25}$$

From (10.17) the time derivative of the entropy per volume is

$$\frac{\partial(\varrho s)}{\partial t} = \frac{1}{T}\frac{\partial(\varrho h)}{\partial t} - \frac{1}{T}\sum_k \frac{\mu_k}{m_k}\frac{\partial \varrho_k}{\partial t} \tag{10.26}$$

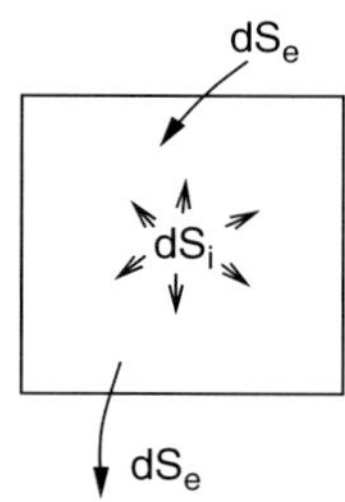

Fig. 10.2. Entropy change. The local entropy production dS_i is always positive, whereas entropy exchange with the environment dS_e can have both signs

and inserting (10.5) and (10.26)

$$\frac{\partial(\varrho s)}{\partial t} = -\frac{1}{T}\text{div}(\mathbf{J}_h) - \frac{1}{T}\sum_k \frac{\mu_k}{m_k}\left(-\text{div}(m_k\mathbf{J}_k) + \sum_{j=1}^r m_k\nu_{kj}r_j\right), \quad (10.27)$$

which can be written as

$$\frac{\partial(\varrho s)}{\partial t} = -\text{div}\left(\frac{1}{T}\mathbf{J}_h + \sum_k \frac{\mu_k}{T}\mathbf{J}_k\right) + \mathbf{J}_h\text{grad}\left(\frac{1}{T}\right)$$
$$- \sum_k \mathbf{J}_k\text{grad}\left(\frac{\mu_k}{T}\right) + \frac{1}{T}\sum_j A_j r_j \qquad (10.28)$$

with the chemical affinities

$$A_j = -\sum_k \mu_k\nu_{kj}. \qquad (10.29)$$

From comparison of (10.28) and (10.24), we find the entropy flux

$$\mathbf{J}_s = \frac{1}{T}\left(\mathbf{J}_h - \sum_k \mu_k\mathbf{J}_k\right) \qquad (10.30)$$

and the entropy production

$$\sigma = \mathbf{J}_h\text{grad}\left(\frac{1}{T}\right) - \sum_k \mathbf{J}_k\text{grad}\left(\frac{\mu_k}{T}\right) + \frac{1}{T}\sum_j A_j r_j$$
$$= \left(\mathbf{J}_h - \sum_k \mu_k\mathbf{J}_k\right)\text{grad}\left(\frac{1}{T}\right) - \sum_k \frac{1}{T}\mathbf{J}_k\text{grad}(\mu_k) + \frac{1}{T}\sum_j A_j r_j. $$
$$(10.31)$$

We define the heat flux as[2]

$$\mathbf{J}_q = T\mathbf{J}_s. \qquad (10.32)$$

The entropy production is a bilinear form

$$\sigma = \mathbf{K}_q\mathbf{J}_q + \sum_k \mathbf{K}_k\mathbf{J}_k + \sum_j K_j r_j \qquad (10.33)$$

of the fluxes

$$\mathbf{J}_q, \quad \mathbf{J}_k, \quad r_j \qquad (10.34)$$

---

[2] The definition of the heat flux is not unique in the case of simultaneous diffusion and heat transport. Our definition is in analogy to the relation $T\text{d}S = \text{d}H - V\text{d}p - \sum_k \mu_k\text{d}N_k$ for an isobaric system. With this convention, the diffusion flux does not depend on the temperature gradient explicitly.

and the thermodynamic forces

$$\mathbf{K}_q = \mathrm{grad}\left(\frac{1}{T}\right),$$

$$\mathbf{K}_k = -\frac{1}{T}\mathrm{grad}(\mu_k),$$

$$K_j = \frac{1}{T}A_j. \tag{10.35}$$

Instead of entropy production, alternatively the rate of energy dissipation can be considered which follows from multiplication of (10.31) with T

$$T\sigma = -\frac{1}{T}\mathbf{J}_q\mathrm{grad}(T) - \sum_k \mathbf{J}_k\mathrm{grad}(\mu_k) + \sum_j A_j r_j. \tag{10.36}$$

10.4 Phenomenological Relations

In equilibrium, all fluxes and thermodynamic forces vanish

$$T = \mathrm{const}, \quad \mu_k = \mathrm{const}, \quad A_j = 0. \tag{10.37}$$

The entropy production is also zero and entropy has its maximum value. If the deviation from equilibrium is small, the fluxes[3] can be approximated as linear functions of the forces:

$$J_\alpha = \sum_\beta L_{\alpha,\beta} K_\beta, \tag{10.38}$$

where the so-called Onsager coefficients $L_{\alpha,\beta}$ are material constants.[4] As Onsager has shown, the matrix of coefficients is symmetric $L_{\alpha,\beta} = L_{\beta,\alpha}$. The entropy production has the approximate form

$$\sigma = \sum_\alpha J_\alpha K_\alpha = \sum_{\alpha,\beta} L_{\alpha,\beta} K_\alpha K_\beta. \tag{10.39}$$

According to the second law of thermodynamics, $\sigma \geq 0$ and therefore the matrix of Onsager coefficients is positive definite.

10.5 Stationary States

Under certain circumstances,[5] stationary states are characterized by a minimum of entropy production, which is compatible with certain external conditions. Consider a system with n independent fluxes and thermodynamic

---

[3] Only a set of independent fluxes should be used here.
[4] Fluxes with different tensorial character are not coupled.
[5] Especially the Onsager coefficients have to be constants.

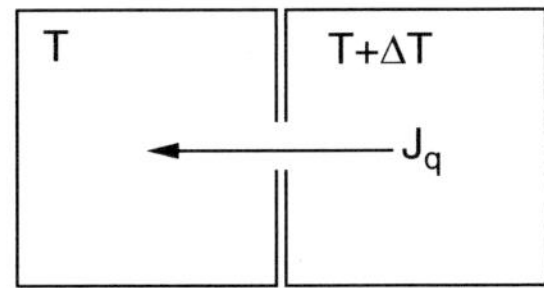

Fig. 10.3. Combined diffusion and heat transport

forces. If the forces $K_1, \ldots, K_m$ are kept fixed and a stationary state is reached, then the fluxes corresponding to the remaining forces vanish:

$$J_\alpha = 0, \quad \alpha = m+1, \ldots, n \tag{10.40}$$

and the entropy production is minimal since

$$\frac{\partial \sigma}{\partial K_\alpha}\bigg|_{K_1, \ldots, K_m} = J_\alpha, \quad \alpha = m+1, \ldots, n. \tag{10.41}$$

A simple example is the coupling of diffusion and heat transport between two reservoirs with fixed temperatures T and $T + \Delta T$ (Fig. 10.3).

In a stationary state, the diffusion current must vanish $\mathbf{J}_d = 0$ since any diffusion would lead to time-dependent concentrations. The entropy production is

$$\sigma = \mathbf{J}_q \mathbf{K}_q + \mathbf{J}_d \mathbf{K}_d. \tag{10.42}$$

If the force

$$\mathbf{K}_q = \operatorname{grad}\left(\frac{1}{T}\right) \approx -\frac{\Delta T}{T^2} \tag{10.43}$$

is fixed by keeping the temperature distribution fixed, then in the stationary state there is only transport of heat but not of mass. The entropy production is

$$\begin{aligned}
\sigma &= L_{qq} K_q^2 + L_{dd} K_d^2 + 2 L_{qd} K_q K_d \\
&= \frac{L_{qq} L_{dd} - L_{qd}^2}{L_{dd}} K_q^2 + L_{dd}\left(K_d + \frac{L_{qd}}{L_{dd}} K_q\right)^2,
\end{aligned} \tag{10.44}$$

where due to the positive definiteness of L_{ij}

$$L_{dd} > 0, \quad L_{qq} > 0, \quad L_{dd} L_{qq} - L_{qd}^2 > 0. \tag{10.45}$$

As a function of K_d, the entropy production is minimal for

$$0 = K_d + \frac{L_{qd}}{L_{dd}} K_q = \frac{1}{L_{dd}} J_d \tag{10.46}$$

hence in the stationary state. Obviously, the chemical potential adjusts such that

$$K_d = -\frac{1}{T}\Delta\mu = \frac{L_{qd}}{L_{dd}}\frac{\Delta T}{T^2},$$ (10.47)

which leads to a concentration gradient[6]:

$$\frac{\Delta c}{c} \approx -\frac{L_{qd}}{L_{dd}}\frac{\Delta T}{k_B T^2}.$$ (10.48)

Problems

10.1. Entropy Production by Chemical Reactions

Consider an isolated homogeneous system with $T = $ const and $\mu_k = $ const but nonzero chemical affinities

$$A_j = -\sum_k \mu_k \nu_{kj}$$

and determine the rate of entropy increase.

---

[6] This is connected to the Seebeck effect.

11

Simple Transport Processes

11.1 Heat Transport

Consider a system without chemical reactions and diffusion and with constant chemical potentials.[1] The thermodynamic force is

$$\mathbf{K}_q = \mathrm{grad}\left(\frac{1}{T}\right) \approx -\frac{1}{T_\mathrm{o}^2}\mathrm{grad}T \tag{11.1}$$

and the phenomenological relation is known as Fourier's law:

$$\mathbf{J}_q = -\kappa\,\mathrm{grad}T. \tag{11.2}$$

The entropy flux is

$$\mathbf{J}_s = \frac{1}{T}\mathbf{J}_q = \frac{1}{T}\mathbf{J}_h \tag{11.3}$$

and the energy dissipation is

$$T\sigma = -\frac{1}{T}J_q\mathrm{grad}T = \frac{\kappa}{T}\left(\mathrm{grad}T\right)^2. \tag{11.4}$$

From the continuity equation for the enthalpy, we find

$$\frac{\partial(\varrho h)}{\partial t} = c_\mathrm{p}\frac{\partial T}{\partial t} = \kappa\Delta T \tag{11.5}$$

and hence the diffusion equation for heat transport

$$\frac{\partial T}{\partial t} = \frac{\kappa}{c_\mathrm{p}}\Delta T. \tag{11.6}$$

For stationary heat transport in one dimension, the temperature gradient is constant, hence also the heat flux. The entropy flux, however, is coordinate dependent due to the local entropy production (Fig. 11.1).

$$\mathrm{div}\mathbf{J}_s = -\mathbf{J}_q\frac{1}{T^2}\mathrm{grad}(T) = \frac{\kappa}{T^2}\left(\mathrm{grad}T\right)^2 = \sigma. \tag{11.7}$$

---

[1] We neglect the temperature dependence of the chemical potential here.

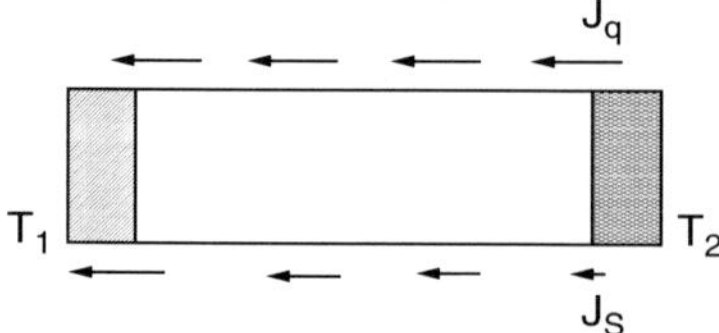

Fig. 11.1. Heat transport

11.2 Diffusion in an External Electric Field

Consider an ionic solution at constant temperature without chemical reactions. The thermodynamic force is

$$\mathbf{K}_k = -\frac{1}{T}\mathrm{grad}(\mu_k).$$ (11.8)

For a dilute solution, we have[2]

$$\mu_k = \mu_k^0 + k_\mathrm{B}T\ln(c_k) + Z_k e\Phi$$ (11.9)

and hence the thermodynamic force

$$\mathbf{K}_k = -\frac{k_\mathrm{B}}{c_k}\mathrm{grad}(c_k) - \frac{1}{T}\mathrm{grad}\left(Z_k e\Phi\right).$$ (11.10)

The entropy production is given by

$$T\sigma = \sum_k \mathbf{J}_k \left(-\frac{k_\mathrm{B}T}{c_k}\mathrm{grad}\, c_k - Z_k\, e\, \mathrm{grad}\,\Phi\right).$$ (11.11)

The phenomenological relations have the general form

$$\mathbf{J}_k = \sum_{k'} L_{k,k'}\mathbf{K}_k,$$ (11.12)

where the interaction mobilities $L_{k,k'}$ vanish for very small ion concentrations, whereas the direct mobilities $L_{k,k}$ are only weakly concentration dependent [39, 40]. Inserting the forces, we find

$$\mathbf{J}_k = -\sum_{k'} L_{k,k'}\frac{k_\mathrm{B}}{c_k}\mathrm{grad} c_{k'} - \frac{1}{T}\mathrm{grad}(\Phi)\sum_{k'} L_{k,k'}Z_{k'}e,$$ (11.13)

which for small ion concentrations can be written more simply by introducing the diffusion constant as

$$\mathbf{J}_k = -D_k\mathrm{grad}\, c_k - \frac{c_k Z_k e D_k}{k_\mathrm{B}T}\mathrm{grad}\,\Phi,$$ (11.14)

---

[2] The concentration (particles per volume) is $c_k = \varrho_k/m_k = \varrho x_k/m_k$.

which is known as the Nernst–Planck equation. This equation can be understood in terms of motion in a viscous medium (7.1). For the motion of a mass point, we have

$$m\dot{\mathbf{v}} = \mathbf{F} - m\gamma\mathbf{v}. \tag{11.15}$$

In a stationary state, the velocity is

$$\mathbf{v} = \frac{1}{m\gamma}\mathbf{F} \tag{11.16}$$

and the particle current

$$\mathbf{J} = c\mathbf{v} = \frac{c}{m\gamma}\mathbf{F} = \frac{cD}{k_{\mathrm{B}}T}\mathbf{F}, \tag{11.17}$$

where we used the Einstein relation

$$D = \frac{k_{\mathrm{B}}T}{m\gamma}. \tag{11.18}$$

From comparison with the Nernst–Planck equation, we find

$$\begin{aligned}
\mathbf{F} &= -\frac{k_{\mathrm{B}}T}{cD}\left(D\mathrm{grad}c + \frac{cZeD}{k_{\mathrm{B}}T}\mathrm{grad}\,\Phi\right) \\
&= -\frac{k_{\mathrm{B}}T}{c}\mathrm{grad}c - \mathrm{grad}\,Ze\Phi.
\end{aligned} \tag{11.19}$$

The charge current is

$$\begin{aligned}
\mathbf{J}_{\mathrm{el}} = Z_k e\mathbf{J}_k &= -\frac{Z_k e c_k D_k}{k_{\mathrm{B}}T}\left(\frac{k_{\mathrm{B}}T}{c_k}\mathrm{grad}c_k + \mathrm{grad}Z_k e\Phi\right) \\
&= -Z_k e D_k \mathrm{grad}\,c_k - \frac{(Z_k e)^2 c_k D_k}{k_{\mathrm{B}}T}\mathrm{grad}\,\Phi.
\end{aligned} \tag{11.20}$$

The prefactor of the potential gradient is connected to the electrical conductivity:

$$G_k = \frac{(Z_k e)^2 c_k D_k}{k_{\mathrm{B}}T}. \tag{11.21}$$

Together with the continuity equation

$$\frac{\partial c_k}{\partial t} = -\mathrm{div}\,\mathbf{J}_k, \tag{11.22}$$

we arrive at a Smoluchowski-type equation

$$\frac{\partial c_k}{\partial t} = \mathrm{div}\left(D_k\mathrm{grad}c_k + \frac{c_k Z_k eD}{k_{\mathrm{B}}T}\mathrm{grad}\Phi\right). \tag{11.23}$$

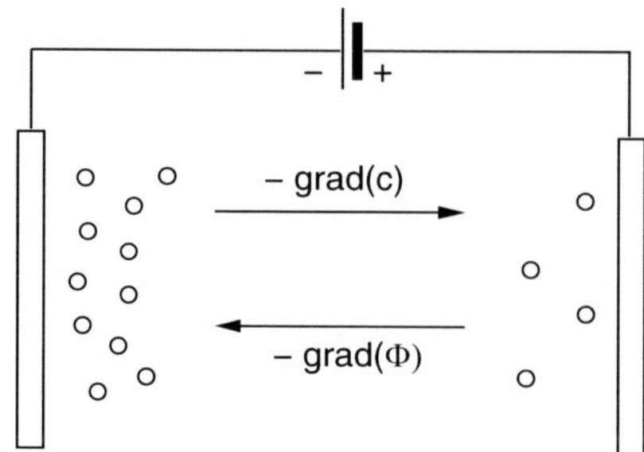

Fig. 11.2. Diffusion in an electric field

In equilibrium, we have

$$\mu_k = \text{const} \tag{11.24}$$

and hence

$$k_{\mathrm{B}}T \ln c_k + Z_k e \Phi = \text{const} \tag{11.25}$$

or

$$c_k = c_k^0 e^{-Z_k e \Phi / k_{\mathrm{B}} T}. \tag{11.26}$$

This is sometimes described as an equilibrium of the currents due to concentration gradient on the one side and to the electrical field on the other side (Fig. 11.2).

The Nernst–Planck equation has to be solved together with the Poisson–Boltzmann equation for the potential

$$\text{div}(\varepsilon \, \text{grad} \, \Phi) = -\sum Z_k e c_k. \tag{11.27}$$

In one dimension, the Poisson–Nernst–Planck equations are

$$J = -D \frac{\mathrm{d}c}{\mathrm{d}x} - \frac{Z e c D}{k_{\mathrm{B}} T} \frac{\mathrm{d}\Phi}{\mathrm{d}x}, \tag{11.28}$$

$$\frac{\mathrm{d}}{\mathrm{d}x} \epsilon \frac{\mathrm{d}}{\mathrm{d}x} \Phi = -\sum Z_k e c_k. \tag{11.29}$$

Problems

11.1. Synthesis of ATP from ADP

In mitochondrial membranes, ATP is synthesized according to

$$\text{ADP} + \text{POH} + 2\text{H}^+_{\text{out}} \rightleftharpoons \text{ATP} + \text{H}_2\text{O} + 2\text{H}^+_{\text{in}}$$

where (in) and (out) refer to inside and outside of the mitochondrial membrane.

Express the equilibrium constant K as a function of the potential difference

$$\Phi_{\mathrm{in}} - \Phi_{\mathrm{out}} \tag{11.30}$$

and the proton concentration ratio

$$c(\mathrm{H}_{\mathrm{in}}^{+})/c(\mathrm{H}_{\mathrm{out}}^{+}). \tag{11.31}$$

Ion Transport Through a Membrane

12.1 Diffusive Transport

We can draw a close analogy between diffusion and reaction on the one side and electronic circuit theory on the other side. We regard the membrane as a resistance for the current of diffusion and the difference in chemical potential as the driving force (Fig. 12.1):

$$\mu = \mu^0 + k_B T \ln c, \tag{12.1}$$

$$c = \exp \frac{\mu - \mu^0}{k_B T}. \tag{12.2}$$

Assuming linear variation of the chemical potential over the thickness a of the membrane, we have

$$\mu(x) = \mu_{\text{inside}} + \frac{x}{a} \Delta \mu. \tag{12.3}$$

The diffusion current is

$$J = -C \frac{\partial \mu}{\partial x} = -C \frac{\Delta \mu}{a} \tag{12.4}$$

and the constant is determined from the diffusion equation in the center of the membrane:

$$0 = \frac{\partial c}{\partial t} = -\frac{\partial}{\partial x}(J) = -\frac{\partial}{\partial x}\left(-C \frac{k_B T}{c} \frac{\partial c}{\partial x}\right) = \frac{\partial}{\partial x}\left(D(x) \frac{\partial}{\partial x} c\right)$$

$$\to C = \frac{\overline{D}}{k_B T} \overline{c} \tag{12.5}$$

with

$$\overline{c} = \exp \frac{\mu_{\text{inside}} + (\Delta \mu / 2)}{k_B T} = \exp \frac{\mu_{\text{outside}} + \mu_{\text{inside}}}{2 k_B T} = \sqrt{c_{\text{inside}} c_{\text{outside}}}. \tag{12.6}$$

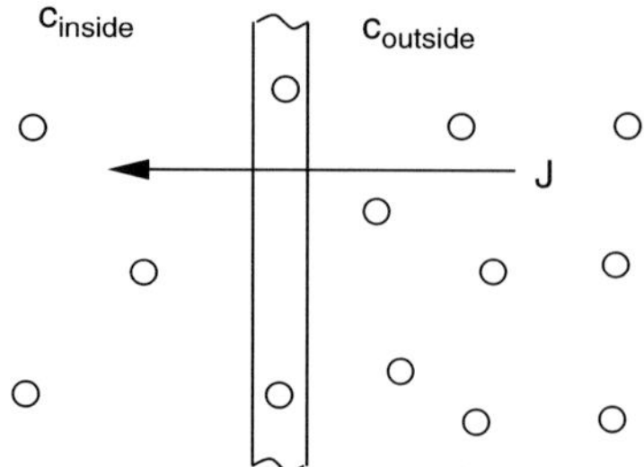

Fig. 12.1. Transport across a membrane

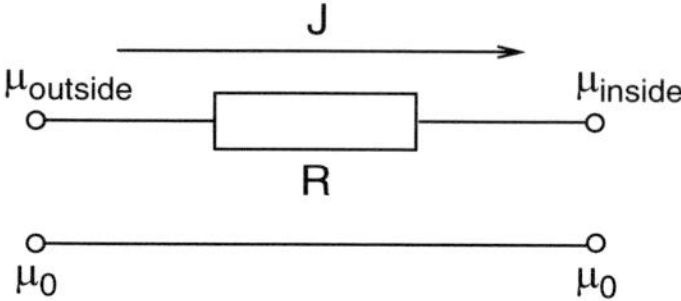

Fig. 12.2. Equivalent circuit for the diffusion through a membrane

Finally, we have (Fig. 12.2)

$$J = -\frac{\overline{D}\overline{c}\Delta\mu}{ak_\mathrm{B}T} = -\frac{\Delta\mu}{R}. \tag{12.7}$$

If we include a gradient of the electrostatic potential, then the equations become

$$\mu = \mu^0 + k_\mathrm{B}T\ln c + Ze\Phi, \tag{12.8}$$

$$c = \exp\frac{\mu - \mu^0 - Ze\Phi}{k_\mathrm{B}T}, \tag{12.9}$$

$$J = -C\frac{\partial\mu}{\partial x} = -C\frac{\Delta\mu}{a} = -C\frac{k_\mathrm{B}T\Delta\ln c + Ze\Delta\Phi}{a}, \tag{12.10}$$

and the electric current is

$$I = ZeJ = -\frac{(Ze)^2}{R}\left(\Delta\frac{k_\mathrm{B}T\ln c}{Ze} + \Delta\phi\right)$$
$$= -g(V_\mathrm{Nernst} - V) \tag{12.11}$$

with the Nernst potential

$$V_\mathrm{Nernst} = \frac{k_\mathrm{B}T}{Ze}\ln\frac{c_\mathrm{outside}}{c_\mathrm{inside}} \tag{12.12}$$

and the membrane potential (Fig. 12.3)

$$V = \Phi_\mathrm{inside} - \Phi_\mathrm{outside}. \tag{12.13}$$

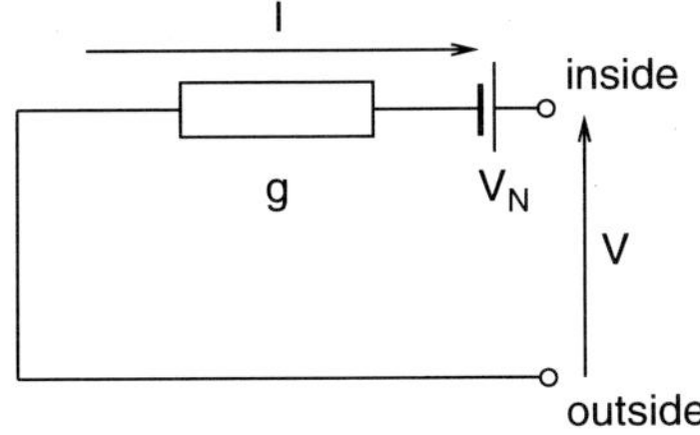

Fig. 12.3. Equivalent circuit for the electric current

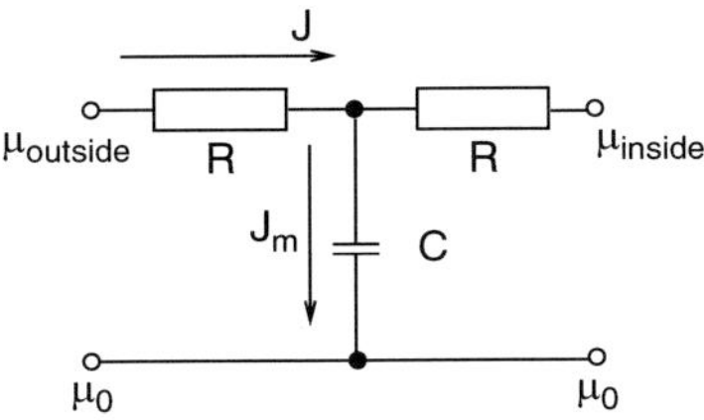

Fig. 12.4. Equivalent circuit with capacity

The transport through biological membranes occurs often via the formation of pore–ligand complexes. The number of these complexes depends on the chemical potential of the ligands and needs time to build up. It may therefore be more realistic to associate also a capacity C_m with the membrane, which we define in analogy to the electronic capacity via the change of the potential (Fig. 12.4)

$$J_\mathrm{m} = C_\mathrm{m}\frac{\mathrm{d}\mu_\mathrm{m}}{\mathrm{d}t}.$$
(12.14)

The change of the potential is

$$\frac{\mathrm{d}\mu_\mathrm{m}}{\mathrm{d}t} = \frac{k_\mathrm{B}T}{c}\frac{\mathrm{d}c}{\mathrm{d}t}$$
(12.15)

and the current is from the continuity equation

$$-\operatorname{div}J = \frac{J_\mathrm{m}}{a} = \frac{\mathrm{d}c}{\mathrm{d}t}.$$
(12.16)

Hence, the capacity is

$$C_\mathrm{m} = \frac{ac}{k_\mathrm{B}T}.$$
(12.17)

12.2 Goldman–Hodgkin–Katz Model

In the rest state of a neuron, the potassium concentration is higher in the interior whereas there are more sodium and chlorine ions outside. The concentration gradients have to be produced by (energy-consuming) ion pumps.

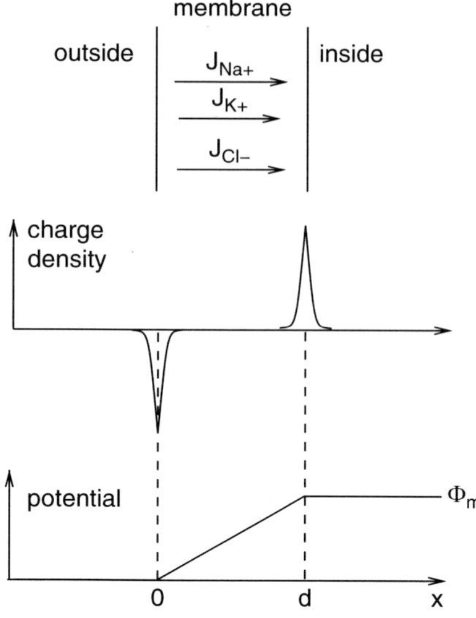

Fig. 12.5. Coupled ionic fluxes

Let us first consider only one type of ions and constant temperature. From (11.25), we have

$$\Phi + \frac{k_B T}{Z_k e} \ln c_k = \text{const.} \tag{12.18}$$

Usually, the potential is defined as zero on the outer side and

$$V_k = \Phi_{\text{inside}} - \Phi_{\text{outside}} = \frac{k_B T}{Z_k e} \ln \frac{c_{k,\text{outside}}}{c_{k,\text{inside}}} \tag{12.19}$$

is the Nernst potential (12.12) of the ion species k.

We now want to calculate the potential for a steady state with more than one ionic species present. If we take the ionic species Na^+, K^+, and Cl^- which are most important in nerve excitation that will define the so-called Goldman–Hodgkin–Katz model [41] (Fig. 12.5).

We want to calculate the dependence of the fluxes J_k on the concentrations $c_{k,\text{in}}$ and $c_{k,\text{out}}$. To that end, we multiply the Nernst–Planck (11.28) equation by e^{y_k} where

$$y_k = Z_k e \frac{\Phi}{k_B T} \tag{12.20}$$

to get

$$J_k e^{y_k} = -D_k e^{y_k} \left(\frac{dc_k}{dx} + c_k \frac{dy_k}{dx} \right) = -D_k \frac{d}{dx} (c_k e^{y_k}). \tag{12.21}$$

This can be integrated over the thickness d of the membrane

$$\int_0^d J_k \mathrm{e}^{y_k}\,\mathrm{d}x = -D\int_0^d \frac{\mathrm{d}}{\mathrm{d}x}(c_k \mathrm{e}^{y_k})\,\mathrm{d}x = -D\left(c_{k,\mathrm{in}}\mathrm{e}^{Z_k e\Phi_\mathrm{m}/k_\mathrm{B}T} - c_{k,\mathrm{out}}\right).$$

$$(12.22)$$

We assume a linear variation of the potential across the membrane

$$\Phi(x) = \frac{x}{d}\Phi_\mathrm{m}. \tag{12.23}$$

This is an approximation which is consistent with the condition of electroneutrality. On the boundary between liquid and membrane, the first derivative of Φ will be discontinuous corresponding to a surface charge distribution.[1] Assuming a stationary state with $\partial J/\partial x = 0$, we can evaluate the left-hand side of (12.22):

$$J_k \frac{k_\mathrm{B}Td}{Z_k e\Phi_\mathrm{m}}\left(\mathrm{e}^{Z_k e\Phi_\mathrm{m}/k_\mathrm{B}T} - 1\right) = -D\left(c_k(d)\mathrm{e}^{Z_k e\Phi_\mathrm{m}/k_\mathrm{B}T} - c_k(0)\right), \tag{12.24}$$

which gives the current

$$J_k = -\frac{D}{d}\frac{Z_k e\Phi_\mathrm{m}}{k_\mathrm{B}T}\frac{c_{k,\mathrm{in}}\mathrm{e}^{Z_k e\Phi_\mathrm{m}/k_\mathrm{B}T} - c_{k,\mathrm{out}}}{\mathrm{e}^{Z_k e\Phi_\mathrm{m}/k_\mathrm{B}T} - 1}. \tag{12.25}$$

In a stationary state, the total charge current has to vanish

$$I = \sum_k Z_k e J_k = 0 \tag{12.26}$$

and we find

$$\frac{e^2\Phi_\mathrm{m}}{k_\mathrm{B}Td}\sum D_k Z_k^2 \frac{c_{k,\mathrm{in}}\mathrm{e}^{Z_k e\Phi_\mathrm{m}/k_\mathrm{B}T} - c_{k,\mathrm{out}}}{\mathrm{e}^{Z_k e\Phi_\mathrm{m}/k_\mathrm{B}T} - 1} = 0. \tag{12.27}$$

With $Z_{\mathrm{Na}^+} = Z_{\mathrm{K}^+} = 1, Z_{\mathrm{Cl}^-} = -1$ and

$$y_\mathrm{m} = \frac{e\Phi_\mathrm{m}}{k_\mathrm{B}T}, \quad b_\mathrm{m} = \mathrm{e}^{y_\mathrm{m}}, \tag{12.28}$$

we have

$$D_{\mathrm{Na}^+}\frac{c_{\mathrm{Na,in}}b_\mathrm{m} - c_{\mathrm{Na,out}}}{b_\mathrm{m} - 1} + D_{\mathrm{K}^+}\frac{c_{\mathrm{K}^+,\mathrm{in}}b_\mathrm{m} - c_{\mathrm{K}^+,\mathrm{out}}}{b_\mathrm{m} - 1} \tag{12.29}$$

$$+D_{\mathrm{Cl}^-}\frac{c_{\mathrm{Cl,in}}b_\mathrm{m}^{-1} - c_{\mathrm{Cl,out}}}{b_\mathrm{m}^{-1} - 1} = 0. \tag{12.30}$$

Multiplication with $(b_\mathrm{m} - 1)$ gives

$$D_{\mathrm{Na}^+}(c_{\mathrm{Na,in}}b_\mathrm{m} - c_{\mathrm{Na,out}}) + D_{\mathrm{K}^+}(c_{\mathrm{K}^+,\mathrm{in}}b_\mathrm{m} - c_{\mathrm{K,out}}) \tag{12.31}$$

$$-b_\mathrm{m}D_{\mathrm{Cl}^-}(C_{\mathrm{Cl,in}}b_\mathrm{m}^{-1} - c_{\mathrm{Cl,out}}) = 0 \tag{12.32}$$

---

[1] We do not consider a possible variation of ϵ here.

and we find

$$b_{\mathrm{m}} = \frac{D_{\mathrm{Na}^+} c_{\mathrm{Na,out}} + D_{\mathrm{K}^+} c_{\mathrm{K,out}} + D_{\mathrm{Cl}^-} c_{\mathrm{Cl,in}}}{D_{\mathrm{Na}^+} c_{\mathrm{Na,in}} + D_{\mathrm{K}^+} c_{\mathrm{K,in}} + D_{\mathrm{Cl}^-} c_{\mathrm{Cl,out}}}, \tag{12.33}$$

and finally

$$\Phi_{\mathrm{m}} = \frac{k_{\mathrm{B}} T}{e} \ln \frac{D_{\mathrm{Na}^+} c_{\mathrm{Na,out}} + D_{\mathrm{K}^+} c_{\mathrm{K,out}} + D_{\mathrm{Cl}^-} c_{\mathrm{Cl,in}}}{D_{\mathrm{Na}^+} c_{\mathrm{Na,in}} + D_{\mathrm{K}^+} c_{\mathrm{K,in}} + D_{\mathrm{Cl}^-} c_{\mathrm{Cl,out}}}. \tag{12.34}$$

This formula should be compared with the Nernst equation:

$$V_k = \Phi_{\mathrm{inside}} - \Phi_{\mathrm{outside}} = \frac{k_{\mathrm{B}} T}{Z_k e} \ln \frac{c_{k,\mathrm{outside}}}{c_{k,\mathrm{inside}}}. \tag{12.35}$$

The ionic contributions appear weighted with their mobilities. The Nernst equation is obtained if the membrane is permeable for only one ionic species.

12.3 Hodgkin–Huxley Model

In 1952, Hodgkin and Huxley won the Nobel prize for their quantitative description of the squid giant axon dynamics [42].

They thought of the axon membrane as an electrical circuit. They assumed independent currents of sodium and potassium, a capacitive current, and a catch-all leak current. The total current is the sum of these (Fig. 12.6):

$$I_{\mathrm{ext}} = I_{\mathrm{C}} + I_{\mathrm{Na}} + I_{\mathrm{K}} + I_{\mathrm{L}}. \tag{12.36}$$

The capacitive current is

$$I_{\mathrm{C}} = C_{\mathrm{m}} \frac{\mathrm{d}V}{\mathrm{d}t}. \tag{12.37}$$

For the ionic currents, we have (12.11)

$$I_k = g_k (V - V_k), \tag{12.38}$$

where g_k is the channel conductance which depends on the membrane potential and on time, and V_k is the specific Nernst potential (12.12). The macroscopic current relates to a large number of ion channels which are controlled

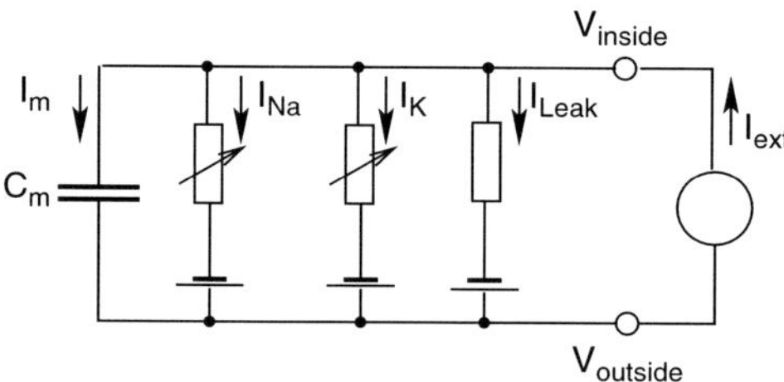

Fig. 12.6. Hodgkin–Huxley model

by gates which can be in either permissive or nonpermissive state. Hodgkin and Huxley considered a simple rate process for the transition between the two states with voltage-dependent rate constants

$$\text{closed} \; \underset{\beta(V)}{\overset{\alpha(V)}{\rightleftharpoons}} \; \text{open.} \tag{12.39}$$

The variable p which denotes the number of open gates obeys a first-order kinetics:

$$\frac{\mathrm{d}p}{\mathrm{d}t} = \alpha(1 - p) - \beta p. \tag{12.40}$$

Hodgkin and Huxley introduced a gating particle n describing the potassium conductance (Fig. 12.7) and two gating particles for sodium (m being activating and h being inactivating). They used the specific equations

$$g_{\mathrm{K}} = \overline{g}_k p_n^4, \tag{12.41}$$

$$g_{\mathrm{Na}} = \overline{g}_{\mathrm{Na}} p_m^3 p_h, \tag{12.42}$$

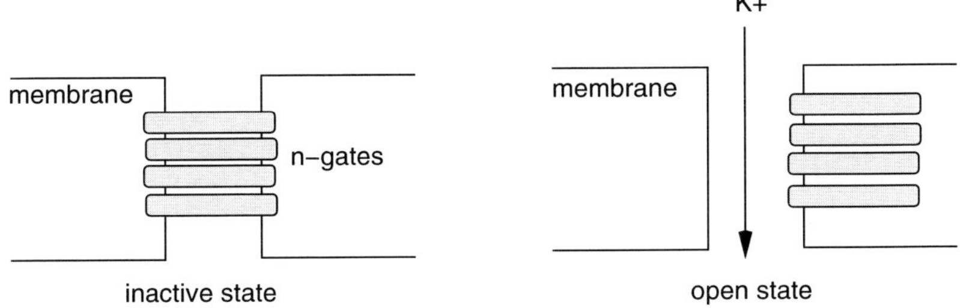

Fig. 12.7. Potassium gates

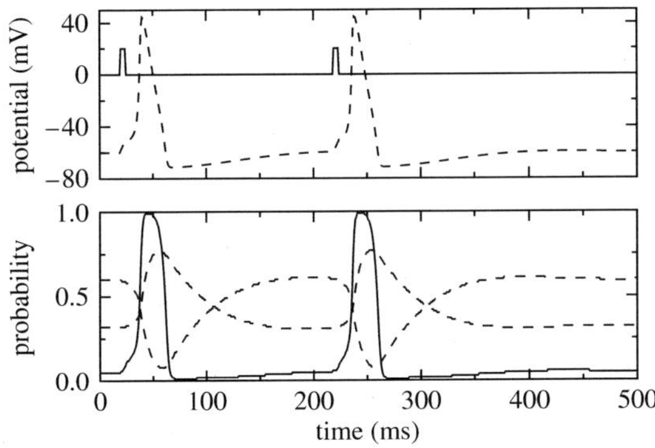

Fig. 12.8. Hodgkin–Huxley model. Equation (12.43) is solved numerically [43]. *Top*: excitation by two short rectangular pulses $I_{\mathrm{ext}}(t)$ (*solid*) and membrane potential $V(t)$ (*dashed*). *Bottom*: fraction of open gates p_m (*full*), p_n (*dashed*), and p_h (*dash-dotted*)

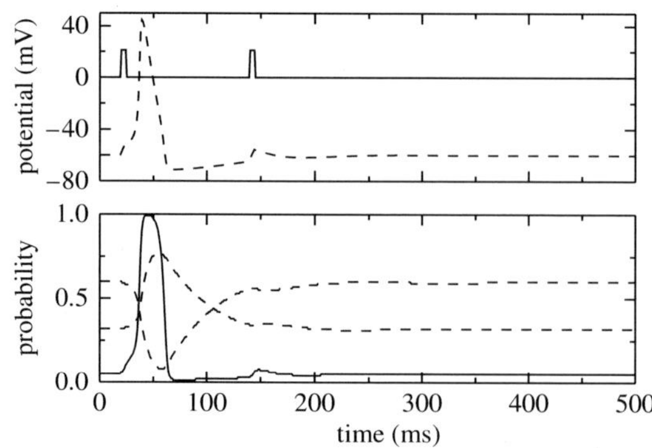

Fig. 12.9. Refractoriness of the Hodgkin–Huxley model. A second pulse cannot excite the system during the refractory period. Details as in Fig. 12.8

which gives finally a highly nonlinear differential equation:

$$I_{\text{ext}}(t) = C_{\text{m}}\frac{\mathrm{d}V}{\mathrm{d}t} + \overline{g}_k p_n^4 (V - V_{\text{K}}) + \overline{g}_{\text{Na}} p_m^3 p_h (V - V_{\text{Na}}) + \overline{g}_{\text{L}} (V - V_{\text{L}}). \quad (12.43)$$

This equation has to be solved numerically. It describes many electrical properties of the neuron-like threshold and refractory behavior, repetitive spiking, and temperature dependence (Figs. 12.8 and 12.9). Later, we will discuss a simplified model with similar properties in more detail (Sect. 13.3).

13

Reaction–Diffusion Systems

Reaction–diffusion systems are described by nonlinear equations and show the formation of structures. Static as well as dynamic patterns found in biology can be very realistically simulated [44].

13.1 Derivation

We consider coupling of diffusive motion and chemical reactions. For constant temperature and total density, we have the continuity equation

$$\frac{\partial}{\partial t} c_k = -\mathrm{div}\,\mathbf{J}_k + \sum_j \nu_{kj} r_j \tag{13.1}$$

together with the phenomenological equation

$$\mathbf{J}_k = -\sum_j L_{kj}\,\mathrm{grad}\,k\ln c_j = -\sum_j D_{kj}\,\mathrm{grad}\,c_j, \tag{13.2}$$

which we combine to get

$$\frac{\partial}{\partial t} c_k = \sum_j D_{kj}\Delta c_j + \sum_j \nu_{kj} r_j, \tag{13.3}$$

where the reaction rates are nonlinear functions of the concentrations:

$$\sum_j \nu_{kj} r_j = F_k(\{c_i\}). \tag{13.4}$$

We assume for the diffusion coefficients the simplest case that the different species diffuse independently (but possibly with different diffusion constants) and that the diffusion fluxes are parallel to the corresponding concentration gradients. We write the diffusion–reaction equation in matrix form

$$\frac{\partial}{\partial t}\begin{pmatrix} c_1 \\ \vdots \\ c_N \end{pmatrix} = \begin{pmatrix} D_1 & & \\ & \ddots & \\ & & D_N \end{pmatrix} \Delta \begin{pmatrix} c_1 \\ \vdots \\ c_N \end{pmatrix} + \begin{pmatrix} F_1(\{c\}) \\ \vdots \\ F_N(\{c\}) \end{pmatrix} \tag{13.5}$$

or briefly

$$\frac{\partial}{\partial t}\mathbf{c} = D\Delta\mathbf{c} + \mathbf{F}(\mathbf{c}). \tag{13.6}$$

13.2 Linearization

Since a solution of the nonlinear equations is generally possible only numerically, we investigate small deviations from an equilibrium solution $\mathbf{c}_0$[1] with

$$\frac{\partial}{\partial t}\mathbf{c}_0 = 0, \tag{13.7}$$

$$\Delta\mathbf{c}_0 = 0. \tag{13.8}$$

Obviously, the equilibrium corresponds to a root

$$\mathbf{F}(\mathbf{c}_0) = 0. \tag{13.9}$$

We linearize the equations by setting

$$\mathbf{c} = \mathbf{c}_0 + \boldsymbol{\xi} \tag{13.10}$$

and expand around the equilibrium

$$\frac{\partial}{\partial t}\boldsymbol{\xi} = D\Delta\boldsymbol{\xi} + \mathbf{F}(\mathbf{c}_0 + \boldsymbol{\xi}) = D\Delta\boldsymbol{\xi} + \left.\frac{\partial \mathbf{F}}{\partial \mathbf{c}}\right|_{\mathbf{c}_0} \boldsymbol{\xi}. \tag{13.11}$$

Denoting the matrix of derivatives by M_0, we can discuss several types of instabilities:

Spatially constant solutions. For a spatially constant solution, we have

$$\frac{\partial}{\partial t}\boldsymbol{\xi} = M_0\boldsymbol{\xi} \tag{13.12}$$

with the formal solution

$$\boldsymbol{\xi} = \boldsymbol{\xi}_0 \exp(M_0 t). \tag{13.13}$$

Depending on the eigenvalues of M, there can be exponentially growing or decaying solutions, oscillating solutions, and exponentially growing or decaying solutions.

---

[1] We assume tacitly that such a solution exists.

Plane waves. Plane waves are solutions of the linearized problem. Using the Ansatz

$$\boldsymbol{\xi} = \boldsymbol{\xi}_0 e^{i(\omega t - \mathbf{k}\mathbf{x})} \tag{13.14}$$

gives

$$i\omega\boldsymbol{\xi} = -k^2 D\boldsymbol{\xi} + M_0\boldsymbol{\xi} = M_k\boldsymbol{\xi}. \tag{13.15}$$

For a stable plane wave solution, $\lambda = i\omega$ is an eigenvalue of

$$M_k = M_0 - k^2 D \tag{13.16}$$

with

$$\Re(\lambda) \le 0. \tag{13.17}$$

If there are purely imaginary eigenvalues for some $\mathbf{k}$, they correspond to stable solutions which are spatially inhomogeneous and lead to formation of certain patterns.

13.3 Fitzhugh–Nagumo Model

The Hodgkin–Huxley model (12.43) describes characteristic properties like the threshold behavior and the refractory period where the membrane is not excitable. Since the system of equations is rather complicated, a simpler model was developed by Fitzhugh (1961) and Nagumo (1962) which involves only two variables u, v (membrane potential and recovery variable):

$$\dot{u} = u - \frac{u^3}{3} - v + I(t), \tag{13.18}$$

$$\dot{v} = \epsilon(u + a - bv). \tag{13.19}$$

The standard parameter values are

$$a = 0.7, \quad b = 0.8, \quad \epsilon = 0.08. \tag{13.20}$$

Nagumo studied an electronic circuit with a tunnel diode which is described by rather similar equations (Fig. 13.1):

$$I = C\dot{V} + I_{\mathrm{diode}}(V) + I_{\mathrm{LR}}, \tag{13.21}$$

$$V = RI_{\mathrm{LR}} + L\dot{I}_{\mathrm{LR}} + E \rightarrow \dot{I}_{\mathrm{LR}} = \frac{V - E - RI_{\mathrm{LR}}}{L}. \tag{13.22}$$

The Fitzhugh–Nagumo model can be used to describe the excitation propagation along the membrane. After rescaling the original equations and adding diffusive terms, we obtain the reaction–diffusion equations:

$$\frac{\partial}{\partial t}\begin{pmatrix} u \\ v \end{pmatrix} = \begin{pmatrix} u - \frac{u^3}{3} - v + I(t) \\ \epsilon(u - bv + a) \end{pmatrix} + \begin{pmatrix} 1 \\ \delta \end{pmatrix} \Delta \begin{pmatrix} u \\ v \end{pmatrix}. \tag{13.23}$$

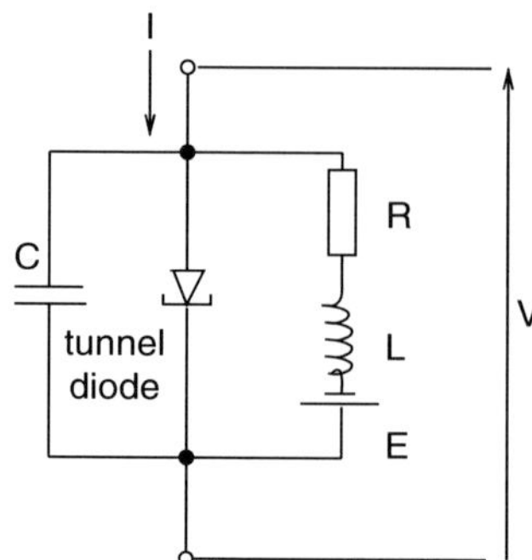

Fig. 13.1. Nagumo circuit

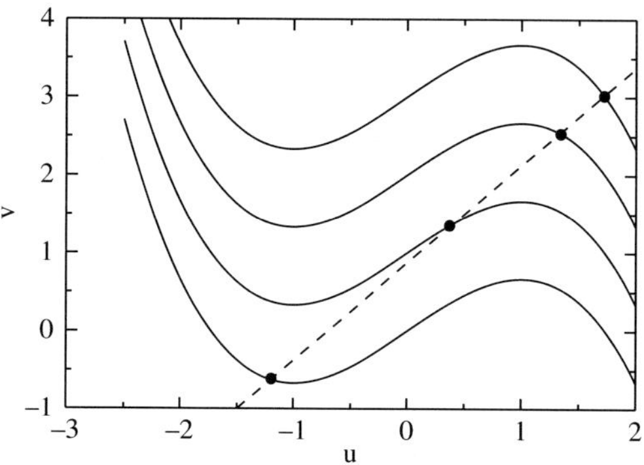

Fig. 13.2. Nullclines of the Fitzhugh–Nagumo equations. The solutions of (13.24) and (13.25) are shown for different values of the current $I = 0, 1, 2, 3$

The evolution of the FN model can be easily represented in a two-dimensional uv-plot. The u-nullcline and the v-nullcline are defined by

$$\dot{u} = 0 \rightarrow v = u - \frac{u^3}{3} + I_0, \tag{13.24}$$

$$\dot{v} = 0 \rightarrow v = \frac{a + u}{b}. \tag{13.25}$$

The intersection of the nullclines is an equilibrium since here $\dot{u} = \dot{v} = 0$ (Fig. 13.2).

Consider small deviations from the equilibrium values

$$u = u_{\mathrm{eq}} + \mu \quad v = v_{\mathrm{eq}} + \eta. \tag{13.26}$$

The linearized equations are

$$\dot{\mu} = (1 - u_{\mathrm{eq}}^2)\mu - \eta, \tag{13.27}$$

$$\dot{\eta} = \epsilon(\mu - b\eta). \tag{13.28}$$

Obviously, instability has to be expected for $u_{\mathrm{eq}}^2 < 1$.

The matrix of derivatives

$$M_0 = \begin{pmatrix} 1 - u_{\text{eq}}^2 & -1 \\ \epsilon & -\epsilon b \end{pmatrix} \qquad (13.29)$$

has the eigenvalues

$$\lambda = \frac{1}{2}\left(1 - u_{\text{eq}}^2 - \epsilon b \pm \sqrt{(1 - u_{\text{eq}}^2 + \epsilon b)^2 - 4\epsilon}\right). \qquad (13.30)$$

The square root is zero for

$$u_{\text{eq}} = \pm\sqrt{1 + \epsilon b \pm 2\sqrt{\epsilon}}. \qquad (13.31)$$

This defines a region around $u_{\text{eq}} = \pm 1$ where the eigenvalues have a nonzero imaginary part (Fig. 13.3).

For standard parameters, we distinguish the regions

$$\begin{aligned} \text{I}: & \quad |u_{\text{eq}}| > \sqrt{1 + \epsilon b + 2\sqrt{M\epsilon}} = 1.277, \\ \text{II}: & \quad 0.706 < |u_{\text{eq}}| < 1.277, \\ \text{III}: & \quad |u_{\text{eq}}| < \sqrt{1 + \epsilon b - 2\sqrt{\epsilon}} = 0.706. \end{aligned} \qquad (13.32)$$

Within region II, the real part of both eigenvalues is (Fig. 13.4)

$$\Re\lambda = \frac{1}{2}(1 - u_{\text{eq}}^2 - \epsilon b) \qquad (13.33)$$

and since

$$1 + \epsilon b - 2\sqrt{\epsilon} < u_{\text{eq}}^2 < 1 + \epsilon b + 2\sqrt{\epsilon}, \qquad (13.34)$$

we have

$$1 - u_{\text{eq}}^2 - \epsilon b - 2\sqrt{\epsilon} + 2\epsilon b < 0 < 1 - u_{\text{eq}}^2 - \epsilon b + 2\sqrt{\epsilon} + 2\epsilon b, \qquad (13.35)$$

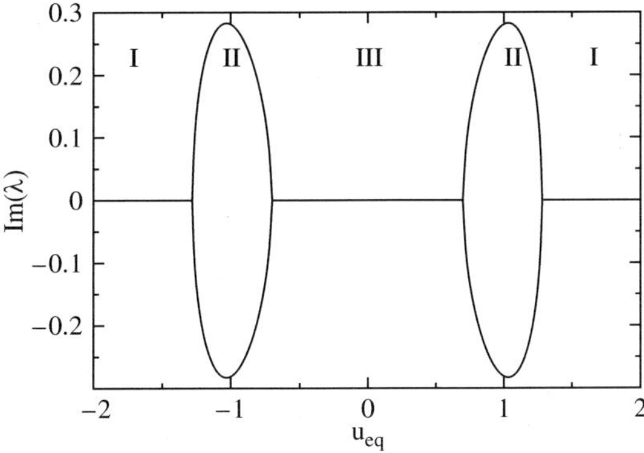

Fig. 13.3. Imaginary part of the eigenvalues

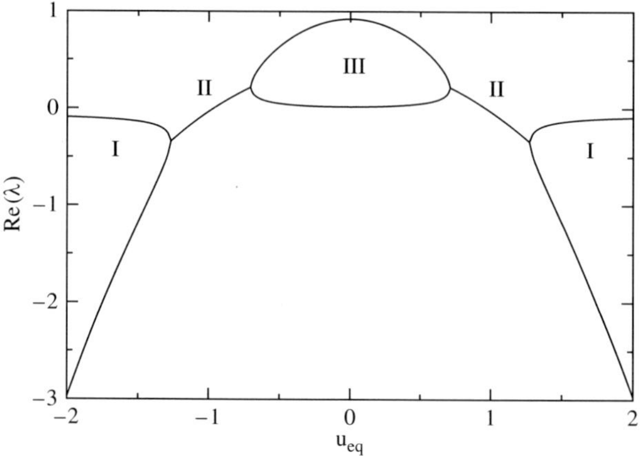

Fig. 13.4. Real part of the eigenvalues

$$\Re\lambda + \epsilon b - \sqrt{\epsilon} < 0 < \Re\lambda + \epsilon b + \sqrt{\epsilon}, \tag{13.36}$$

$$|\Re\lambda + \epsilon b| < \sqrt{\epsilon}. \tag{13.37}$$

For standard parameters,

$$-\,0.35 < \Re\lambda < 0.22. \tag{13.38}$$

Oscillating instabilities appear in region II if

$$\sqrt{\epsilon} < \epsilon b \rightarrow \epsilon < b, \tag{13.39}$$

which is the case for standard parameters. Outside region II instabilities appear if at least the larger of the two real eigenvalues is positive or

$$\sqrt{(1 - u_{\mathrm{eq}}^2 + \epsilon b)^2 - 4\epsilon} > u_{\mathrm{eq}}^2 - 1 + \epsilon b. \tag{13.40}$$

This is the case if the right-hand side is negative or

$$u_{\mathrm{eq}}^2 < 1 - \epsilon b \tag{13.41}$$

hence in the center of region III if $\epsilon b < 1$. For standard parameters $1 - \epsilon b = 0.936$ and according to (13.32), the whole region III is instable. If the right-hand side of (13.40) is positive, we can take the square

$$(1 - u_{\mathrm{eq}}^2 + \epsilon b)^2 - 4\epsilon > (u_{\mathrm{eq}}^2 - 1 + \epsilon b)^2, \tag{13.42}$$

which simplifies to

$$u_{\mathrm{eq}}^2 < 1 - \frac{1}{b}. \tag{13.43}$$

This is false for $b < 1$ and hence for standard parameters the whole region I is stable.

Part V

Reaction Rate Theory

14

Equilibrium Reactions

In this chapter, we study chemical equilibrium reactions. In thermal equilibrium of forward and backward reactions, the overall reaction rate vanishes and the ratio of the rate constants gives the equilibrium constant, which usually shows an exponential dependence on the inverse temperature.[1]

14.1 Arrhenius Law

Reaction rate theory goes back to Arrhenius who in 1889 investigated the temperature-dependent rates of inversion of sugar in the presence of acids. Empirically, a temperature dependence is often observed of the form

$$k(T) = A e^{-E_{\mathrm{a}}/k_{\mathrm{B}}T} \tag{14.1}$$

with the activation energy E_{a}. Considering a chemical equilibrium (Fig. 14.1)

$$A \underset{k'}{\overset{k}{\rightleftharpoons}} B, \tag{14.2}$$

this gives for the equilibrium constant

$$K = \frac{k}{k'} \tag{14.3}$$

and

$$\ln K = \ln k - \ln k' = \ln A - \ln A' - \frac{E_{\mathrm{a}} - E_{\mathrm{a}}'}{k_{\mathrm{B}}T}. \tag{14.4}$$

---

[1] An overview over the development of rate theory during the past century is given in [45].

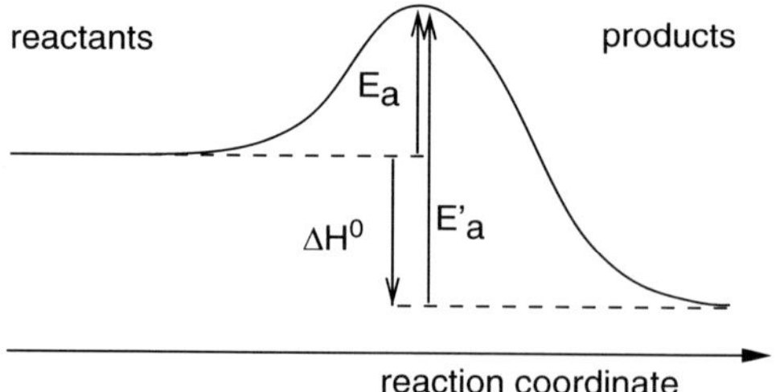

Fig. 14.1. Arrhenius law

In equilibrium, the thermodynamic forces vanish

$$T = \text{const}, \tag{14.5}$$

$$A = \sum_k \mu_k \nu_k = 0. \tag{14.6}$$

For dilute solutions with

$$\mu_k = \mu_k^0 + k_B T \ln c_k, \tag{14.7}$$

we have

$$\sum_k \mu_k^0 \nu_k + k_B T \sum_k \nu_k \ln c_k = 0, \tag{14.8}$$

which gives the van't Hoff relation for the equilibrium constant

$$\ln(K_c) = \sum_k \nu_k \ln c_k = -\frac{\sum_k \mu_k^0 \nu_k}{k_B T} = -\frac{\Delta G^0}{k_B T}. \tag{14.9}$$

The standard reaction free energy can be divided into an entropic and an energetic part:

$$-\frac{\Delta G^0}{k_B T} = \frac{-\Delta H^0}{k_B T} + \frac{\Delta S^0}{k}. \tag{14.10}$$

Since volume changes are not important at atmospheric pressure, the free reaction enthalpy gives the activation energy difference

$$E_a - E'_a = \Delta H^0. \tag{14.11}$$

A catalyst can only change the activation energies but never the difference ΔH^0.

14.2 Statistical Interpretation of the Equilibrium Constant

The chemical potential can be obtained as

$$\mu_k = \left(\frac{\partial F}{\partial N_k}\right)_{T,V,N_k'} = -k_{\mathrm{B}}T\left(\frac{\partial \ln Z}{\partial N_k}\right)_{T,V,N_k'}. \tag{14.12}$$

Using the approximation of the ideal gas, we have

$$Z = \prod \frac{z_k^{N_k}}{N_k!} \tag{14.13}$$

and

$$\ln Z \approx \sum_k N_k \ln z_k - N_k \ln N_k + N_k, \tag{14.14}$$

which gives the chemical potential

$$\mu_k = -k_{\mathrm{B}}T \ln \frac{z_k}{N_k}. \tag{14.15}$$

Let us consider a simple isomerization reaction

$$A \rightleftharpoons B. \tag{14.16}$$

The partition functions for the two species are (Fig. 14.2)

$$z_{\mathrm{A}} = \sum_{n=0,1,\dots} e^{-\epsilon_n(\mathrm{A})/k_{\mathrm{B}}T}, \quad z_{\mathrm{B}} = \sum_{n=0,1,\dots} e^{-\epsilon_n(\mathrm{B})/k_{\mathrm{B}}T}. \tag{14.17}$$

In equilibrium,

$$\mu_{\mathrm{B}} - \mu_{\mathrm{A}} = 0, \tag{14.18}$$

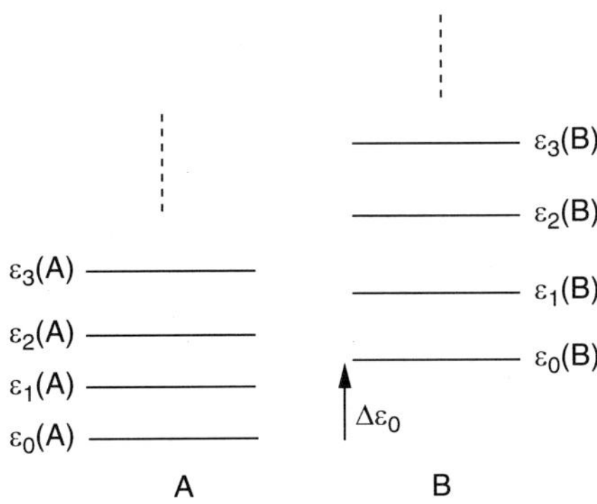

Fig. 14.2. Statistical interpretation of the equilibrium constant

$$- k_{\mathrm{B}} T \ln \frac{z_{\mathrm{B}}}{N_{\mathrm{B}}} = -k_{\mathrm{B}} T \ln \frac{z_{\mathrm{A}}}{N_{\mathrm{A}}}, \tag{14.19}$$

$$\frac{z_{\mathrm{B}}}{z_{\mathrm{A}}} = \frac{N_{\mathrm{B}}}{N_{\mathrm{A}}} = (N_{\mathrm{B}}/V)(N_{\mathrm{A}}/V)^{-1} = K_{\mathrm{c}}, \tag{14.20}$$

$$K_{\mathrm{c}} = \frac{\sum_{n=0,1,\dots} e^{-\epsilon_n(\mathrm{B})/k_{\mathrm{B}}T}}{\sum_{n=0,1,\dots} e^{-\epsilon_n(\mathrm{A})/k_{\mathrm{B}}T}} = \frac{\sum_{n=0,1,\dots} e^{-(\epsilon_n(\mathrm{B})-\epsilon_0(\mathrm{B}))/k_{\mathrm{B}}T}}{\sum_{n=0,1,\dots} e^{-(\epsilon_n(\mathrm{A})-\epsilon_0(\mathrm{A}))/k_{\mathrm{B}}T}} e^{-\Delta\epsilon/k_{\mathrm{B}}T}.$$

$$\tag{14.21}$$

This is nothing but a thermal distribution over all energy states of the system.

15

Calculation of Reaction Rates

The activated behavior of the reaction rate can be understood from a simple model of two colliding atoms (Fig. 15.1). In this chapter, we discuss the connection to transition state theory, which takes into account the internal degrees of freedom of larger molecules and explains not only the activation energy, but also the prefactor of the Arrhenius law [27, 28, 46].

15.1 Collision Theory

The Arrhenius expression consists of two factors. The exponential gives the number of reaction partners with enough energy and the prefactor gives the collision frequency times the efficiency.

We consider the collision of two spherical particles A and B in a coordinate system, where B is fixed and A is moving with the relative velocity

$$\mathbf{v}_r = \mathbf{v}_A - \mathbf{v}_B. \tag{15.1}$$

The two particles collide if the distance between their centers is smaller than the sum of the radii:

$$d = r_A + r_B. \tag{15.2}$$

During a time interval dt, the number of possible collision partners is

$$c_B \pi d^2 v_r dt \tag{15.3}$$

and the number of collisions within a volume V is

$$N_{\text{coll}}/V = c_A c_B \pi (r_A + r_B)^2 v_r dt. \tag{15.4}$$

Assuming independent Maxwell distributions for the velocities

$$f(\mathbf{v}_a \mathbf{v}_b) d^3 v_a d^3 v_b = \left(\frac{m}{2\pi k_B T} \right)^3 \exp\left\{ -\frac{m_a v_a^2}{2k_B T} - \frac{m_b v_b^2}{2k_B T} \right\} d^3 v_a d^3 v_b, \tag{15.5}$$

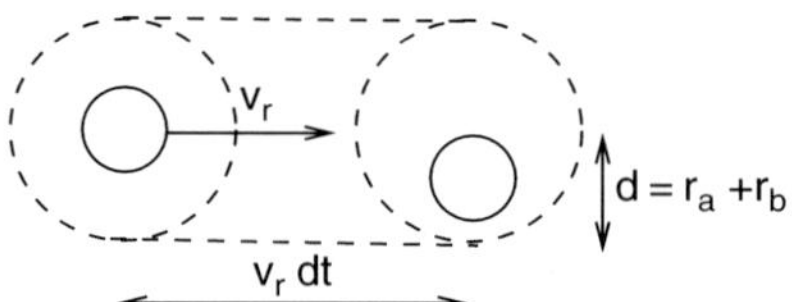

Fig. 15.1. Collision of two particles

we have a Maxwell distribution also for the relative velocities

$$f(\mathbf{v}_\mathrm{r})d^3v_\mathrm{r} = \left(\frac{\mu}{2\pi k_\mathrm{B}T}\right)^{3/2} \exp\left\{-\frac{\mu v_\mathrm{r}^2}{2k_\mathrm{B}T}\right\} d^3 v_\mathrm{r} \tag{15.6}$$

with the reduced mass

$$\mu = \frac{M_\mathrm{A} M_\mathrm{B}}{M_\mathrm{A} + M_\mathrm{B}}. \tag{15.7}$$

The average relative velocity then is

$$\overline{v_\mathrm{r}} = \sqrt{\frac{8k_\mathrm{B}T}{\pi\mu}} \tag{15.8}$$

and the collision frequency is

$$\frac{\mathrm{d}N_\mathrm{coll}}{V\mathrm{d}t} = c_\mathrm{a} c_\mathrm{b} \pi d^2 \sqrt{\frac{8k_\mathrm{B}T}{\pi\mu}}. \tag{15.9}$$

If both particles are of the same species, we have instead

$$\frac{\mathrm{d}N_\mathrm{coll}}{V\mathrm{d}t} = \frac{1}{2} c_\mathrm{a}^2 \pi d^2 \sqrt{\frac{16k_\mathrm{B}T}{\pi M}}, \tag{15.10}$$

since each collision involves two molecules. Since each collision will not lead to a chemical reaction, we introduce the reaction cross section $\sigma(E)$ which depends on the kinetic energy of the relative velocity. As a simple approximation, we use the reaction cross section of hard spheres:

$$\sigma(E) = \begin{cases} 0 & \text{if } E < E_0, \\ \pi(r_\mathrm{a} + r_\mathrm{b})^2 & \text{if } E > E_0, \end{cases} \tag{15.11}$$

where E_0 is the minimum energy necessary for a reaction to occur (Fig. 15.2).

The distribution of relative kinetic energy can be determined as follows. From the one-dimensional Maxwell distribution

$$f(v)\mathrm{d}v = \sqrt{\frac{m}{2\pi k_\mathrm{B}T}} e^{-mv^2/2k_\mathrm{B}T} \mathrm{d}v, \tag{15.12}$$

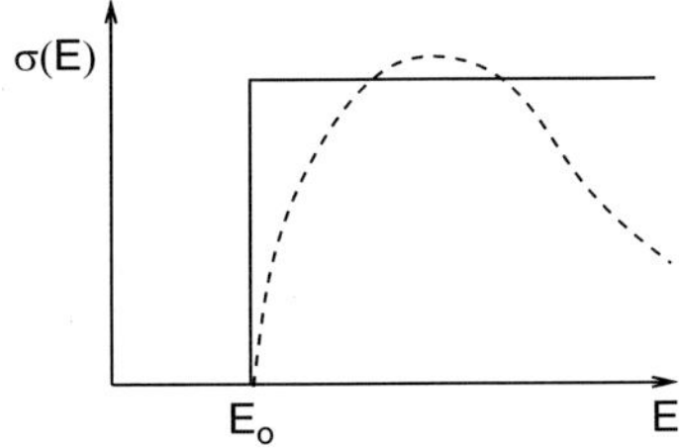

Fig. 15.2. Hard-sphere approximation to the reaction cross section

we find the distribution of relative kinetic energy for one particle as[1]

$$f(E_\mathrm{a})\mathrm{d}E_\mathrm{a} = 2\sqrt{\frac{m_\mathrm{a}}{2\pi k_\mathrm{B}T}}\mathrm{e}^{-E_\mathrm{a}/k_\mathrm{B}T}\frac{\mathrm{d}E_\mathrm{a}}{\sqrt{2mE_\mathrm{a}}} = \frac{1}{\sqrt{\pi k_\mathrm{B}TE_\mathrm{a}}}\mathrm{e}^{-E_\mathrm{a}/k_\mathrm{B}T}\mathrm{d}E_\mathrm{a}.$$

$$(15.13)$$

The joint distribution for the two particles is

$$f(E_\mathrm{a}, E_\mathrm{b})\mathrm{d}E_\mathrm{a}\mathrm{d}E_\mathrm{b}$$
$$= \left(\frac{1}{\sqrt{\pi k_\mathrm{B}TE_\mathrm{a}}}\mathrm{e}^{-E_\mathrm{a}/k_\mathrm{B}T}\mathrm{d}E_\mathrm{a}\right)\left(\frac{1}{\sqrt{\pi k_\mathrm{B}TE_\mathrm{b}}}\mathrm{e}^{-E_\mathrm{b}/k_\mathrm{B}T}\mathrm{d}E_\mathrm{b}\right) \quad (15.14)$$

and the distribution of the total kinetic energy is given by

$$f(E) = \int_0^E f(E_\mathrm{a}, E - E_\mathrm{a})\mathrm{d}E_\mathrm{a}$$
$$= \frac{1}{\pi k_\mathrm{B}T}\int_0^E \frac{1}{\sqrt{E_\mathrm{a}(E - E_\mathrm{a})}}\mathrm{e}^{-E_\mathrm{a}/k_\mathrm{B}T-(E-E_\mathrm{a})/k_\mathrm{B}T}\mathrm{d}E_\mathrm{a}$$
$$= \frac{1}{\pi k_\mathrm{B}T}\mathrm{e}^{-E/k_\mathrm{B}T}\int_0^E \frac{\mathrm{d}E_\mathrm{a}}{\sqrt{E_\mathrm{a}(E - E_\mathrm{a})}}. \quad (15.15)$$

This integral can be evaluated by substitution

$$E_\mathrm{a} = E\sin^2\theta, \quad \mathrm{d}E_\mathrm{a} = 2E\sin\theta\cos\theta\,\mathrm{d}\theta,$$

$$\int_0^E \frac{\mathrm{d}E_\mathrm{a}}{\sqrt{E_\mathrm{a}(E - E_\mathrm{a})}} = \int_0^{\pi/2}\frac{2E\sin\theta\cos\theta\mathrm{d}\theta}{\sqrt{E\sin^2\theta E\cos^2\theta}} = \int_0^{\pi/2}2\mathrm{d}\theta = \pi \quad (15.16)$$

and we find

$$f(E) = \frac{1}{k_\mathrm{B}T}\mathrm{e}^{-E/k_\mathrm{B}T}. \quad (15.17)$$

Now

$$\int_0^\infty f(E)\sigma(E)\mathrm{d}E = \int_{E_0}^\infty f(E)\pi d^2\mathrm{d}E = \pi d^2\mathrm{e}^{-E_0/k_\mathrm{B}T} \quad (15.18)$$

---

[1] There is a factor of 2 since $v^2 = (-v)^2$.

and we have finally

$$r = c_\mathrm{a} c_\mathrm{b} \pi d^2 \sqrt{\frac{8 k_\mathrm{B} T}{\pi \mu}}\, \mathrm{e}^{-E_0/k_\mathrm{B} T}. \tag{15.19}$$

For nonspherical molecules, the reaction rate depends on the relative orientation. Therefore, a so-called steric factor p is introduced:

$$r = c_\mathrm{a} c_\mathrm{b} p \pi d^2 \sqrt{\frac{8 k_\mathrm{B} T}{\pi \mu}}\, \mathrm{e}^{-E_0/k_\mathrm{B} T}. \tag{15.20}$$

There is no general way to calculate p and often it has to be estimated. Together, p and d^2 give an effective molecular diameter[2]

$$d_\mathrm{eff}^2 = p d^2. \tag{15.21}$$

15.2 Transition State Theory

According to transition state theory [47, 48], the reactants form an equilibrium amount of an activated complex which decomposes to yield the products (Figs. 15.3 and 15.4).

The reaction path [49] leads over a saddle point called the transition state. According to TST, the reaction of two substances A and B is written

$$\begin{array}{ccccc} & K & & k_1 & \\ \mathrm{A+B} & \rightleftharpoons & [\mathrm{A-B}]^{\ddagger} & \rightarrow & \mathrm{C+D} \\ \text{reactants} & & \text{activated complex} & & \text{products} \end{array} \quad . \tag{15.22}$$

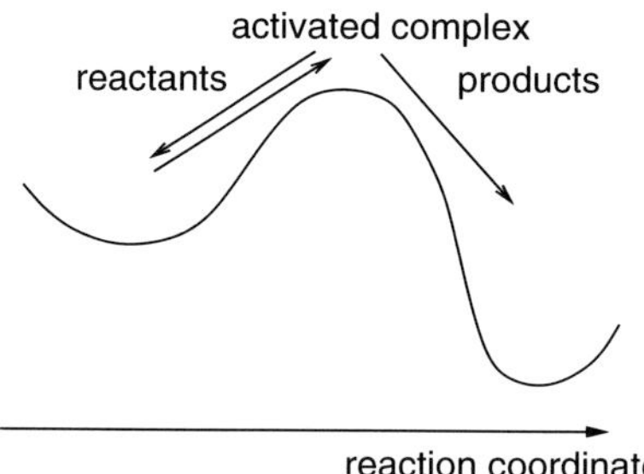

Fig. 15.3. Transition state theory. The reactants are in chemical equilibrium with the activated complex which can decompose into the products

---

[2] The quantity $\pi d^2/4$ is equivalent to the collision cross section used in connection with nuclear reactions.

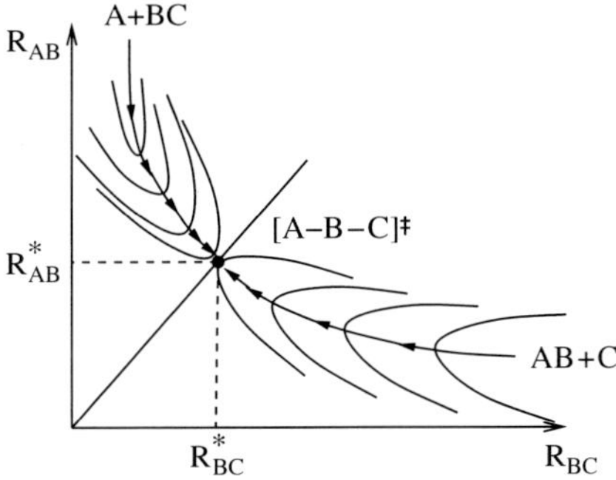

Fig. 15.4. Potential energy contour diagram. The potential energy for the reaction $A+BC \rightarrow [A-B-C]^{\ddagger} \rightarrow AB+C$ is shown schematically as a function of the distances R_{AB} and R_{BC}. In general, three coordinates are needed to give the relative positions of the nuclei, but if atom A approaches the molecule BC, the direction of minimum potential energy is along the line of centers for the reaction. *Arrows* indicate the reaction coordinate. The activated complex corresponds to a saddle point of the potential energy surface

If all molecules behave as ideal gases, the equilibrium constant K may be calculated as

$$K_{\mathrm{c}} = \frac{c_{[A-B]^{\ddagger}}}{c_A c_B} = \frac{z_{[A-B]^{\ddagger}}}{z_A z_B} e^{-\Delta E_0/k_B T}, \tag{15.23}$$

where ΔE_0 is the difference in zero-point energies of the reactants and the activated complex. The activated complex may be considered a normal molecule, except that one of its vibrational modes is so loose that it allows immediate decomposition into the products. The corresponding contribution to the partition function is

$$z_{\omega \to 0} = \frac{1}{1 - e^{-\hbar\omega/k_B T}} \to \frac{k_B T}{\hbar\omega}. \tag{15.24}$$

Thus, the equilibrium constant becomes

$$K = \frac{k_B T}{\hbar\omega} \frac{z^{\ddagger}}{z_A z_B} e^{-\Delta E_0/k_B T}, \tag{15.25}$$

where $z^{\ddagger}$ differs from $z_{[A-B]^{\ddagger}}$ by the contribution of the unique vibrational mode.

If $\omega/2\pi$ is considered the decomposition frequency, then the rate of decomposition is

$$r = \frac{dc_C}{dt} = \frac{\omega}{2\pi} c_{[A-B]^{\ddagger}} = \frac{k_B T}{2\pi\hbar} \frac{z^{\ddagger}}{z_A z_B} e^{-\Delta E_0/k_B T} c_A c_B = k_2 c_A c_B. \tag{15.26}$$

If not every activated complex decays into products, the expression for the rate constant has to be modified by a transmission coefficient κ:

$$k_2 = \kappa \frac{k_\mathrm{B}T}{2\pi\hbar} \frac{z^{\ddagger}}{z_\mathrm{A}z_\mathrm{B}} \mathrm{e}^{-\Delta E_0/k_\mathrm{B}T}. \tag{15.27}$$

In most cases, κ is close to 1. Important exceptions are the formation of a diatomic molecule, since the excess energy can only be eliminated by third-body collisions or radiation and reactions that involve changes from one type of electronic state to another – for instance in certain *cis–trans* isomerizations.

15.3 Comparison Between Collision Theory and Transition State Theory

For comparison, we apply transition state theory to the reaction of two spherical particles. For the reactants, only translational degrees of freedom are relevant giving the partition function

$$z_\mathrm{T} = \frac{(2\pi m k_\mathrm{B}T)^{3/2}}{h^3}. \tag{15.28}$$

For the activated complex, we consider the rotation around the center of mass. The moment of inertia is

$$I = (r_\mathrm{A} + r_\mathrm{B})^2 \frac{m_\mathrm{A}m_\mathrm{B}}{m_\mathrm{A} + m_\mathrm{B}} \tag{15.29}$$

and the partition function is

$$z_\mathrm{R} = \frac{8\pi^2 I k_\mathrm{B}T}{h^2}. \tag{15.30}$$

The rate expression from TST is

$$\begin{aligned}
k &= \frac{k_\mathrm{B}T}{h} \frac{(2\pi(m_\mathrm{A} + m_\mathrm{B})k_\mathrm{B}T)^{3/2}h^3}{(2\pi k_\mathrm{B}T)^3 m_\mathrm{A}^{3/2} m_\mathrm{B}^{3/2}} 8\pi^2 k_\mathrm{B}T \\
&\quad \times \frac{(r_\mathrm{A} + r_\mathrm{B})^2}{h^2} \frac{m_\mathrm{A}m_\mathrm{B}}{m_\mathrm{A} + m_\mathrm{B}} \mathrm{e}^{-\Delta E_0/k_\mathrm{B}T} \\
&= \sqrt{8\pi k_\mathrm{B}T} \sqrt{\frac{m_\mathrm{A} + m_\mathrm{B}}{m_\mathrm{A}m_\mathrm{B}}} (r_\mathrm{A} + r_\mathrm{B})^2 \mathrm{e}^{-\Delta E_0/k_\mathrm{B}T},
\end{aligned} \tag{15.31}$$

which is the same result as obtained from collision theory.

We consider now a bimolecular reaction. Each reactant has three translational, three rotational, and $3N - 6$ vibrational[3] degrees of freedom. The

---

[3] $3N - 5$ for a collinear molecule

partition functions are

$$
\begin{aligned}
z_{\mathrm{A}} &= z_{\mathrm{T}}^3 z_{\mathrm{R}}^3 z_{\mathrm{V}}^{3N_{\mathrm{A}}-6}, \\
z_{\mathrm{B}} &= z_{\mathrm{T}}^3 z_{\mathrm{R}}^3 z_{\mathrm{V}}^{3N_{\mathrm{B}}-6}, \\
z^{\ddagger} &= z_{\mathrm{T}}^3 z_{\mathrm{R}}^3 z_{\mathrm{V}}^{3N_{\mathrm{A}}+3N_{\mathrm{B}}-7}.
\end{aligned}
\tag{15.32}
$$

The reaction rate from TST is

$$
k_{\mathrm{TST}} = \frac{k_{\mathrm{B}}T}{h}\frac{z^{\ddagger}}{z_{\mathrm{A}}z_{\mathrm{B}}}\mathrm{e}^{-E_0/k_{\mathrm{B}}T} \approx \frac{k_{\mathrm{B}}T}{h}\frac{z_{\mathrm{V}}^5}{z_{\mathrm{T}}^3 z_{\mathrm{R}}^3}\mathrm{e}^{-E_0/k_{\mathrm{B}}T}.
\tag{15.33}
$$

For the rigid sphere model

$$
z_{\mathrm{A}} = z_{\mathrm{B}} = z_{\mathrm{T}}^3, \quad z^{\ddagger} = z_{\mathrm{T}}^3 z_{\mathrm{R}}^2
\tag{15.34}
$$

hence the rate constant is

$$
k_{\mathrm{rigid}} = \frac{k_{\mathrm{B}}T}{h}\frac{z_{\mathrm{R}}^2}{z_{\mathrm{T}}^3}\mathrm{e}^{-E_0/k_{\mathrm{B}}T}.
\tag{15.35}
$$

From comparison, we see that the steric factor

$$
p \approx \left(\frac{z_{\mathrm{V}}}{z_{\mathrm{R}}}\right)^5
\tag{15.36}
$$

which is typically in the order of 10^{-5}.

15.4 Thermodynamical Formulation of TST

Consider again the reaction

$$
\mathrm{A+B} \rightleftharpoons [\mathrm{A-B}]^{\ddagger} \rightarrow \text{products}
\tag{15.37}
$$

with the equilibrium constant

$$
K_{\mathrm{c}} = \frac{c_{\mathrm{AB}^{\ddagger}}}{c_{\mathrm{A}}c_{\mathrm{B}}}.
\tag{15.38}
$$

The TST rate expression (15.25) gives[4]

$$
k_2 = \frac{k_{\mathrm{B}}T}{2\pi\hbar}K_{\mathrm{c}}
\tag{15.39}
$$

---

[4] The activated complex is treated as a normal molecule with the exception of the special mode.

and with (14.9) we have for an ideal solution

$$k_2 = \frac{k_{\mathrm{B}}T}{2\pi\hbar}\mathrm{e}^{-\Delta G^{0\ddagger}/k_{\mathrm{B}}T} = \frac{k_{\mathrm{B}}T}{2\pi\hbar}\mathrm{e}^{-\Delta H^{0\ddagger}/k_{\mathrm{B}}T}\mathrm{e}^{\Delta S^{0\ddagger}/k_{\mathrm{B}}}. \tag{15.40}$$

The temperature dependence of the rate constant is

$$\frac{\mathrm{d}\ln k_2}{\mathrm{d}T} = \frac{1}{T} - \frac{\partial}{\partial T}\frac{\Delta G^{0\ddagger}}{k_{\mathrm{B}}T}. \tag{15.41}$$

Now for ideal gases or solutions, the chemical potential has the form

$$\mu = k_{\mathrm{B}}T\ln c - k_{\mathrm{B}}T\ln\frac{z}{V} \tag{15.42}$$

hence

$$\Delta G^{0\ddagger} = -k_{\mathrm{B}}T\sum\nu_i\ln\frac{z_i}{V} \tag{15.43}$$

and

$$\frac{\partial}{\partial T}\frac{\Delta G^{0\ddagger}}{k_{\mathrm{B}}T} = -\sum\nu_i\frac{\partial}{\partial T}\ln z_i = \sum\nu_i\frac{U_i}{k_{\mathrm{B}}T^2} = \Delta U^{0\ddagger}. \tag{15.44}$$

Comparison with the Arrhenius law

$$\frac{\mathrm{d}}{\mathrm{d}T}\left(\ln A - \frac{E_{\mathrm{a}}}{k_{\mathrm{B}}T}\right) = \frac{E_{\mathrm{a}}}{k_{\mathrm{B}}T^2} \tag{15.45}$$

shows that the activation energy is

$$E_{\mathrm{a}} = k_{\mathrm{B}}T + \Delta U^{0\ddagger} \approx k_{\mathrm{B}}T + \Delta H^{0\ddagger} \quad \text{(ideal solutions).} \tag{15.46}$$

For a bimolecular reaction in solution, we find from

$$r_{\mathrm{TST}} = \frac{k_{\mathrm{B}}T}{h}\mathrm{e}^{\Delta S^{0\ddagger}/k_{\mathrm{B}}}\mathrm{e}^{-\Delta H^{0\ddagger}/k_{\mathrm{B}}T} = \frac{k_{\mathrm{B}}T}{h}\mathrm{e}^{\Delta S^{0\ddagger}/k_{\mathrm{B}}}\mathrm{e}^{-(E_{\mathrm{a}}-k_{\mathrm{B}}T)/k_{\mathrm{B}}T} \tag{15.47}$$

that the steric factor is determined by the entropy of formation of the activated complex.

15.5 Kinetic Isotope Effects

There are two origins of the kinetic isotope effect. The first is quantum mechanical tunneling through the potential barrier.

It is only important at very low temperatures and for reactions involving very light atoms. For the simplest model case of a square barrier, the tunneling probability is approximately

$$P = \exp\left\{-2\sqrt{\frac{2m(V_0 - E)}{\hbar^2}}\,a\right\}. \tag{15.48}$$

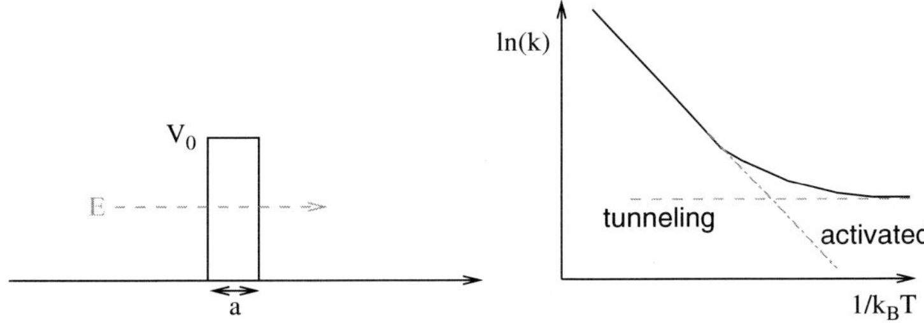

Fig. 15.5. Tunneling through a rectangular barrier. *Left*: the tunneling probability depends on the height and width of the barrier. *Right*: at low temperatures, tunneling dominates (*dashed*). At higher temperatures, activated behavior is observed (*dash-dotted*)

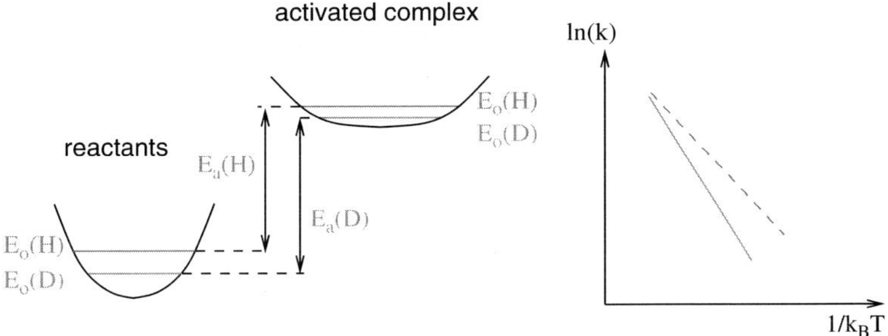

Fig. 15.6. Isotope dependence of the activation energy

The tunneling probability depends on the mass m but is independent of temperature (Fig. 15.5).

The second origin of isotope effects is the difference in activation energy for reactions involving different isotopes. Vibrational frequencies and therefore vibrational zero-point energies depend on the reduced mass μ of the vibrating system:

$$\hbar\omega = \frac{\hbar}{2\pi}\sqrt{\frac{f}{\mu}}.$$ (15.49)

Since the transition state is usually more loosely bound than the reactants, the vibrational levels are more closely spaced. Therefore, the activation energy is higher for the heavier isotopomer (a normal isotope effect). The maximum isotope effect is obtained when the bond involving the isotope is completely broken in the transition state. Then, the difference in activation energies is simply the difference in zero-point energies of the reactants (Fig. 15.6).

15.6 General Rate Expressions

The potential energy surface has a saddle point in the transition state region. The surface S^*, passing through that saddle point along the direction of steepest ascent, defines the border separating the reactants from the products in quite a natural way. We introduce the so-called intrinsic reaction coordinate (IRC) q_r as the path of steepest descent from the cole connecting reactants and products. We set $q_r = 0$ for the points on the surface S^*. The instantaneous rate $r(t)$ is given by the flux of systems that cross the surface S^* at time t and are on the product side at $t \to \infty$. This definition allows for multiple crossings of the surface S^*. In general, however, it will depend on how exactly the surface S^* is chosen. Only if the fluxes become stationary, this dependence vanishes.

15.6.1 The Flux Operator

Classically, the flux is the product of the number of systems passing through the surface S^* and their velocity v_r normal to that surface. As we defined the reaction coordinate q_r to be normal to the surface S^*, the velocity is

$$v_r = \frac{dq_r}{dt} \tag{15.50}$$

and the classical flux is

$$\langle F \rangle = \int d^{n-1}q \int d^n p \, v_r W(q,p,t) = \int d^n q \int d^n p \, \delta(q_r) \frac{p_r}{m_r} W(q,p,t). \tag{15.51}$$

Here, $W(q,p,t)$ is the phase-space distribution and m_r is the reduced mass corresponding to the coordinate q_r. We may therefore define the classical flux operator

$$F(q,p) = \delta(q_r) \frac{p_r}{m_r}. \tag{15.52}$$

In the quantum mechanical case, we have to modify this definition since p and q do not commute. It can be shown that we have to use the symmetrized product

$$F = \frac{1}{2m_r}\left(p_r \delta(q_r) + \delta(q_r) p_r\right) = \frac{1}{2m_r}\{p_r, \delta(q_r)\} \tag{15.53}$$

with

$$p_r = \frac{\hbar}{i} \frac{\partial}{\partial q_r}. \tag{15.54}$$

The projection operator on to the systems on the product side is simply given by

$$P = \theta(q_r). \tag{15.55}$$

However, we need the projector $P(t)$ on that states that at some time t in the future are on the product side. If H is the system Hamiltonian, then

$$P(t) = \mathrm{e}^{\mathrm{i}Ht/\hbar}\,\theta(q_\mathrm{r})\mathrm{e}^{\mathrm{i}Ht/\hbar} \tag{15.56}$$

and that part of the density matrix $\rho(t)$ that will end in the product side in the far future is

$$\lim_{t'\to\infty} P(t'-t)\rho(t)P(t'-t). \tag{15.57}$$

The rate is given by

$$r(t) = \lim_{t'\to\infty}\,\mathrm{tr}\left(P(t'-t)\rho(t)P(t'-t)F\right). \tag{15.58}$$

The time dependence of the density matrix is given by the van Neumann equation

$$\mathrm{i}\hbar\frac{\mathrm{d}\rho(t)}{\mathrm{d}t} = [H,\rho(t)]. \tag{15.59}$$

In thermal equilibrium,

$$\rho_\mathrm{eq} = Q^{-1}\exp\left(-\beta H\right), \quad Q = \mathrm{tr}\left(\exp(-\beta H)\right) \tag{15.60}$$

is time independent as $[\exp(-\beta H), H] = 0$. It can be shown that for $t' \to \infty$, the density matrix ρ_eq and the projector $P(t)$ commute[5]:

$$\lim_{t'\to\infty} P(t'-t)\rho_\mathrm{eq}P(t'-t) = \lim_{t'\to\infty} P(t')\rho_\mathrm{eq}. \tag{15.61}$$

The rate will then be also independent of t:

$$r = Q^{-1}\lim_{t'\to\infty}\,\mathrm{tr}\left(\mathrm{e}^{-\beta H}\mathrm{e}^{\mathrm{i}Ht'/\hbar}\,\theta(q_\mathrm{r})\,\mathrm{e}^{-\mathrm{i}Ht'/\hbar}\,F\right). \tag{15.62}$$

This expression can be modified such that the limit operation does not appear explicitly any more. To that end, we note that the trace vanishes for $t'=0$[6] and

$$r = Q^{-1}\mathrm{tr}\left(\mathrm{e}^{-\beta H}\mathrm{e}^{\mathrm{i}Ht'/\hbar}\,\theta(q_\mathrm{r})\,\mathrm{e}^{-\mathrm{i}Ht'/\hbar}\,F\right)\bigg|_{t'=0}^{t'=\infty}$$

$$= Q^{-1}\int_0^\infty \frac{\mathrm{d}}{\mathrm{d}t'}\left(\mathrm{e}^{-\beta H}\mathrm{e}^{\mathrm{i}Ht'/\hbar}\,\theta(q_\mathrm{r})\,\mathrm{e}^{-\mathrm{i}Ht'/\hbar}\,F\right)\mathrm{d}t'. \tag{15.63}$$

---

[5] Asymptotically, the states on the product side will leave the reaction zone with positive momentum p_r, so that we may replace $\theta(q_\mathrm{r})$ with $\theta(p_\mathrm{r})$. Again asymptotically, these states are eigenfunctions of the Hamiltonian, so that $P(t = \infty)$ and $\exp(-\beta H)$ commute.

[6] This follows from time inversion symmetry: the trace has to be symmetric with respect to that operation. Time inversion changes $p_\mathrm{r} \to -p_\mathrm{r}$. The operators ρ and q_r are not affected. So the trace has to be symmetric and antisymmetric at the same time.

For the derivative, we find

$$\frac{d}{dt} e^{iHt'/\hbar}\, \theta(q_r)\, e^{-iHt'/\hbar} = \frac{i}{\hbar} e^{iHt'/\hbar}\, [H, \theta(q_r)]\, e^{-iHt'/\hbar}. \tag{15.64}$$

Since only the kinetic energy $p_r^2/2m_r$ does not commute with $\theta(q_r)$, the commutator is

$$\begin{aligned}
[H, \theta(q_r)] &= \left[\frac{p_r^2}{2m_r}, \theta(q_r)\right] = \frac{\hbar}{i}\left(\frac{\partial}{\partial q_r}\frac{p_r}{2m_r}\theta(q_r) - \theta(q_r)\frac{\partial}{\partial q_r}\frac{p_r}{2m_r}\right)\\
&= \frac{\hbar}{i}\left(\frac{p_r}{2m_r}\delta(q_r) + \frac{p_r}{2m_r}\theta(q_r)\frac{\partial}{\partial q_r} - \theta(q_r)\frac{\partial}{\partial q_r}\frac{p_r}{2m_r}\right)\\
&= \frac{\hbar}{i}\left(\frac{p_r}{2m_r}\delta(q_r) + \delta(q_r)\frac{p_r}{2m_r}\right) = \frac{\hbar}{i}F. \tag{15.65}
\end{aligned}$$

Therefore, we can express the reaction rate as an integral over the flux–flux correlation function

$$r = Q^{-1}\int_0^\infty \mathrm{tr}\left(e^{-\beta H}F(t)F(0)\right)\mathrm{d}t = \int_0^\infty \langle F(t)F(0)\rangle \mathrm{d}t. \tag{15.66}$$

Ultimately, this expression for the reaction rate has been used a lot in numerical computations. It is quite general and may easily be extended to nonadiabatic reactions.

Problems

15.1. Transition State Theory

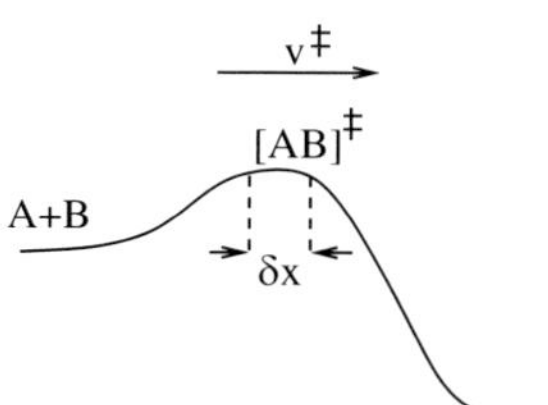

Instead of using a vibrational partition function to describe the motion of the activated complex over the reaction barrier, we can also use a translational partition function. We consider all complexes lying within a distance δx of the barrier to be activated complexes. Use the translational partition function for a particle of mass m in a box of length δx to obtain the TST rate constant. Assume that the average velocity of the particles moving over the barrier is[7]

$$v^\ddagger = \frac{1}{2}\sqrt{\frac{2k_B T}{\pi m^\ddagger}}.$$

---

[7] The particle moves to the left or right side with equal probability.

15.2. Harmonic Transition State Theory

For systems such as solids, which are well described as harmonic oscillators around stationary points, the harmonic form of TST is often a good approximation, which can be used to evaluate the TST rate constant, which can then be written as the product of the probability of finding the system in the transition state and the average velocity at the transition state:

$$k_{\text{TST}} = \langle v\delta(x - x^{\ddagger}) \rangle.$$

For a one-dimensional model, assume that the transition state is an exit point from the parabola at some position $x = x^{\ddagger}$ with energy $\Delta E = m\omega^2 x^{\ddagger 2}/2$ and evaluate the TST rate constant

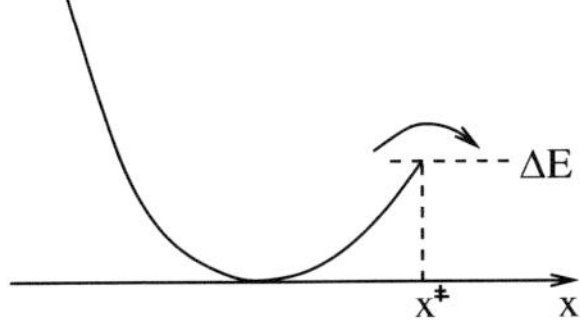

16

Marcus Theory of Electron Transfer

In 1992, Rudolph Marcus received the Nobel Prize for chemistry. His theory is currently the dominant theory of electron transfer in chemistry. Originally, Marcus explained outer-sphere electron transfer, introducing reorganization energy and electronic coupling to relate the thermodynamic transition state to nonequilibrium fluctuations of the medium polarization [50–52].

16.1 Phenomenological Description of ET

We want to describe the rate of electron transfer (ET) between two species, D and A, in solution. If D and A are neutral, this is a charge separation process:

$$D + A \rightarrow D^+ + A^-. \tag{16.1}$$

If the charge is transferred from one molecule to the other, this is a charge resonance process

$$D^- + A \rightarrow D + A^- \tag{16.2}$$

or

$$D + A^+ \rightarrow D^+ + A. \tag{16.3}$$

In the following, we want to treat all these kinds of processes together. Therefore from now on, the symbols D and A implicitly have the meaning D^{Z_D} and A^{Z_A} and the general reaction scheme

$$D^{Z_D} + A^{Z_A} \rightarrow D^{Z_D+1} + A^{Z_A-1} \tag{16.4}$$

will be simply denoted as

$$D + A \rightarrow D^+ + A^-. \tag{16.5}$$

Marcus theory is a phenomenological theory, in which the assumption is made, that ET proceeds in the following three steps:

1. The two species approach each other by diffusive motion and form the so-called encounter complex:

$$D + A \underset{k_{-\mathrm{dif}}}{\overset{k_{\mathrm{dif}}}{\rightleftharpoons}} [D - A]^{\ddagger}. \tag{16.6}$$

2. In the activated complex, ET takes place

$$[D - A]^{\ddagger} \underset{k_{-\mathrm{et}}}{\overset{k_{\mathrm{et}}}{\rightleftharpoons}} [D^{+}A^{-}]^{\ddagger}. \tag{16.7}$$

The actual ET is much faster than the motion of the nuclei. The nuclear conformation of the activated complex and the solvent polarization do not change. According to the Franck–Condon principle, the two states $[D - A]^{\ddagger}$ and $[D^{+}A^{-}]^{\ddagger}$ have to be isoenergetic.

3. After the transfer, the polarization returns to a new equilibrium and the products separate

$$[D^{+}A^{-}]^{\ddagger} \overset{k_{\mathrm{sep}}}{\rightarrow} D^{+} + A^{-}. \tag{16.8}$$

The overall reaction rate is

$$r = \frac{\mathrm{d}}{\mathrm{d}t} c_{A^-} = k_{\mathrm{sep}} c_{[D^+A^-]^{\ddagger}}. \tag{16.9}$$

We want to calculate the observed quenching rate

$$r = -\frac{\mathrm{d}}{\mathrm{d}t} c_{D} = k_{\mathrm{q}} c_{D} c_{A}. \tag{16.10}$$

We consider the stationary state

$$0 = \frac{\mathrm{d}}{\mathrm{d}t} c_{[D-A]^{\ddagger}} = \frac{\mathrm{d}}{\mathrm{d}t} c_{[D^+A^-]^{\ddagger}}. \tag{16.11}$$

We deduce that

$$0 = k_{\mathrm{dif}} c_{D} c_{A} + k_{-\mathrm{et}} c_{[D^+A^-]^{\ddagger}} - (k_{-\mathrm{dif}} + k_{\mathrm{et}}) c_{[D-A]^{\ddagger}}, \tag{16.12}$$

$$0 = k_{\mathrm{et}} c_{[D-A]^{\ddagger}} - (k_{-\mathrm{et}} + k_{\mathrm{sep}}) c_{[D^+A^-]^{\ddagger}}. \tag{16.13}$$

Eliminating the concentration of the encounter complex from these two equations, we find

$$\begin{aligned}
c_{[D^+A^-]^{\ddagger}} &= \frac{k_{\mathrm{et}}}{k_{-\mathrm{et}} + k_{\mathrm{sep}}} c_{[D-A]^{\ddagger}} \\
&= \frac{k_{\mathrm{et}} k_{\mathrm{dif}}}{k_{-\mathrm{dif}} k_{-\mathrm{et}} + k_{-\mathrm{dif}} k_{\mathrm{sep}} + k_{\mathrm{et}} k_{\mathrm{sep}}} c_{D} c_{A}
\end{aligned} \tag{16.14}$$

and the observed quenching rate is

$$k_{\mathrm{q}} = \frac{k_{\mathrm{dif}}}{1 + \frac{k_{-\mathrm{dif}}}{k_{\mathrm{et}}} + \frac{k_{-\mathrm{dif}}k_{-\mathrm{et}}}{k_{\mathrm{et}}k_{\mathrm{sep}}}} . \tag{16.15}$$

The reciprocal quenching rate can be written as

$$\frac{1}{k_{\mathrm{q}}} = \frac{1}{k_{\mathrm{dif}}} + \frac{k_{-\mathrm{dif}}}{k_{\mathrm{dif}}k_{\mathrm{et}}} + \frac{k_{-\mathrm{dif}}k_{-\mathrm{et}}}{k_{\mathrm{dif}}k_{\mathrm{et}}k_{\mathrm{sep}}} . \tag{16.16}$$

There are some interesting limiting cases:

- If $k_{\mathrm{et}} \gg k_{-\mathrm{dif}}$ and $k_{\mathrm{sep}} \gg k_{-\mathrm{et}}$, then the observed quenching rate is given by the diffusion rate $k_{\mathrm{q}} = k_{\mathrm{dif}}$ and the ET rate is limited by the formation of the activated complex.
- If the probability of ET is very small $k_{\mathrm{et}} \ll k_{-\mathrm{dif}}$ and $k_{-\mathrm{et}} \ll k_{\mathrm{dif}}$, then the quenching rate is given by $k_{\mathrm{q}} = k_{\mathrm{et}}k_{\mathrm{dif}}/k_{-\mathrm{dif}} = k_{\mathrm{et}}K_{\mathrm{dif}}$.
- If diffusion is not important as in solids or proteins and in the absence of further decay channels, the quenching rate is given by $k_{\mathrm{q}} \approx k_{\mathrm{et}}$.

16.2 Simple Explanation of Marcus Theory

The TST expression for the ET rate is

$$k_{\mathrm{et}} = \frac{\omega_{\mathrm{N}}}{2\pi}\kappa_{\mathrm{et}}\mathrm{e}^{-\Delta G^{\ddagger}/k_{\mathrm{B}}T}, \tag{16.17}$$

where κ_{et} describes the probability of a transition between the two electronic states DA and $\mathrm{D^+A^-}$, and ω_{N} is the effective frequency of the nuclear motion. We consider a collective reaction coordinate Q which summarizes the polarization coordinates of the system (Fig. 16.1).

Initially, the polarization is in equilibrium with the reactants and for small polarization fluctuations, series expansion of the Gibbs free energy gives

$$G_{\mathrm{r}}(Q) = G_{\mathrm{r}}^{(0)} + \frac{a}{2}Q^2 . \tag{16.18}$$

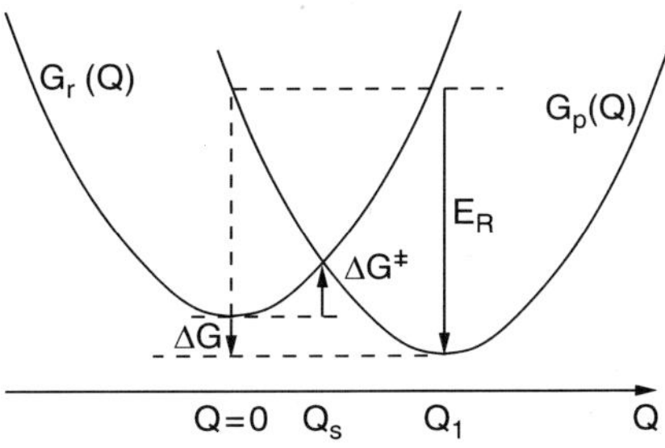

Fig. 16.1. Displaced oscillator model

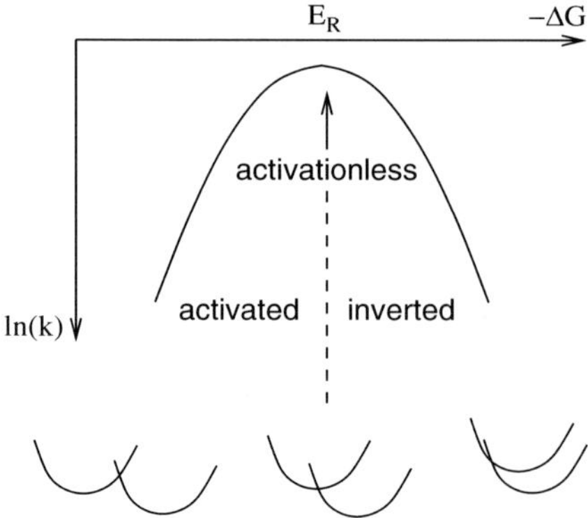

Fig. 16.2. Marcus parabola

If the polarization change induced by the ET is not too large (no dielectric saturation), the potential curve for the final state will have the same curvature but is shifted to a new equilibrium value Q_1:

$$G_p(Q) = G_p^{(0)} + \frac{a}{2}(Q - Q_1)^2 = G_r^{(0)} + \Delta G + \frac{a}{2}(Q - Q_1)^2. \qquad (16.19)$$

In the new equilibrium, the final state is stabilized by the reorganization energy[1]

$$E_R = -(G_p(Q_1) - G_p(0)) = \frac{a}{2}Q_1^2. \qquad (16.20)$$

The transition between the two states is only possible if due to some thermal fluctuations the crossing point is reached which is at

$$Q_s = \frac{\frac{a}{2}Q_1^2 + \Delta G}{aQ_1} = \frac{E_R + \Delta G}{\sqrt{2aE_R}}. \qquad (16.21)$$

The corresponding activation energy is

$$\Delta G^{\ddagger} = \frac{a}{2}Q_s^2 = \frac{(\Delta G + E_R)^2}{4E_R} \qquad (16.22)$$

hence the rate expression becomes

$$k_{et} = \frac{\omega_N}{2\pi}\kappa_{et}\exp\left(-\frac{(\Delta G + E_R)^2}{4E_R k_B T}\right). \qquad (16.23)$$

If we plot the logarithm of the rate as a function of the driving force ΔG, we obtain the famous Marcus parabola which has a maximum at $\Delta G = -E_R$ (Fig. 16.2).

---

[1] Sometimes the reorganization energy is defined as the negative quantity $G_p(Q_1) - G_p(0)$.

16.3 Free Energy Contribution of the Nonequilibrium Polarization

In Marcus theory, the solvent is treated in a continuum approximation. The medium is characterized by its static dielectric constant $\varepsilon_{\mathrm{st}}$ and its optical dielectric constant $\varepsilon_{\mathrm{op}} = \varepsilon_0 n^2$. The displacement $\mathbf{D}$ is generated by the charge distribution ϱ:

$$\mathrm{div}\,\mathbf{D} = \varrho. \tag{16.24}$$

In equilibrium, it is related to the electric field by

$$\mathbf{D} = \varepsilon_0\mathbf{E} + \mathbf{P} = \varepsilon\mathbf{E}. \tag{16.25}$$

Here $\mathbf{P}$ is the polarization, which is nothing else but the influenced dipole density. It consists of two parts [53]:

1. The optical or electronical polarization $\mathbf{P}_{\mathrm{op}}$. It is due to the induced dipole moments due to the response of the molecular electrons to the applied field. At optical frequencies, the electrons still move fast enough to follow the electric field adiabatically.
2. The inertial polarization $\mathbf{P}_{\mathrm{in}}$ which is due to the reorganization of the solvent molecules (reorientation of the dipoles, changes of molecular geometry). This part of the polarization corresponds to the static dielectric constant. The nuclear motion of the solvent molecules cannot follow the rapid oscillations at optical frequencies.

The total frequency-dependent polarization is

$$\mathbf{P}(\omega) = \mathbf{P}_{\mathrm{op}}(\omega) + \mathbf{P}_{\mathrm{in}}(\omega) = \mathbf{D}(\omega) - \varepsilon_0\mathbf{E}(\omega) = (\varepsilon(\omega) - \varepsilon_0)\mathbf{E}(\omega). \tag{16.26}$$

In the static limit, this becomes

$$\mathbf{P}_{\mathrm{op}} + \mathbf{P}_{\mathrm{in}} = (\varepsilon_{\mathrm{st}} - \varepsilon_0)\mathbf{E} \tag{16.27}$$

and at optical frequencies

$$\mathbf{P}_{\mathrm{op}} = (\varepsilon_{\mathrm{op}} - \varepsilon_0)\mathbf{E}. \tag{16.28}$$

Hence the contribution of the inertial polarization is in the static limit

$$\mathbf{P}_{\mathrm{in}} = (\varepsilon_{\mathrm{st}} - \varepsilon_{\mathrm{op}})\mathbf{E}. \tag{16.29}$$

In general, $\mathrm{rot}\,\mathbf{D} \neq 0$ and the calculation of the field for given charge distribution is difficult. Therefore, we use a model of spherical ions behaving as ideal conductors with the charge distributed over the surface. Then in our case, the displacement is normal to the surface and $\mathrm{rot}\,\mathbf{D} = 0$ and the displacement is the same as it would be generated by the charge distribution ϱ in vacuum. The corresponding electric field in vacuum would be $\varepsilon_0^{-1}\mathbf{D}$ and the field in the medium can be expressed as

$$\mathbf{E} = \frac{1}{\varepsilon_0}(\mathbf{D} - \mathbf{P}) = \frac{1}{\varepsilon_0}\mathbf{D} - \frac{1}{\varepsilon_0}(\mathbf{P}_{\mathrm{op}} + \mathbf{P}_{\mathrm{in}})$$

$$= \frac{1}{\varepsilon_0}(\mathbf{D} - \mathbf{P}_{\mathrm{in}}) - \frac{1}{\varepsilon_0}(\varepsilon_{\mathrm{op}} - \varepsilon_0)\mathbf{E},$$

$$\mathbf{E} = \frac{1}{\varepsilon_{\mathrm{op}}}(\mathbf{D} - \mathbf{P}_{\mathrm{in}}), \tag{16.30}$$

where we assumed that the optical polarization reacts instantaneously to changes of the charge distribution or to fluctuations of the inertial polarization. In equilibrium, (16.29) gives

$$\mathbf{E} = \frac{1}{\varepsilon_{\mathrm{op}}}(\mathbf{D} - (\varepsilon_{\mathrm{st}} - \varepsilon_{\mathrm{op}})\mathbf{E}),$$

$$\mathbf{E} = \frac{1}{\varepsilon_{\mathrm{st}}}\mathbf{D}. \tag{16.31}$$

If the polarization is in equilibrium with the field $(1/\varepsilon_0)\mathbf{D}$ produced by the charge distribution of either AB or A^+B^-, the free energy G of the two configurations will, in general, not be the same. We designate these two configurations as I and II, respectively. Marcus calculated the free energy of a nonequilibrium polarization for which the free energies of the reaction complexes $[AB]^{\ddagger}$ and $[A^+B^-]^{\ddagger}$ become equal. The contribution to the free energy is

$$G_{\mathrm{el}} = \int \mathrm{d}^3r \int \mathbf{E}\,\mathrm{d}\mathbf{D} = \frac{1}{\varepsilon_{\mathrm{op}}} \int \mathrm{d}^3r \int (\mathbf{D} - \mathbf{P}_{\mathrm{in}})\mathrm{d}\mathbf{D}, \tag{16.32}$$

which can be divided into two contributions:

1. The energy without inertial polarization

$$G_{\mathrm{el},0} = \frac{1}{\varepsilon_{\mathrm{op}}} \int \mathrm{d}^3r \frac{D^2}{2}. \tag{16.33}$$

2. The contribution of the inertial polarization

$$- \frac{1}{\varepsilon_{\mathrm{op}}} \int \mathrm{d}^3r \mathbf{P}_{\mathrm{in}}\mathrm{d}\mathbf{D}. \tag{16.34}$$

In equilibrium, we have

$$\mathbf{P}_{\mathrm{in}} = (\varepsilon_{\mathrm{st}} - \varepsilon_{\mathrm{op}})\mathbf{E} = (\varepsilon_{\mathrm{st}} - \varepsilon_{\mathrm{op}})\frac{1}{\varepsilon_{\mathrm{st}}}\mathbf{D} = \varepsilon_{\mathrm{op}}\left(\frac{1}{\varepsilon_{\mathrm{op}}} - \frac{1}{\varepsilon_{\mathrm{st}}}\right)\mathbf{D} \tag{16.35}$$

and we can express the free energy as a function of P_{in}

$$G_{\mathrm{el}}^{\mathrm{eq}} = G_{\mathrm{el},0} - \frac{1}{\frac{1}{\varepsilon_{\mathrm{op}}} - \frac{1}{\varepsilon_{\mathrm{st}}}} \int \mathrm{d}^3r \frac{1}{2}\left(\frac{\mathbf{P}_{\mathrm{in}}}{\varepsilon_{\mathrm{op}}}\right)^2. \tag{16.36}$$

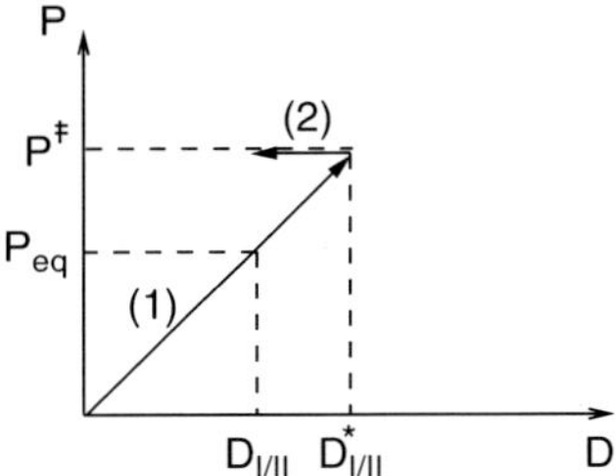

Fig. 16.3. Calculation of the free energy change

Now, we have to calculate the free energies of the activated complexes. Consider a thermal fluctuation of the inertial polarization. We will do this in two steps (Fig. 16.3).

In the first step, it is assumed that the polarization is always in equilibrium with the charge distribution. However, this charge density will be chosen such that the nonequilibrium polarization $\mathbf{P}_{\mathrm{in}}^{\ddagger}$ of the activated complex is generated. The electric fields are $\mathbf{D}^*$ and $\mathbf{E}^*$:

$$\frac{1}{\varepsilon_{\mathrm{op}}}\mathbf{P}_{\mathrm{in}}^{\ddagger} = \left(\frac{1}{\varepsilon_{\mathrm{op}}} - \frac{1}{\varepsilon_{\mathrm{st}}}\right)\mathbf{D}^*. \tag{16.37}$$

The corresponding free energy is

$$\Delta G_{\mathrm{el}}^{1} = \frac{1}{\varepsilon_{\mathrm{op}}}\int \mathrm{d}^3 r \int_{D}^{D^*} \mathbf{D}\,\mathrm{d}\mathbf{D} - \frac{1}{\varepsilon_{\mathrm{op}}}\int \mathrm{d}^3 r \int_{D}^{D^*} \mathbf{P}_{\mathrm{in}}\mathrm{d}\mathbf{D}$$

$$= G_{\mathrm{el},0}^{*} - G_{\mathrm{el},0} - \frac{1}{\frac{1}{\varepsilon_{\mathrm{op}}} - \frac{1}{\varepsilon_{\mathrm{st}}}}\int \mathrm{d}^3 r \frac{\mathbf{P}_{\mathrm{in}}^{\ddagger 2} - \mathbf{P}_{\mathrm{in}}^2}{2\varepsilon_{\mathrm{op}}^2}. \tag{16.38}$$

In the second step, the electric field of the charge distribution will be reduced to the equilibrium value while keeping the nonequilibrium polarization fixed.[2] Then this fixed polarization acts as an additional displacement:

$$\mathbf{E} = \frac{1}{\varepsilon_0}(\mathbf{D} - \mathbf{P}_{\mathrm{in}}^{\ddagger}) - \frac{1}{\varepsilon_0}\mathbf{P}_{\mathrm{op}}, \tag{16.39}$$

where the optical polarization depends on the electric field according to (16.28)

$$\mathbf{E} = \frac{1}{\varepsilon_0}(\mathbf{D} - \mathbf{P}_{\mathrm{in}}^{\ddagger}) - \left(\frac{\varepsilon_{\mathrm{op}}}{\varepsilon_0} - 1\right)\mathbf{E}. \tag{16.40}$$

---

[2] We tacitly assume that this is possible. In general, the polarization $P_{\mathrm{in}}^{\ddagger}$ will also modify the charge distribution on the reactants. If, however, the distance between the two ions is large in comparison with the radii, these changes can be neglected. Marcus calls this the point-charge approximation.

Hence, if we treat the optical polarization as following instantaneously, the slow fluctuations of the inertial polarization

$$\mathbf{E} = \frac{1}{\varepsilon_{\mathrm{op}}}(\mathbf{D} - \mathbf{P}_{\mathrm{in}}^{\ddagger}), \tag{16.41}$$

which means the inertial polarization is shielded by the optical polarization it creates. We may now calculate the change in free energy as

$$\Delta G_{\mathrm{el}}^{2} = \frac{1}{\varepsilon_{\mathrm{op}}} \int \mathrm{d}^{3}r \int_{D^{*}}^{D} (\mathbf{D} - \mathbf{P}_{\mathrm{in}}^{\ddagger})\mathrm{d}\mathbf{D}$$
$$= G_{\mathrm{el},0} - G_{\mathrm{el},0}^{*} - \int \mathrm{d}^{3}r \frac{1}{\varepsilon_{\mathrm{op}}}\mathbf{P}_{\mathrm{in}}^{\ddagger}(\mathbf{D} - \mathbf{D}^{*}). \tag{16.42}$$

If we substitute $\mathbf{D}^{*}$ from (16.37), we get

$$\frac{1}{\varepsilon_{\mathrm{op}}}\mathbf{P}_{\mathrm{in}}^{\ddagger}(\mathbf{D} - \mathbf{D}^{*}) = \frac{1}{\frac{1}{\varepsilon_{\mathrm{op}}} - \frac{1}{\varepsilon_{\mathrm{st}}}} \frac{\mathbf{P}_{\mathrm{in}}^{\ddagger}}{\varepsilon_{\mathrm{op}}} \left(\frac{\mathbf{P}_{\mathrm{in}} - \mathbf{P}_{\mathrm{in}}^{\ddagger}}{\varepsilon_{\mathrm{op}}} \right) \tag{16.43}$$

and the total free energy change due to the polarization fluctuation is

$$G_{\mathrm{el}}^{\ddagger} - G_{\mathrm{el}}^{\mathrm{eq}} = \Delta G_{\mathrm{el}}^{1} + \Delta G_{\mathrm{el}}^{2}$$
$$= -\frac{1}{\frac{1}{\varepsilon_{\mathrm{op}}} - \frac{1}{\varepsilon_{\mathrm{st}}}} \int \mathrm{d}^{3}r \frac{\mathbf{P}_{\mathrm{in}}^{\ddagger 2} - \mathbf{P}_{\mathrm{in}}^{2}}{2\varepsilon_{\mathrm{op}}^{2}}$$
$$- \frac{1}{\frac{1}{\varepsilon_{\mathrm{op}}} - \frac{1}{\varepsilon_{\mathrm{st}}}} \int \mathrm{d}^{3}r \frac{1}{\varepsilon_{\mathrm{op}}^{2}}\mathbf{P}_{\mathrm{in}}^{\ddagger}(\mathbf{P}_{\mathrm{in,eq}} - \mathbf{P}_{\mathrm{in}}^{\ddagger})$$
$$= \frac{1}{\frac{1}{\varepsilon_{\mathrm{op}}} - \frac{1}{\varepsilon_{\mathrm{st}}}} \int \mathrm{d}^{3}r \frac{1}{2} \left(\frac{\mathbf{P}_{\mathrm{in}}^{\ddagger} - \mathbf{P}_{\mathrm{in,eq}}}{\varepsilon_{\mathrm{op}}} \right)^{2}. \tag{16.44}$$

16.4 Activation Energy

The free energy $G_{\mathrm{el}}^{\ddagger}$ is a functional of the polarization fluctuation

$$\delta\mathbf{P} = \mathbf{P}_{\mathrm{in}}^{\ddagger} - \mathbf{P}_{\mathrm{in}}. \tag{16.45}$$

If we minimize the free energy

$$\delta G_{\mathrm{el}}^{\ddagger} = 0, \tag{16.46}$$

we find that

$$\delta\mathbf{P} = 0, \quad G_{\mathrm{el}}^{\ddagger} = G_{\mathrm{el}}^{\mathrm{eq}}. \tag{16.47}$$

Marcus asks now, for which polarizations $\mathbf{P}_{\text{in}}^{\ddagger}$ of the two states I and II become isoenergetic, i.e.

$$\Delta U^{\ddagger} = U_{\text{II}}^{\ddagger} - U_{\text{I}}^{\ddagger} = 0. \tag{16.48}$$

We want to rephrase the last condition in terms of the free energy

$$G = U - TS + pV \approx U - TS. \tag{16.49}$$

We conclude that

$$\Delta G^{\ddagger} = G_{\text{II}}^{\ddagger} - G_{\text{I}}^{\ddagger} = -T\Delta S^{\ddagger}. \tag{16.50}$$

Here (Fig. 16.4)

$$G_{\text{I}}^{\ddagger} = \frac{1}{\frac{1}{\varepsilon_{\text{op}}} - \frac{1}{\varepsilon_{\text{st}}}} \frac{1}{2} \int \mathrm{d}^3 r \left(\frac{\delta \mathbf{P}_{\text{I}}}{\varepsilon_{\text{op}}}\right)^2 + G_{\text{el,I}}^{\text{eq}}(R) - G_{\text{el,I}}^{\text{eq}}(\infty) + G_{\text{I}}^{\infty},$$

$$G_{\text{II}}^{\ddagger} = \frac{1}{\frac{1}{\varepsilon_{\text{op}}} - \frac{1}{\varepsilon_{\text{st}}}} \frac{1}{2} \int \mathrm{d}^3 r \left(\frac{\delta \mathbf{P}_{\text{II}}}{\varepsilon_{\text{op}}}\right)^2 + G_{\text{el,II}}^{\text{eq}}(R) - G_{\text{el,II}}^{\text{eq}}(\infty) + G_{\text{II}}^{\infty}. \tag{16.51}$$

The quantity

$$\Delta G_{\infty} = G_{\text{II}}^{\infty} - G_{\text{I}}^{\infty} = \Delta G_{\text{vac}} + \Delta G_{\text{solv},\infty} \tag{16.52}$$

is the difference of free energies for an ideal solution of A and B on the one side and A^+ and B^- on the other side at infinite distance $R \to \infty$. Usually, it is determined experimentally. It contains all the information on the internal structure of the ions. The entropy difference is the difference of internal entropies of the reactants since the contribution from the polarization drops out due to $\mathbf{P}_{\text{in,I}}^{\ddagger} = \mathbf{P}_{\text{in,II}}^{\ddagger}$. Usually, this entropy difference is small. The electrostatic energies $G_{\text{el,I/II}}^{\text{eq}}$ are composed of the polarization contribution (the solvation free energy) and the mutual interaction of the reactants which will be considered later. The reaction free energy is

$$\Delta G_{\text{eq}} = \Delta G_{\infty} + \Delta G_{\text{el}}^{\text{eq}}(R) - \Delta G_{\text{el}}^{\text{eq}}(\infty). \tag{16.53}$$

It can be easily seen that the condition $\Delta G^{\ddagger} = 0$ will be fulfilled for infinitely many polarizations $\mathbf{P}_{\text{in}}^{\ddagger}$. We will therefore look for that one which minimizes simultaneously the free energy of the transition state $G_{\text{I,II}}^{\ddagger}$. We introduce a Lagrange parameter m to impose the isoenergeticity condition:

$$\delta \left[G_{\text{I}}^{\ddagger}(\mathbf{P}_{\text{in}}^{\ddagger}) + m(G_{\text{II}}^{\ddagger}(\mathbf{P}_{\text{in}}^{\ddagger}) - G_{\text{I}}^{\ddagger}(\mathbf{P}_{\text{in}}^{\ddagger}))\right] = 0. \tag{16.54}$$

The variation is with respect to $\mathbf{P}_{\text{in}}^{\ddagger}$ or, equivalently with respect to $\delta \mathbf{P}_{\text{I}}$ for fixed change of the inertial polarization:

$$\Delta \mathbf{P}_{\text{in}} = \delta \mathbf{P}_{\text{I}} - \delta \mathbf{P}_{\text{II}} = (\mathbf{P}_{\text{in}}^{\ddagger} - \mathbf{P}_{\text{I,in}}) - (\mathbf{P}_{\text{in}}^{\ddagger} - \mathbf{P}_{\text{II,in}}) = \mathbf{P}_{\text{II,in}} - \mathbf{P}_{\text{I,in}}. \tag{16.55}$$

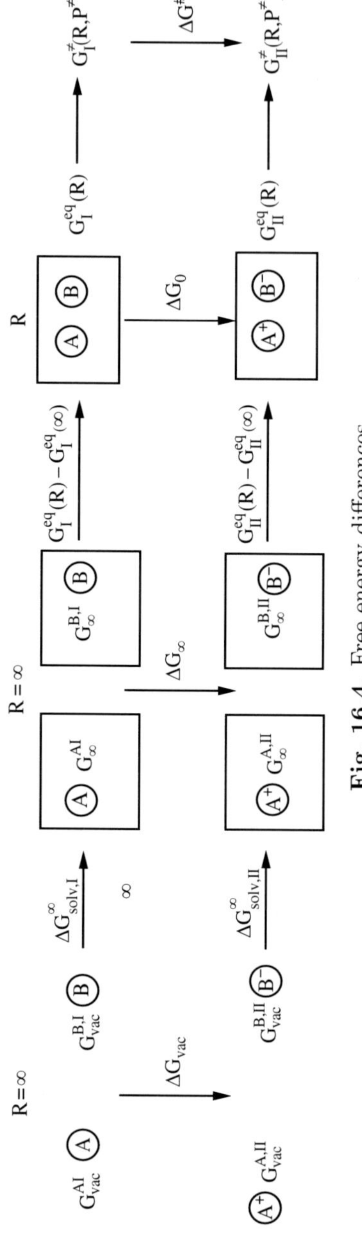

Fig. 16.4. Free energy differences

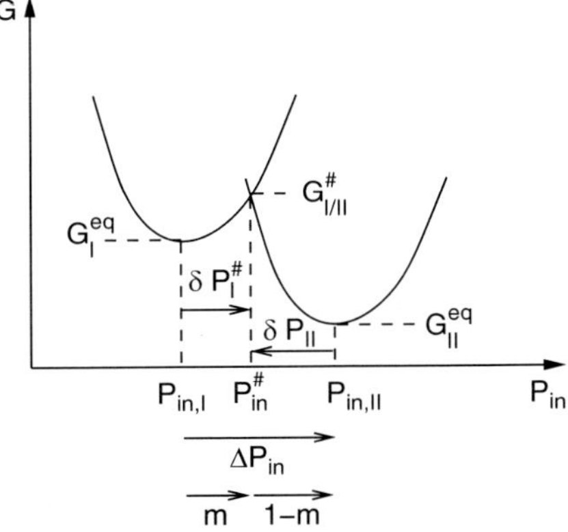

Fig. 16.5. Variation of the nonequilibrium polarization

We have to solve

$$\delta\left[(1-m)G_I^{\ddagger}(\mathbf{P}_{I,in}+\delta\mathbf{P}_I)+mG_{II}^{\ddagger}(\mathbf{P}_{II,in}-\Delta\mathbf{P}_{in}+\delta\mathbf{P}_{I,in})\right]=0. \quad (16.56)$$

Due to the quadratic dependence of the free energies, we find the condition

$$[(1-m)\delta\mathbf{P}_I+m(\delta\mathbf{P}_I-\Delta\mathbf{P}_{in})]=0 \quad (16.57)$$

and hence the solution (Fig. 16.5)

$$\delta\mathbf{P}_I^{\ddagger}=m\Delta\mathbf{P}_{in}. \quad (16.58)$$

The Lagrange parameter m is determined by inserting this solution into the isoenergicity condition:

$$\begin{aligned}
-T\Delta S &= \Delta G^{\ddagger}(\mathbf{P}_{in}^{\ddagger}) \\
&= G_{II}^{\ddagger}(\mathbf{P}_{in,II}+(m-1)\Delta P_{in})-G_I^{\ddagger}(\mathbf{P}_{in,I}+m\Delta P_{in}) \\
&= \Delta G_{eq}+((m-1)^2-m^2)\frac{1}{\frac{1}{\varepsilon_{op}}-\frac{1}{\varepsilon_{st}}}\frac{1}{2}\int d^3r\left(\frac{\Delta\mathbf{P}_{in}}{\varepsilon_{op}}\right)^2 \\
&= \Delta G_{eq}+(1-2m)\lambda, \quad (16.59)
\end{aligned}$$

where

$$\begin{aligned}
\lambda &= G_{II}(\mathbf{P}_{in,II}-\Delta\mathbf{P}_{in})-G_{II}(\mathbf{P}_{in,II}) \\
&= \frac{1}{\frac{1}{\varepsilon_{op}}-\frac{1}{\varepsilon_{st}}}\frac{1}{2}\int d^3r\left(\frac{\Delta\mathbf{P}_{in}}{\varepsilon_{op}}\right)^2=\left(\frac{1}{\varepsilon_{op}}-\frac{1}{\varepsilon_{st}}\right)\frac{1}{2}\int d^3r\,(\Delta\mathbf{D})^2
\end{aligned}$$

$$(16.60)$$

is the reorganization energy. Solving for m we find

$$m = \frac{\Delta G_{\text{eq}} + T\Delta S^{\ddagger} + \lambda}{2\lambda}.$$

(16.61)

The free energy of activation is given by

$$\Delta G_{\text{a}} = G_{\text{I}}^{\ddagger} - G_{\text{I}}^{\text{eq}} = m^2 \frac{1}{\frac{1}{\varepsilon_{\text{op}}} - \frac{1}{\varepsilon_{\text{st}}}} \frac{1}{2} \int d^3 r \left(\frac{\Delta \mathbf{P}_{\text{in}}}{\varepsilon_{\text{op}}} \right)^2$$

$$= m^2 \lambda = \frac{(\Delta G_{\text{eq}} + T\Delta S^{\ddagger} + \lambda)^2}{4\lambda}.$$

(16.62)

Finally, we insert this into the TST rate expression to find

$$k_{\text{et}} = \frac{\omega_{\text{N}}}{2\pi} \kappa_{\text{et}} e^{-\Delta G_{\text{a}}/k_{\text{B}}T} = \frac{\omega_{\text{N}}}{2\pi} \kappa_{\text{et}} \exp\left\{ -\frac{(\Delta G_{\text{eq}} + T\Delta S^{\ddagger} + \lambda)^2}{4\lambda k_{\text{B}}T} \right\}.$$

(16.63)

16.5 Simple Model Systems

We will now calculate reorganization energy and free energy differences explicitly for a model system consisting of two spherically reactants. We assume that the charge distribution of the reactants is not influenced by mutual Coulombic interaction or changes of the medium polarization (we apply the point-charge approximation again) (Fig. 16.6).

We have already calculated the solvation energy at large distances 43

$$W_{\text{elec}}^{(1)} = \left(-\frac{Q_{\text{A}}^2}{8\pi a_{\text{A}}} - \frac{Q_{\text{B}}^2}{8\pi a_{\text{B}}} \right) \left(\frac{1}{\varepsilon_0} - \frac{1}{\varepsilon_{\text{st}}} \right)$$

(16.64)

from which we find the polarization energy at large distances

$$G_{\text{el}}^{\text{eq}}(R = \infty) = \frac{1}{8\pi\varepsilon_{\text{st}}} \left(\frac{Q_{\text{A}}^2}{a_{\text{A}}} + \frac{Q_{\text{B}}^2}{a_{\text{B}}} \right).$$

(16.65)

We now consider a finite distance R. If we take the origin of the coordinate system to coincide with the center of the reactant A, then the dielectric displacement is

$$\mathbf{D}_{\text{I/II}}(\mathbf{r}) = \frac{1}{4\pi} \left(Q_{\text{A,I/II}} \frac{\mathbf{r}}{r^3} + Q_{\text{B,I/II}} \frac{\mathbf{r} - \mathbf{R}}{|\mathbf{r} - \mathbf{R}|^3} \right)$$

(16.66)

Fig. 16.6. Simple donor–acceptor geometry

and the polarization free energy

$$G^{\text{eq}}_{\text{el,I/II}} = \int \mathrm{d}^3 r \frac{D^2}{2\varepsilon_{\text{st}}} \tag{16.67}$$

contains two types of integrals. The first is

$$I_1 = \int \frac{\mathrm{d}^3 r}{r^4}. \tag{16.68}$$

The range of this integral is outside the spheres A and B. As the integrand falls off rapidly with distance, we may extend the integral over the volume of the other sphere. Introducing polar coordinates, we find

$$I_1 \approx \int_a^\infty 4\pi \frac{\mathrm{d}r}{r^2} = \frac{4\pi}{a}. \tag{16.69}$$

The second kind of integral is

$$I_2 = \int \mathrm{d}^3 r \frac{\mathbf{r}(\mathbf{r} - \mathbf{R})}{r^3 |\mathbf{r} - \mathbf{R}|^3}. \tag{16.70}$$

Again we extend the integration range and use polar coordinates

$$I_2 \approx 2\pi \int_0^\pi \sin\theta\,\mathrm{d}\theta \int_0^\infty r^2 \mathrm{d}r \frac{r^2 - rR\cos\theta}{r^3(r^2 + R^2 - 2rR\cos\theta)^{3/2}}. \tag{16.71}$$

The integral over r gives

$$\int_0^\infty \frac{r - R\cos\theta}{(r^2 + R^2 - 2rR\cos\theta)^{3/2}}\mathrm{d}r = -\left.\frac{1}{\sqrt{r^2 + R^2 - 2rR\cos\theta}}\right|_0^\infty = \frac{1}{R} \tag{16.72}$$

and the integral over θ then gives

$$I_2 = \int_0^\pi \frac{2\pi\sin\theta}{R}\mathrm{d}\theta = \frac{4\pi}{R}. \tag{16.73}$$

Together, we have the polarization free energy

$$G^{\text{eq}}_{\text{el}} = \frac{1}{2\varepsilon_{\text{st}}} \int \mathrm{d}^3 r D^2_\varrho = \frac{1}{8\pi\varepsilon_{\text{st}}}\left(\frac{Q^2_{\text{A}}}{a_{\text{A}}} + \frac{Q^2_{\text{B}}}{a_{\text{B}}}\right) + \frac{1}{8\pi\varepsilon_{\text{st}}}\frac{2Q_{\text{A}}Q_{\text{B}}}{R} \tag{16.74}$$

$$= G^{\text{eq}}(\infty) + \frac{Q_{\text{A}}Q_{\text{B}}}{4\pi\varepsilon_{\text{st}} R}$$

and the difference

$$G^{\text{eq}}_{\text{el,II}} - G^{\text{eq}}_{\text{el,I}} = \frac{e^2}{8\pi\varepsilon_{\text{st}}}\left(\frac{1 + 2Z_{\text{A}}}{a_{\text{A}}} + \frac{1 - 2Z_{\text{B}}}{a_{\text{B}}} + 2\frac{Z_{\text{B}} - Z_{\text{A}} - 1}{R}\right). \tag{16.75}$$

Similarly, the reorganization energy is

$$\lambda = \frac{1}{2}\left(\frac{1}{\varepsilon_{\mathrm{op}}} - \frac{1}{\varepsilon_{\mathrm{st}}}\right)\int \mathrm{d}^3r\,(\Delta\mathbf{D})^2$$

$$= \frac{1}{8\pi}\left(\frac{1}{\varepsilon_{\mathrm{op}}} - \frac{1}{\varepsilon_{\mathrm{st}}}\right)\left(\frac{\Delta Q_{\mathrm{A}}^2}{a_{\mathrm{A}}} + \frac{\Delta Q_{\mathrm{B}}^2}{a_{\mathrm{B}}} + \frac{2\Delta Q_{\mathrm{A}}\Delta Q_{\mathrm{B}}}{R}\right). \tag{16.76}$$

Now since $\Delta Q_{\mathrm{A}} = -\Delta Q_{\mathrm{B}} = e$, we can write this as

$$\lambda = \frac{e^2}{8\pi}\left(\frac{1}{\varepsilon_{\mathrm{op}}} - \frac{1}{\varepsilon_{\mathrm{st}}}\right)\left(\frac{1}{a_{\mathrm{A}}} + \frac{1}{a_{\mathrm{B}}} - \frac{2}{R}\right). \tag{16.77}$$

We consider here the most important special cases.

16.5.1 Charge Separation

If an ion pair is created during the ET reaction ($Q_{\mathrm{A,I}} = Q_{\mathrm{B,I}} = 0$, $Q_{\mathrm{A,II}} = e$, $Q_{\mathrm{B,II}} = -e$), we have

$$\frac{Q_{\mathrm{A}}^2}{a_{\mathrm{A}}} + \frac{Q_{\mathrm{B}}^2}{a_{\mathrm{B}}} + \frac{2Q_{\mathrm{A}}Q_{\mathrm{B}}}{R} = \begin{cases} 0 & \text{(I)} \\ \frac{1}{a_{\mathrm{A}}} + \frac{1}{a_{\mathrm{B}}} - \frac{2}{R} & \text{(II)} \end{cases}. \tag{16.78}$$

The free energy is

$$G_{\mathrm{el,I}}^{\mathrm{eq}} = 0,$$

$$G_{\mathrm{el,II}}^{\mathrm{eq}} = \frac{e^2}{8\pi\varepsilon_{\mathrm{st}}}\left(\frac{1}{a_{\mathrm{A}}} + \frac{1}{a_{\mathrm{B}}} - \frac{2}{R}\right) = G_{\mathrm{el,II}}^{\mathrm{eq}}(\infty) - \frac{e^2}{4\pi\varepsilon_{\mathrm{st}}R} \tag{16.79}$$

and the activation energy becomes

$$\Delta G_{\mathrm{a}} = \frac{(\Delta G_\infty + T\Delta S^\ddagger + \lambda - \frac{e^2}{4\pi\varepsilon_{\mathrm{st}}R})^2}{4\lambda}. \tag{16.80}$$

If the process is photoinduced, the free energy at large distance can be expressed in terms of the ionization potential of the donor, the electron affinity of the acceptor, and the energy of the optical transition (Fig. 16.7)

$$\Delta G_\infty = \mathrm{EA(A)} - (\mathrm{IP(D)} - \hbar\omega). \tag{16.81}$$

16.5.2 Charge Shift

For a charge shift reaction of the type $Q_{\mathrm{A,I}} = Q_{\mathrm{B,II}} = -e$, $Q_{\mathrm{B,I}} = Q_{\mathrm{A,II}} = 0$,

$$G_{\mathrm{I}}^{\mathrm{eq}} = \frac{e^2}{8\pi\varepsilon_{\mathrm{st}}a_{\mathrm{A}}}, \quad G_{\mathrm{II}}^{\mathrm{eq}} = \frac{e^2}{8\pi\varepsilon_{\mathrm{st}}a_{\mathrm{B}}}, \tag{16.82}$$

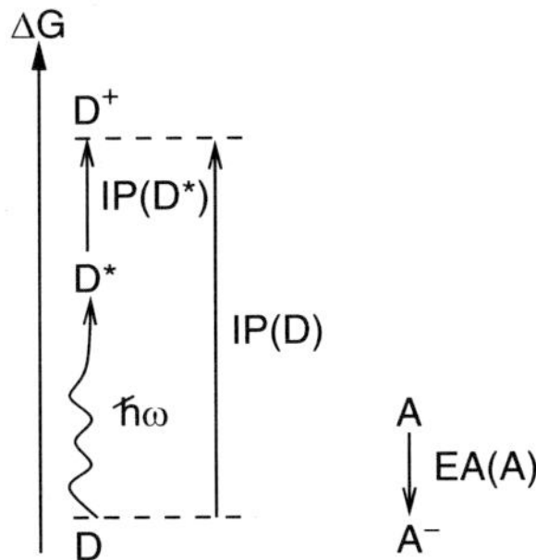

Fig. 16.7. Photoinduced charge separation. At large distance, the energy difference ΔG_∞ is given by $\mathrm{EA(A)} - \mathrm{IP(D^*)} = \mathrm{EA(A)} - (\mathrm{IP(D)} - \hbar\omega)$

and for the second type $Q_{\mathrm{A,I}} = Q_{\mathrm{B,II}} = 0$, $Q_{\mathrm{B,I}} = Q_{\mathrm{A,II}} = e$,

$$G_{\mathrm{I}}^{\mathrm{eq}} = \frac{e^2}{8\pi\varepsilon_{\mathrm{st}}a_{\mathrm{B}}}, \quad G_{\mathrm{II}}^{\mathrm{eq}} = \frac{e^2}{8\pi\varepsilon_{\mathrm{st}}a_{\mathrm{A}}}. \tag{16.83}$$

The activation energy now does not contain a Coulombic term

$$\Delta G_{\mathrm{a}} = \frac{(\Delta G_\infty + T\Delta S^\ddagger + \lambda)^2}{4\lambda}. \tag{16.84}$$

16.6 The Energy Gap as the Reaction Coordinate

We want to construct one single reaction coordinate which summarizes the polarization fluctuations. To that end, we consider the energy gap

$$\begin{aligned}
\Delta U^\ddagger &= \Delta G^\ddagger + T\Delta S^\ddagger \approx G_{\mathrm{II}}^\ddagger(\mathbf{P}_{\mathrm{in}}^\ddagger) - G_{\mathrm{I}}^\ddagger(\mathbf{P}_{\mathrm{in}}^\ddagger) \\
&= G_{\mathrm{II}}^\ddagger(\mathbf{P}_{\mathrm{II,in}} - \Delta\mathbf{P}_{\mathrm{in}} + \delta\mathbf{P}_{\mathrm{I,in}}) - G_{\mathrm{I}}^\ddagger(\mathbf{P}_{\mathrm{I,in}} + \delta\mathbf{P}_{\mathrm{I,in}}) \\
&= \Delta G_{\mathrm{eq}} + \frac{1}{\frac{1}{\varepsilon_{\mathrm{op}}} - \frac{1}{\varepsilon_{\mathrm{st}}}} \frac{1}{2} \int \mathrm{d}^3 r \left(\frac{(\delta\mathbf{P}_{\mathrm{I,in}} - \Delta\mathbf{P}_{\mathrm{in}})}{\varepsilon_{\mathrm{op}}} \right)^2 \\
&\quad - \frac{1}{\frac{1}{\varepsilon_{\mathrm{op}}} - \frac{1}{\varepsilon_{\mathrm{st}}}} \frac{1}{2} \int \mathrm{d}^3 r \left(\frac{\delta\mathbf{P}_{\mathrm{I,in}}}{\varepsilon_{\mathrm{op}}} \right)^2 \\
&= \Delta G_{\mathrm{eq}} + \frac{1}{\frac{1}{\varepsilon_{\mathrm{op}}} - \frac{1}{\varepsilon_{\mathrm{st}}}} \frac{1}{2} \int \mathrm{d}^3 r \frac{\Delta\mathbf{P}_{\mathrm{in}}^2 - 2\delta\mathbf{P}_{\mathrm{I,in}}\Delta\mathbf{P}_{\mathrm{in}}}{\varepsilon_{\mathrm{op}}^2} \\
&= \Delta G_{\mathrm{eq}} + \lambda - \delta U(t) \tag{16.85}
\end{aligned}$$

with the thermally fluctuating energy

$$\delta U(t) = \frac{1}{\frac{1}{\varepsilon_{\mathrm{op}}} - \frac{1}{\varepsilon_{\mathrm{st}}}} \int \mathrm{d}^3 r \frac{\delta\mathbf{P}_{\mathrm{I,in}}\Delta\mathbf{P}_{\mathrm{in}}}{\varepsilon_{\mathrm{op}}^2}. \tag{16.86}$$

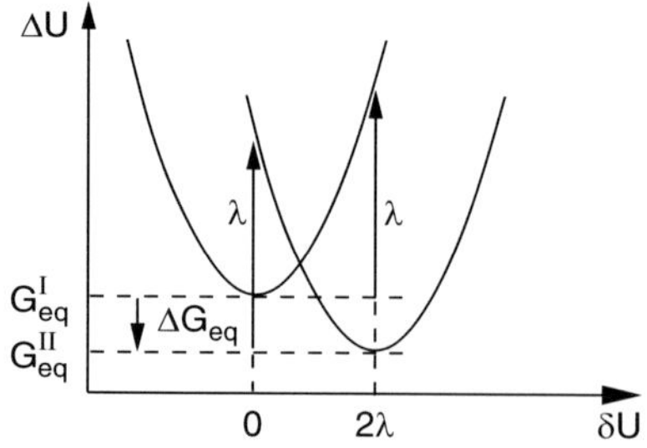

Fig. 16.8. Energy gap as reaction coordinate

In the equilibrium of the reactants,

$$\delta U = 0, \quad \Delta U = \Delta G_{eq} + \lambda \,(\mathrm{I}) \tag{16.87}$$

and in the equilibrium of the products (Fig. 16.8)

$$\delta U = 2\lambda, \quad \Delta U = \Delta G_{eq} - \lambda \,(\mathrm{II}). \tag{16.88}$$

The free energies now take the form

$$G_{\mathrm{I}} = G_{\mathrm{I}}^{eq} + \frac{1}{4\lambda}(\delta U)^2, \tag{16.89}$$

$$G_{\mathrm{II}} = G_{\mathrm{II}}^{eq} + \frac{1}{4\lambda}(\delta U - 2\lambda)^2. \tag{16.90}$$

The free energy fluctuations in the reactant state are given by

$$\langle G_{\mathrm{I}} - G_{\mathrm{I}}^{eq} \rangle = \frac{1}{4\lambda}\langle \delta U^2 \rangle. \tag{16.91}$$

If we identify this with the thermal average $k_{\mathrm{B}}T/2$ of a one-dimensional harmonic motion, we have

$$\langle \delta U^2 \rangle = 2k_{\mathrm{B}}T\lambda. \tag{16.92}$$

The fluctuations of the energy gap can for instance be investigated with molecular dynamics methods. They can be modeled by diffusive motion in a harmonic potential $U(Q) = Q^2/4\lambda$:

$$\frac{\partial}{\partial t}W(Q,t) = -D_{\mathrm{E}}\left\{\frac{\partial^2}{\partial Q^2}W + \frac{1}{k_{\mathrm{B}}T}\frac{\partial}{\partial Q}\left(\frac{\partial U}{\partial Q}W\right)\right\} - \kappa\delta(Q - \lambda)W, \tag{16.93}$$

where the diffusion constant is

$$D_{\mathrm{E}} = \frac{2k_{\mathrm{B}}T\lambda}{\tau_{\mathrm{L}}} \tag{16.94}$$

and τ_{L} is the longitudinal relaxation time of the medium.

16.7 Inner-Shell Reorganization

Intramolecular modes can also contribute to the reorganization. Low-frequency modes ($\hbar\omega < k_\mathrm{B}T$) which can be treated classically can be taken into account by adding an inner-shell contribution to the reorganization energies:

$$\lambda = \lambda_\mathrm{outer} + \lambda_\mathrm{inner}. \tag{16.95}$$

Often C–C stretching modes at $\hbar\omega \approx 1{,}400\,\mathrm{cm}^{-1}$ change their equilibrium positions during the ET process. They can accept the excess energy in the inverted region and reduce the activation energy. Often the Franck–Condon progression of one dominant mode $\hbar\omega_v$ is used to analyze experimental data in the inverted region. Since $\hbar\omega_v \gg k_\mathrm{B}T$, thermal occupation is negligible. For a harmonic oscillator, the relative intensity of the $0 \to n$ transition[3] is

$$\mathrm{FC}(0, n) = \frac{g_\mathrm{r}^{2n}}{n!}\mathrm{e}^{-g_\mathrm{r}^2} \tag{16.96}$$

and the sum over all transitions gives the rate expression

$$k = \frac{\omega}{2\pi}\kappa_\mathrm{el} \sum_{n=0}^{\infty} \mathrm{e}^{-g^2}\frac{g^{2n}}{n!} \exp\left(-\frac{(\Delta G + E_\mathrm{R} + n\hbar\omega_v)^2}{4 E_\mathrm{R} k_\mathrm{B}T}\right), \tag{16.97}$$

where $g^2\hbar\omega_v$ is the partial reorganization energy of the stretching mode (Figs. 16.9 and 16.10).

16.8 The Transmission Coefficient for Nonadiabatic Electron Transfer

The transmission coefficient κ will be considered later in more detail. For adiabatic electron transfer, it approaches unity. For nonadiabatic transfer (small crossing probability), it depends on the electronic coupling matrix element V_et. The quantum mechanical rate expression for nonadiabatic transfer becomes in the high-temperature limit ($\hbar\omega \ll k_\mathrm{B}T$ for all coupling modes)

$$k = \frac{2\pi V_\mathrm{et}^2}{\hbar}\frac{\mathrm{e}^{-(\Delta G + \lambda)^2/4\lambda k_\mathrm{B}T}}{\sqrt{4\pi\lambda k_\mathrm{B}T}}. \tag{16.98}$$

---

[3] A more detailed discussion follows later.

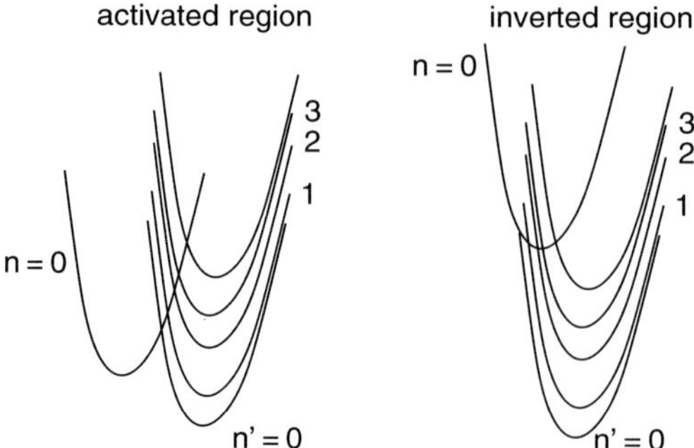

Fig. 16.9. Accepting mode. In the inverted region, vibrations can accept the excess energy

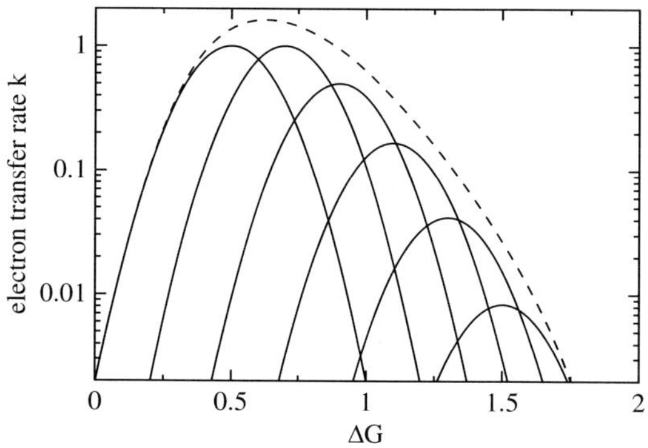

Fig. 16.10. Inclusion of a high-frequency mode progression. Equation (16.97) is evaluated for typical values of $\hbar\omega_v = 0.2\,\mathrm{eV}$, $g = 1$, and $\lambda = 0.5\,\mathrm{eV}$ (*broken curve*). The *full curves* show the contributions of different $0 \to n$ transitions

Problems

16.1. Marcus Cross Relation

(a) Calculate the activation energy for the self-exchange reaction

$$\mathrm{A}^- + \mathrm{A} \quad \overset{k_{\mathrm{AA}}}{\to} \quad \mathrm{A} + \mathrm{A}^-$$

in the harmonic model

$$G_{\mathrm{R}}(Q) = \frac{a}{2}Q^2, \quad G_{\mathrm{P}}(Q) = \frac{a}{2}(Q - Q_1)^2.$$

(b) Show that for the cross reaction

$$A + D \quad \xrightarrow{k} \quad A^- + D^+$$

the reorganization energy is given by the average of the reorganization energies for the two self-exchange reactions:

$$\lambda = \frac{\lambda_{AA} + \lambda_{DD}}{2}$$

and the rate k can be expressed as

$$k = \sqrt{k_{AA} k_{DD} K_{eq} f}$$

where k_{AA} and k_{DD} are the rate constants of the self-exchange reactions, K_{eq} is the equilibrium constant of the cross reaction and f is a factor, which is usually close to unity.

Part VI

Elementary Photophysics

Molecular States

Biomolecules have a large number of vibrational degrees of freedom, which are more or less strongly coupled to transitions between different electronic states. In this chapter, we discuss the vibronic states on the basis of the Born–Oppenheimer separation and the remaining nonadiabatic coupling of the adiabatic states. In the following, we do not consider translational or rotational motion which is essentially hindered for Biomolecules in the condensed phase.

17.1 Born–Oppenheimer Separation

In molecular physics, usually the Born–Oppenheimer separation of electronic (r) and nuclear motion (Q) is used. The molecular Hamiltonian (without considering spin or relativistic effects) can be written as

$$H = T_{\mathrm{N}}(Q) + T_{\mathrm{el}}(r) + V_{\mathrm{N}}(Q) + V_{\mathrm{eN}}(Q, r) + V_{\mathrm{ee}}(r) \tag{17.1}$$

with kinetic energy

$$T_{\mathrm{N}} = \sum_j -\frac{\hbar^2}{2m_j}\frac{\partial^2}{\partial Q_j^2}, \quad T_{\mathrm{el}} = \sum -\frac{\hbar^2}{2m_{\mathrm{e}}}\frac{\partial^2}{\partial r_k^2} \tag{17.2}$$

and Coulombic interaction

$$\begin{aligned}
V &= V_{\mathrm{N}}(Q) + V_{\mathrm{eN}}(Q, r) + V_{\mathrm{ee}}(r) \\
&= \sum_{j,j'} \frac{Z_j Z_{j'} \mathrm{e}^2}{4\pi\varepsilon |R_j - R_{j'}|} - \sum_{j,k} \frac{Z_j \mathrm{e}^2}{4\pi\varepsilon |R_j - r_k|} + \sum_{k,k'} \frac{\mathrm{e}^2}{4\pi\varepsilon |r_k - r_{k'}|}.
\end{aligned} \tag{17.3}$$

The Born–Oppenheimer wave function is a product

$$\psi(r, Q)\chi(Q), \tag{17.4}$$

where the electronic part depends parametrically on the nuclear coordinates. The nuclear masses are much larger than the electronic mass. Therefore, the kinetic energy of the nuclei is neglected for the electronic motion. The electronic wave function is obtained approximately, from the eigenvalue problem,

$$(T_{\text{el}} + V)\psi_s(r, Q) = E_s(Q)\psi_s(r, Q), \tag{17.5}$$

which has to be solved separately for each set of nuclear coordinates. Using now the Born–Oppenheimer product ansatz, we have

$$\begin{aligned}
H\psi_s(r, Q)\chi_s(Q) &= T_{\text{N}}\psi_s(r, Q)\chi_s(Q) + E_s(Q)\psi_s(r, Q)\chi_s(Q) \\
&= \psi_s(r, Q)(T_{\text{N}} + E_s(Q))\chi_s(Q) \\
&\quad - \sum_j \frac{\hbar^2}{2m_j}\left(\chi_s(Q)\frac{\partial^2\psi_s(r, Q)}{\partial Q_j^2} + \frac{\partial\chi_s(Q)}{\partial Q_j}\frac{\partial\psi_s}{\partial Q_j}\right).
\end{aligned} \tag{17.6}$$

The sum constitutes the so-called nonadiabatic interaction V_{nad}. If it is neglected in lowest order, the nuclear wave function is a solution of the eigenvalue problem:

$$(T_{\text{N}} + E_s(Q))\chi_s(Q) = E\chi_s(Q). \tag{17.7}$$

Often the potential energy $E_s(Q)$ can be expanded around the equilibrium configuration Q_{0s}[1]

$$E_s(Q) = E_s(Q_{0s}) + \frac{1}{2}\sum_{j,j'}(Q_j - Q_{j0s})(Q_{j'} - Q_{j'0s})\frac{\partial^2}{\partial Q_j\partial Q_{j'}}E_s + \cdots \tag{17.8}$$

Within the harmonic approximation, the matrix of mass-weighted second derivatives

$$D_{jj'} = \frac{1}{\sqrt{m_j m_{j'}}}\frac{\partial^2}{\partial Q_j\partial Q_{j'}} \tag{17.9}$$

is diagonalized

$$\sum_{j'} D_{jj'}u_{j'}^r = \omega_r^2 u_j^r \tag{17.10}$$

and the nuclear motion becomes a superposition of independent normal mode vibrations with amplitudes q_r and frequencies ω_r[2]

$$Q_j - Q_{j0s} = \frac{1}{\sqrt{m_j}}\sum_r q_r u_j^r. \tag{17.11}$$

After transformation to normal mode coordinates, the eigenvalue problem (17.7) decouples according to

---

[1] which will be different for different electronic states s in general.

[2] These quantities will be different for different electronic states s of course.

$$E_s + T_{\mathrm{N}} = E_s^{(0)} + \sum_r \left(\frac{\omega_r^2}{2} q_r^2 - \frac{\hbar^2}{2} \frac{\partial^2}{\partial q_r^2} \right) \qquad (17.12)$$

and the nuclear wave function factorizes

$$\chi_s = \prod_r \chi_{s,r,n(r)}(q_r). \qquad (17.13)$$

The total energy of a molecular state is

$$E_s^0 + \sum_r \hbar \omega_r^{(s)} \left(n(r) + \frac{1}{2} \right). \qquad (17.14)$$

The vibrations form a very dense manifold of states for biological molecules. This is often schematically represented in a Jablonsky diagram[3] (Figs. 17.1–17.4)

17.2 Nonadiabatic Interaction

The nonadiabatic interaction couples the adiabatic electronic states.[4] If we expand the wave function as a linear combination of the adiabatic wave functions, which form a complete basis at each configuration Q

$$\Psi = \sum_s C_s \psi_s(r, Q) \chi_s(Q), \qquad (17.15)$$

we have to consider the nonadiabatic matrix elements[5]

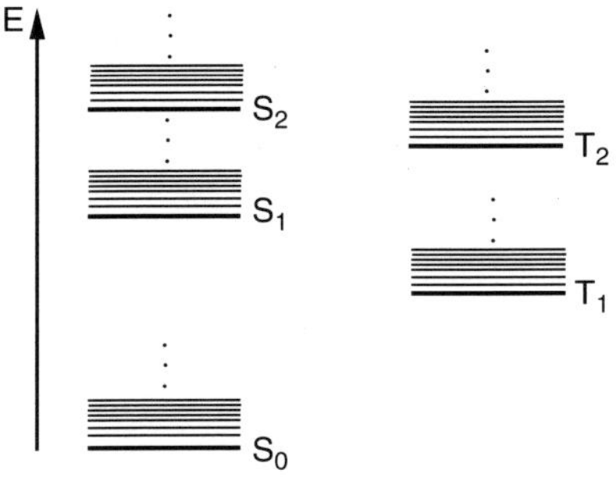

Fig. 17.1. Jablonsky diagram

---

[3] which usually also shows electronic states of higher multiplicity.
[4] Sometimes called channels.
[5] Without a magnetic field, the molecular wave functions can be chosen real-valued.

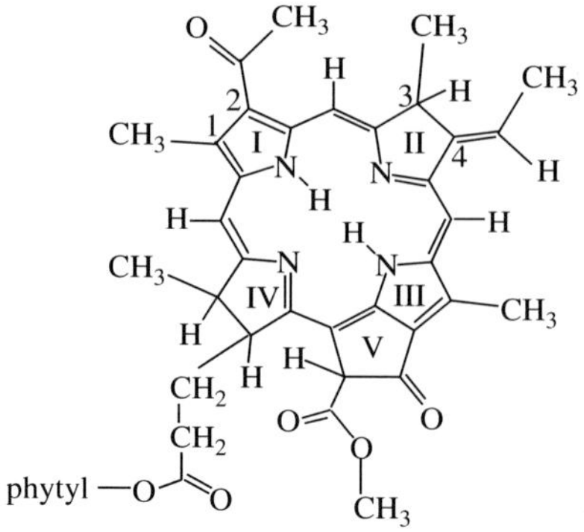

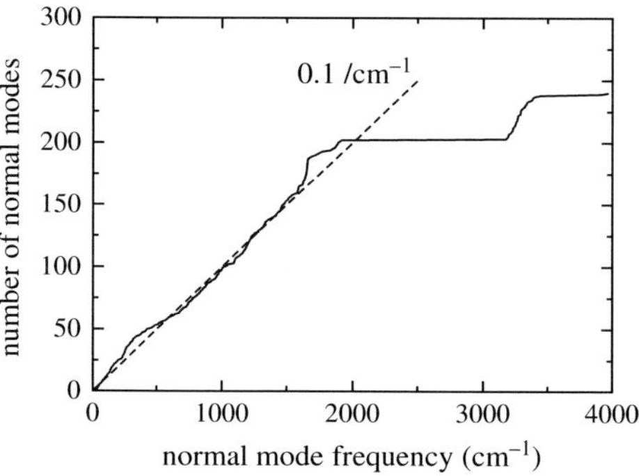

Fig. 17.2. Bacteriopheophytine-b

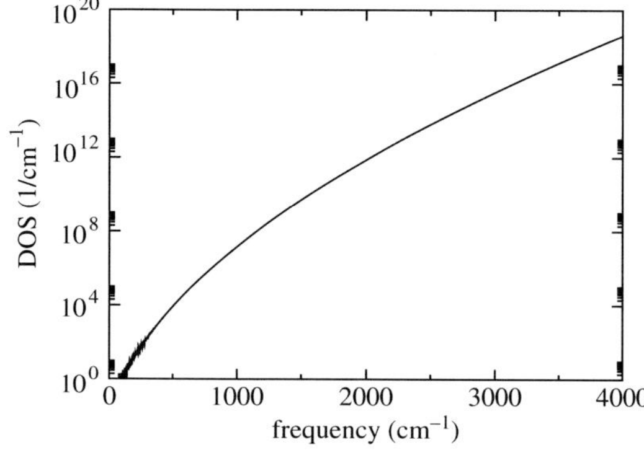

Fig. 17.3. Distribution of normal mode frequencies. Normal modes for bacteriopheophytine-b were calculated quantum chemically. The phytyl tail was replaced by a methyl group. The cumulative frequency distribution (*solid*) is almost linear at small frequencies corresponding to a constant density of $0.1\,\mathrm{cm}^{-1}$ normal modes (*dashed*)

Fig. 17.4. Density of vibrational states. The density of vibrational states including overtones and combination tones was evaluated for bacteriopheophytine-b with a simple counting algorithm

$$
V_{s,s'}^{\mathrm{nad}} = \int \mathrm{d}r\,\mathrm{d}Q\,\chi_{s'}(Q)\psi_{s'}(r,Q)\,V_{\mathrm{nad}}\,\psi_s(r,Q)\chi_s(Q)
$$

$$
= -\sum_j \frac{\hbar^2}{2m_j}\int \mathrm{d}Q\,\chi_{s'}(Q)\,\chi_s(Q)\int \mathrm{d}r\,\psi_{s'}(r,Q)\left(\frac{\partial^2}{\partial Q_j^2}\psi_s(r,Q)\right)
$$

$$
-\sum_j \frac{\hbar^2}{2m_j}\int \mathrm{d}Q\,\chi_{s'}(Q)\left(\frac{\partial}{\partial Q_j}\chi_s(Q)\right)\int \mathrm{d}r\,\psi_{s'}(r,Q)\left(\frac{\partial}{\partial Q_j}\psi_s(r,Q)\right).
\tag{17.16}
$$

The first term is generally small. If we evaluate the second at an equilibrium configuration $Q^{(0)}$ of the state s and neglect the dependence on Q[6] we have

$$
V_{s,s'}^{\mathrm{nad}} \approx -V_{s,s'}^{\mathrm{nad,el}}\sum_j \frac{\hbar^2}{2m_j}\int \mathrm{d}Q\,\chi_{s'}(Q)\left(\frac{\partial}{\partial Q_j}\chi_s(Q)\right),
\tag{17.17}
$$

$$
V_{s,s'}^{\mathrm{nad,el}} = \int \mathrm{d}r\,\psi_{s'}(r,Q^{(0)})\left(\frac{\partial}{\partial Q_j}\psi_s(r,Q^{(0)})\right).
\tag{17.18}
$$

Within the harmonic approximation, the gradient operator can be expressed as

$$
\frac{1}{\sqrt{m_j}}\frac{\partial}{\partial Q_j} = \sum_r u_j^r\frac{\partial}{\partial q_r} = \sum_r u_j^r\sqrt{\frac{\omega_r}{2\hbar}}(b_r^+ - b_r)
\tag{17.19}
$$

and the nonadiabatic matrix element becomes

$$
V_{s',n(r'),s,n(r)}^{\mathrm{nad}} = -V_{s,s'}^{\mathrm{nad,el}}\sum \frac{\hbar^2}{2}\sqrt{\frac{\omega_r}{2\hbar}}\int \mathrm{d}q_1\mathrm{d}q_2\cdots\left(\prod_{r'}\chi_{s'r'n(r')}\right)
$$

$$
\times \left(\prod_{r''\neq r}\chi_{s,r'',n(r'')}\right)\left(\sqrt{n(r)+1}\,\chi_{s,r,n(r)+1} + \sqrt{n(r)}\,\chi_{s,r,n(r)-1}\right).
\tag{17.20}
$$

This expression simplifies considerably if the mixing of normal modes in the state s' can be neglected (the so-called parallel mode approximation). Then the overlap integral factorizes into a product of Franck–Condon factors:

$$
V_{s',n(r'),s,n'(r)}^{\mathrm{nad}} = -V_{s,s'}^{\mathrm{nad,el}}\sum \frac{\hbar^2}{2}\sqrt{\frac{\omega_r}{2\hbar}}\left(\prod_{r'\neq r}\mathrm{FC}_{r'}^{s's}(n',n)\right)
$$

$$
\times \left(\sqrt{n_r+1}\,\mathrm{FC}_r^{s's}(n_r',n_r+1) + \sqrt{n_r}\,\mathrm{FC}_r^{s's}(n_r',n_r-1)\right)
\tag{17.21}
$$

with

$$
\mathrm{FC}_r^{s's}(n_r',n_r) = \int \mathrm{d}q_r\,\chi_{s'r,n'(r)}(q_r)\,\chi_{s,r,n(r)}(q_r).
\tag{17.22}
$$

---

[6] This is known as the Condon approximation.

18

Optical Transitions

We consider allowed optical transitions between two electronic states of a molecule, for instance the ground state (g) and an excited singlet state (e). Within the adiabatic approximation (17.13), we take into account transitions between the manifolds of Born–Oppenheimer vibronic states

$$\psi_{\mathrm{g}}(r;Q)\chi_{\mathrm{g}}(Q) \to \psi_{\mathrm{e}}(r;Q)\chi_{\mathrm{e}}(Q). \tag{18.1}$$

Assuming that the occupation of initial states is given by a canonical distribution $P(\chi_{\mathrm{g}})$, the probability for dipole transitions in an oscillating external field

$$\mathbf{E}_0 \exp(\mathrm{i}\omega_0 t) \tag{18.2}$$

is given by the golden rule expression

$$k = \frac{2\pi}{\hbar} \sum_{\chi_{\mathrm{g}}} P(\chi_{\mathrm{g}}) \left| \langle \psi_{\mathrm{g}}\chi_{\mathrm{g}} | \mathbf{E}_0 er | \psi_{\mathrm{e}}\chi_{\mathrm{e}} \rangle \right|^2 \delta(\hbar\omega_{\mathrm{eg}} - \hbar\omega_0). \tag{18.3}$$

18.1 Dipole Transitions in the Condon Approximation

The matrix element of the dipole operator er can be simplified by performing the integration over the electronic coordinates:

$$\int \mathrm{d}Q \chi_{\mathrm{g}}^*(Q)\chi_{\mathrm{e}}(Q) \int \mathrm{d}r \psi_{\mathrm{g}}^*(r;Q) er \psi_{\mathrm{e}}(r;Q)$$

$$= \int \mathrm{d}Q \chi_{\mathrm{g}}^*(Q)\chi_{\mathrm{e}}(Q) \, \mathbf{M}_{\mathrm{ge}}(Q). \tag{18.4}$$

The dipole moment function $\mathbf{M}_{\mathrm{eg}}(Q)$ is expanded as a series with respect to the nuclear coordinates around the equilibrium configuration

$$\mathbf{M}_{\mathrm{eg}}(Q) = \mathbf{M}_{\mathrm{eg}}(Q_{\mathrm{eq}}) + \frac{\partial \mathbf{M}_{\mathrm{eg}}}{\partial Q}(Q - Q_{\mathrm{eq}}) + \cdots \tag{18.5}$$

If its equilibrium value does not vanish for symmetry reasons, the dipole moment can be approximated by neglecting all higher-order terms (this is known as Condon approximation)

$$\mathbf{M}_{\mathrm{eg}}(Q) \approx \boldsymbol{\mu}_{\mathrm{eg}} = \mathbf{M}_{\mathrm{eg}}(Q_{\mathrm{eq}}). \tag{18.6}$$

The transition rate becomes a product of an electronic factor and an overlap integral of the nuclear wave functions which is known as Franck–Condon factor (in fact, the Franck–Condon-weighted density of states)

$$k = \frac{2\pi}{\hbar} |\mathbf{E}_0 \boldsymbol{\mu}_{\mathrm{eg}}|^2 \sum_{\chi_{\mathrm{g}}, \chi_{\mathrm{e}}} P(\chi_{\mathrm{g}}) \, |\langle \chi_{\mathrm{g}}(Q)|\chi_{\mathrm{e}}(Q)\rangle|^2 \, \delta(\hbar\omega_{\mathrm{eg}} - \hbar\omega_0) \tag{18.7}$$

$$= \frac{2\pi}{\hbar} |\mathbf{E}_0 \boldsymbol{\mu}_{\mathrm{eg}}|^2 \, \mathrm{FCF}(\hbar\omega_0). \tag{18.8}$$

18.2 Time Correlation Function Formalism

In the following, we rewrite the transition rate within the time correlation function (TCF) formalism.

The unperturbed molecular Hamiltonian is

$$H_0 = |\psi_{\mathrm{e}}\rangle (H_{\mathrm{e}} + \hbar\omega_{00})\langle\psi_{\mathrm{e}}| + |\psi_{\mathrm{g}}\rangle H_{\mathrm{g}}\langle\psi_{\mathrm{g}}|, \tag{18.9}$$

where we introduced the Hamiltonians for the nuclear motion in the two states:

$$H_{\mathrm{g(e)}} = T_{\mathrm{N}} + E_{\mathrm{g(e)}}(Q) \tag{18.10}$$

with

$$H_{\mathrm{g(e)}}\chi_{\mathrm{g(e)}} = \hbar\omega_{\mathrm{g(e)}}\chi_{\mathrm{g(e)}} \tag{18.11}$$

and $\hbar\omega_{00}$ is the pure electronic, so-called "0–0" transition energy.[1] The energy of the transition

$$|\psi_{\mathrm{g}}\chi_{\mathrm{g}}\rangle \rightarrow |\psi_{\mathrm{e}}\chi_{\mathrm{e}}\rangle \tag{18.12}$$

is

$$\hbar\omega_{\mathrm{eg}} = \hbar\omega_{00} + \hbar\omega_{\mathrm{e}} - \hbar\omega_{\mathrm{g}}. \tag{18.13}$$

The interaction between molecule and radiation field is within the Condon approximation[2]

$$H' = |\psi_{\mathrm{e}}\rangle(-\mathbf{E}_0\boldsymbol{\mu}_{\mathrm{eg}})\langle\psi_{\mathrm{g}}| + \mathrm{h.c.} \tag{18.14}$$

After replacing the δ-function by a Fourier integral

$$\delta(\hbar\omega) = \frac{1}{2\pi\hbar} \int_{-\infty}^{\infty} \mathrm{e}^{\mathrm{i}\omega t}\mathrm{d}t, \tag{18.15}$$

---

[1] Including the zero-point energy difference.
[2] We consider here only the nondiagonal term.

(18.7) becomes

$$k = \frac{|\mathbf{E}_0\boldsymbol{\mu}_{\mathrm{eg}}|^2}{\hbar^2} \int \mathrm{d}t \sum_{\chi_{\mathrm{g}}\chi_{\mathrm{e}}} P(\chi_{\mathrm{g}})|\langle\chi_{\mathrm{g}}|\chi_{\mathrm{e}}\rangle|^2 e^{\mathrm{i}(\omega_{\mathrm{e}}-\omega_{\mathrm{g}}+\omega_{00}-\omega_0)t}, \tag{18.16}$$

which can be written as a thermal average over the vibrations of the ground state

$$k = \frac{|\mathbf{E}_0\boldsymbol{\mu}_{\mathrm{eg}}|^2}{\hbar^2} \int \mathrm{d}t \sum_{\chi_{\mathrm{g}}\chi_{\mathrm{e}}} \left\langle \chi_{\mathrm{g}} \frac{e^{-\beta H_{\mathrm{g}}}}{Q_{\mathrm{g}}} e^{-\mathrm{i}\omega_{\mathrm{g}}t}\Big|\chi_{\mathrm{e}}\right\rangle e^{\mathrm{i}\omega_{\mathrm{e}}t}\langle\chi_{\mathrm{e}}|\chi_{\mathrm{g}}\rangle e^{\mathrm{i}(\omega_{00}-\omega_0)t}$$

$$= \frac{|\mathbf{E}_0\boldsymbol{\mu}_{\mathrm{eg}}|^2}{\hbar^2} \int \mathrm{d}t\, e^{-\mathrm{i}(\omega_0-\omega_{00})t} \left\langle e^{-\mathrm{i}t H_{\mathrm{g}}/\hbar} e^{\mathrm{i}t H_{\mathrm{e}}/\hbar}\right\rangle_{\mathrm{g}} \tag{18.17}$$

$$= \frac{|\mathbf{E}_0\boldsymbol{\mu}_{\mathrm{eg}}|^2}{\hbar^2} \int \mathrm{d}t\, e^{-\mathrm{i}(\omega_0-\omega_{00})t} F(t) \tag{18.18}$$

with the correlation function

$$F(t) = \left\langle e^{-\mathrm{i}t H_{\mathrm{g}}/\hbar} e^{\mathrm{i}t H_{\mathrm{e}}/\hbar}\right\rangle_{\mathrm{g}} \tag{18.19}$$

which will be evaluated for some simple models in the following chapter. From comparison with (18.8), we find that the Franck–Condon-weighted density of states is given by

$$\mathrm{FCF}(\hbar\omega) = \frac{1}{2\pi\hbar} \int \mathrm{d}t\, e^{-\mathrm{i}(\omega_0-\omega_0 0)} F(t). \tag{18.20}$$

Problems

18.1. Absorption Spectrum

Within the Born–Oppenheimer approximation, the rate for optical transitions of a molecule from its ground state to an electronically excited state is proportional to

$$\alpha(\hbar\omega) = \sum_{i,f} P_i|\langle i|\mu|f\rangle|^2 \delta(+\hbar\omega_f - \hbar\omega_i - \hbar\omega).$$

Show that this golden rule expression can be formulated as the Fourier integral of the dipole moment correlation function

$$\frac{1}{2\pi\hbar} \int \mathrm{d}t\, e^{\mathrm{i}\omega t}\langle\mu(0)\mu(t)\rangle$$

and that within the Condon approximation, this reduces to the correlation function of the nuclear motion.

The Displaced Harmonic Oscillator Model

We would like now to discuss a more specific model for the transition between the vibrational manifolds. We apply the harmonic approximation for the nuclear motion which is described by Hamiltonians

$$H_{\mathrm{g,e}} = \sum_r \hbar \omega_r^{(\mathrm{g,e})} b_r^{(\mathrm{g,e})+} b_r^{(\mathrm{g,e})} \tag{19.1}$$

and discuss the model of displaced oscillators.

19.1 The Time Correlation Function in the Displaced Harmonic Oscillator Approximation

In a very simplified model, we neglect mixing of the normal modes (parallel mode approximation) and frequency changes (Fig. 19.1).

This leads to the "displaced harmonic oscillator" model (DHO) with

$$H_{\mathrm{g}} = \sum_r \hbar \omega_r b_r^+ b_r, \tag{19.2}$$

$$H_{\mathrm{e}} = \sum_r \hbar \omega_r (b_r^+ + g_r)(b_r + g_r) = H_{\mathrm{g}} + \sum_r g_r \hbar \omega_r (b_r^\dagger + b_r) + \sum_r g_r^2 \hbar \omega_r. \tag{19.3}$$

In this approximation, the normal mode frequencies and eigenvectors are the same for both electronic states but the equilibrium positions are shifted relative to each other. The correlation function (18.19)

$$F(t) = \left\langle \mathrm{e}^{-(\mathrm{i}t/\hbar)H_{\mathrm{g}}} \mathrm{e}^{(\mathrm{i}t/\hbar)H_{\mathrm{e}}} \right\rangle_{\mathrm{g}} = Q^{-1} \mathrm{tr}\left(\mathrm{e}^{-H_{\mathrm{g}}/k_{\mathrm{B}}T} \mathrm{e}^{-(\mathrm{i}t/\hbar)H_{\mathrm{g}}} \mathrm{e}^{(\mathrm{i}t/\hbar)H_{\mathrm{e}}} \right) \tag{19.4}$$

with

$$Q = \mathrm{tr}(\mathrm{e}^{-H_{\mathrm{g}}/k_{\mathrm{B}}T}) \tag{19.5}$$

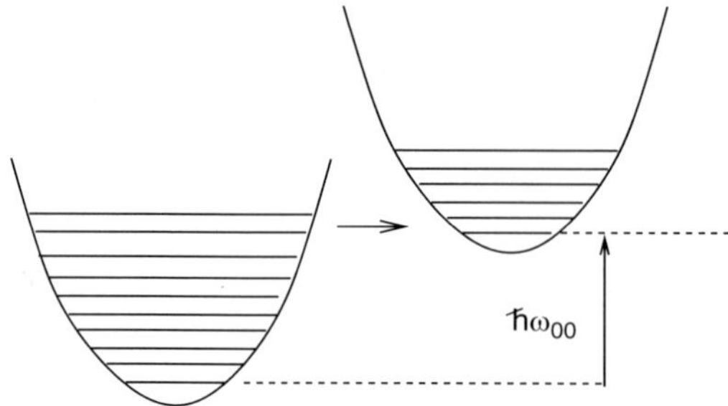

Fig. 19.1. Displaced oscillator model

factorizes in the parallel mode approximation:

$$F(t) = \prod_r F_r(t) \tag{19.6}$$

$$F_r(t) = Q_r^{-1}\mathrm{tr}\left(\mathrm{e}^{-\hbar\omega_r b_r^\dagger b_r/k_\mathrm{B}T}\mathrm{e}^{-\mathrm{i}t\omega_r b_r^\dagger b_r}\mathrm{e}^{\mathrm{i}t\omega_r(b_r^\dagger+g_r)(b_r+g_r)}\right)$$
$$= \left\langle \mathrm{e}^{-\mathrm{i}t\omega_r b_r^\dagger b_r}\mathrm{e}^{\mathrm{i}t\omega_r(b_r^\dagger+g_r)(b_r+g_r)}\right\rangle. \tag{19.7}$$

As shown in the appendix, this can be evaluated as

$$F_r(t) = \exp\left(g_r^2(\mathrm{e}^{\mathrm{i}\omega t}-1)(\overline{n_r}+1)+(\mathrm{e}^{-\mathrm{i}\omega t}-1)\overline{n_r}\right) \tag{19.8}$$

with the average phonon numbers

$$\overline{n_r} = \frac{1}{\mathrm{e}^{\hbar\omega_r/k_\mathrm{B}t}-1}. \tag{19.9}$$

Expression (19.7) contains phonon absorption (positive frequencies) and emission processes (negative frequencies). We discuss two important limiting cases.

19.2 High-Frequency Modes

In the limit $\hbar\omega_r \gg kT$, the average phonon number

$$\overline{n_r} = \frac{1}{\mathrm{e}^{\hbar\omega/k_\mathrm{B}T}-1} \tag{19.10}$$

is small and the correlation function becomes

$$F_r(t) \rightarrow \exp\left(g_r^2(\mathrm{e}^{\mathrm{i}\omega t}-1)\right). \tag{19.11}$$

Expansion of $F_r(t)$ as a power series of g_r^2 gives

$$F_r(t) = \sum_j \frac{g_r^{2j}}{j!}\mathrm{e}^{-g_r^2}\mathrm{e}^{\mathrm{i}(j\omega_r)t} \tag{19.12}$$

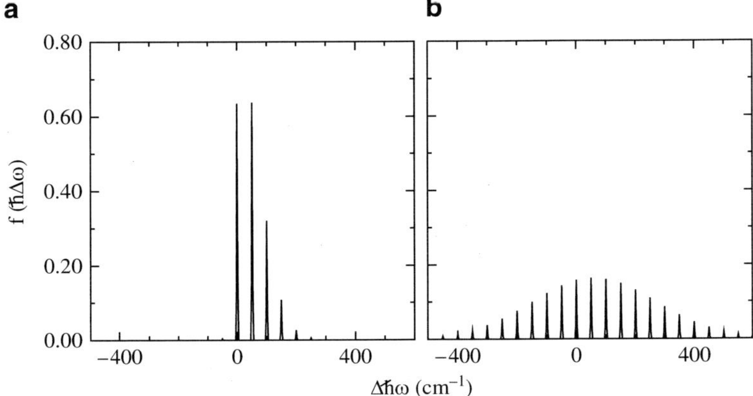

Fig. 19.2. Progression of a low-frequency mode. The Fourier transform of (19.8) is shown for typical values of $\hbar\omega = 50\,\mathrm{cm}^{-1}$, 1, $E_r = 50\,\mathrm{cm}^{-1}$ and (**a**) $kT = 10\,\mathrm{cm}^{-1}$, (**b**) $kT = 200\,\mathrm{cm}^{-1}$. A small damping was introduced to obtain finite linewidths

which corresponds to a progression of transitions $0 \rightarrow j\,\omega_r$ with Franck-Condon factors (Fig. 19.2)

$$FC(0,j) = \frac{g_r^{2j}}{j!}\mathrm{e}^{-g_r^2}. \tag{19.13}$$

19.3 The Short-Time Approximation

To obtain a smooth approximation to the spectrum

$$f(\hbar\Delta\omega) = \frac{1}{2\pi\hbar}\int_{-\infty}^{\infty}\mathrm{d}t\,\mathrm{e}^{-\mathrm{i}\Delta\omega t}\prod_r F_r(t), \tag{19.14}$$

we expand the oscillating functions

$$\mathrm{e}^{\mathrm{i}\omega_r t} = 1 + \mathrm{i}\omega_r t - \frac{\omega_r^2 t^2}{2} + \cdots \tag{19.15}$$

and find

$$f(\hbar\Delta\omega) = \frac{1}{2\pi\hbar}\int_{-\infty}^{\infty}\mathrm{d}t\,\mathrm{e}^{-\mathrm{i}\Delta\omega}\exp\left\{\sum_r g_r^2\mathrm{i}\omega_r t - \sum_r g_r^2(2\bar{n}_r+1)\frac{\omega_r^2 t^2}{2}\right\}$$

$$= \frac{1}{2\pi\hbar}\sqrt{\frac{2\pi}{\sum_r g_r^2\omega_r^2(2\bar{n}_r+1)}}\exp\left\{-\frac{(\Delta\omega - \sum_r g_r^2\omega_r)^2}{2\left(\sum_r g_r^2\omega_r^2(2\bar{n}_r+1)\right)}\right\}. \tag{19.16}$$

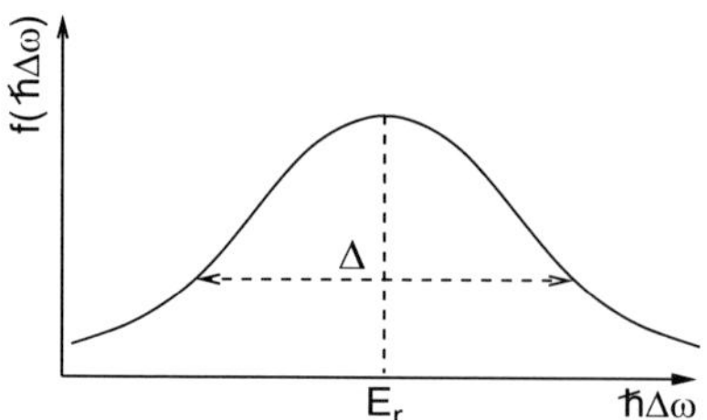

Fig. 19.3. Gaussian envelope

With the definition of the reorganization energy

$$E_r = \sum_r g_r^2 \hbar\omega_r \tag{19.17}$$

and the width

$$\Delta^2 = \sum_r g_r^2 (\hbar\omega_r)^2 (2\bar{n}_r + 1), \tag{19.18}$$

it takes the form

$$f(\hbar\Delta\omega) = \frac{1}{\Delta\sqrt{2\pi}} \exp\left\{ -\frac{(\hbar\Delta\omega - E_r)^2}{2\Delta^2} \right\}. \tag{19.19}$$

In the high-temperature limit (all modes can be treated classically), the width becomes (Fig. 19.3)

$$\Delta^2 \to \sum_r 2g_r^2 \hbar\omega_r k_{\mathrm{B}} T = 2E_r k_{\mathrm{B}} T \tag{19.20}$$

and

$$f(\hbar\Delta\omega) \to \frac{1}{\sqrt{4\pi E_r k_{\mathrm{B}} T}} \exp\left(-\frac{(\hbar\Delta\omega - E_r)^2}{4 E_r k_{\mathrm{B}} T} \right). \tag{19.21}$$

20

Spectral Diffusion

Electronic excitation energies of a chromophore within a protein environment are not static quantities but fluctuate in time. This can be directly observed with the methods of single molecule spectroscopy. If instead an ensemble average is measured, then the relative time scales of measurement and fluctuations determine if an inhomogeneous distribution is observed or if the fluctuations lead to a homogeneous broadening. In this chapter, we discuss simple models [54–56], which are capable of describing the transition between these two limiting cases.

20.1 Dephasing

We study a semiclassical model of a two-state system. Due to interaction with the environment, the energies of the two states are fluctuating quantities. The system is described by a time-dependent Hamiltonian

$$H_0 = \begin{pmatrix} E_1(t) & 0 \\ 0 & E_2(t) \end{pmatrix} \tag{20.1}$$

and the time evolution of the unperturbed system

$$\psi(t) = \exp\left\{ \frac{1}{i\hbar} \int_0^t H_0 \mathrm{d}t \right\} \psi(0) = U_0(t)\psi(0) \tag{20.2}$$

can be described with the help of a propagator

$$U_0(t) = \exp\left\{ \frac{1}{i\hbar} \int_0^t H_0 \mathrm{d}t \right\} = \begin{pmatrix} e^{(1/i\hbar)\int_0^t E_1(t)\mathrm{d}t} & \\ & e^{(1/i\hbar)\int_0^t E_2(t)\mathrm{d}t} \end{pmatrix}. \tag{20.3}$$

Optical transitions are induced by the interaction operator

$$H' = \begin{pmatrix} 0 & \mu e^{i\omega t} \\ \mu e^{-i\omega t} & 0 \end{pmatrix}. \tag{20.4}$$

We make the following ansatz

$$\psi(t) = U_0(t)\phi(t) \tag{20.5}$$

and find

$$(H_0 + H')\psi = i\hbar\frac{d}{dt}\psi = H_0 U_0(t)\phi + i\hbar U_0(t)\frac{d}{dt}\phi. \tag{20.6}$$

Hence

$$i\hbar\frac{d}{dt}\phi = U_0(-t)H'U_0(t)\phi. \tag{20.7}$$

The product of the operators gives

$$U_0(-t)H'U_0(t) = \begin{pmatrix} 0 & \mu e^{i\omega t - (i/\hbar)\int_0^t (E_2 - E_1)dt} \\ \mu e^{-i\omega t + (i/\hbar)\int_0^t (E_2 - E_1)dt} & 0 \end{pmatrix} \tag{20.8}$$

and (20.7) can be written in the form

$$i\hbar\frac{d}{dt}\phi = \begin{pmatrix} 0 & \mu(t)e^{i\omega t} \\ \mu(t)^* e^{-i\omega t} & 0 \end{pmatrix}, \tag{20.9}$$

where the time-dependent dipole moment $\mu(t)$ obeys the differential equation

$$\begin{aligned} \frac{d}{dt}\mu(t) &= \frac{1}{i\hbar}(E_2(t) - E_1(t))\mu(t) \\ &= -i\omega_{21}(t)\mu(t) = -i(\langle\omega_{21}\rangle + \delta\omega(t))\mu(t). \end{aligned} \tag{20.10}$$

Starting from the initial condition

$$\phi(t = 0) = \begin{pmatrix} 1 \\ 0 \end{pmatrix}, \tag{20.11}$$

we find in lowest order

$$\phi(t) = \begin{pmatrix} 1 \\ \frac{1}{i\hbar}\int_0^t dt\, e^{-i\omega t}\mu^*(t) \end{pmatrix} \tag{20.12}$$

and the transition probability is given by

$$\begin{aligned} P(t) &= \frac{1}{\hbar^2}\int_0^t dt'' \int_0^t dt' e^{i\omega(t'' - t')}\mu(t'')\mu^*(t') \\ &= \frac{|\mu_0|^2}{\hbar^2}\int_0^t dt'' \int_0^t dt' e^{i\omega(t'' - t')}e^{-i\langle\omega_{21}\rangle(t'' - t')}e^{-i\int_{t'}^{t''}\delta\omega(t''')dt'''}. \end{aligned} \tag{20.13}$$

The ensemble average gives for stationary fluctuations

$$P(t) = \frac{|\mu_0|^2}{\hbar^2}\int_0^t \int_0^t dt''\, dt'\, e^{i\omega(t'' - t')}e^{-i\langle\omega_{21}\rangle(t'' - t')}F(t'' - t') \tag{20.14}$$

with the dephasing function

$$F(t) = \left\langle \exp\left(-\mathrm{i}\int_0^t \delta\omega(t')\mathrm{d}t'\right) \right\rangle. \tag{20.15}$$

With the help of the Fourier transformation

$$F(t) = \int_{-\infty}^{\infty} \mathrm{e}^{-\mathrm{i}\omega' t}\hat{F}(\omega)\mathrm{d}\omega',$$

the transition probability becomes

$$P(t) = \frac{|\mu_0|^2}{\hbar^2}\int_{-\infty}^{\infty}\mathrm{d}\omega'\int_0^t\int_0^t\mathrm{d}t''\,\mathrm{d}t'\,\mathrm{e}^{\mathrm{i}(\omega-\omega')(t''-t')}\mathrm{e}^{-\mathrm{i}\langle\omega_{21}\rangle(t''-t')}\hat{F}(\omega)$$

$$= \frac{|\mu_0|^2}{\hbar^2}\int_{-\infty}^{\infty}\mathrm{d}\omega'\hat{F}(\omega)\frac{2(1-\cos((\omega-\omega'-\langle\omega_{21}\rangle)t))}{(\omega-\omega'-\langle\omega_{21}\rangle)^2}, \tag{20.16}$$

where the quotient approximates a δ-function for longer times

$$\frac{2(1-\cos((\omega-\omega'-\langle\omega_{21}\rangle)t))}{(\omega-\omega'-\langle\omega_{21}\rangle)^2} \to 2\pi t\delta(\omega-\omega'-\langle\omega_{21}\rangle) \tag{20.17}$$

and finally the golden rule expression is obtained in the form

$$\frac{P(t)}{t} \to \frac{2\pi|\mu_0|^2}{\hbar^2}\hat{F}(\omega-\langle\omega_{21}\rangle). \tag{20.18}$$

20.2 Gaussian Fluctuations

For Gaussian fluctuations, the dephasing function can be approximated by a cumulant expansion. Consider the expression $\ln(\langle\mathrm{e}^{\mathrm{i}\lambda x}\rangle)$. Series expansion with respect to λ gives

$$\ln(\langle\mathrm{e}^{\mathrm{i}\lambda x}\rangle) = \lambda\frac{\langle\mathrm{i}x\mathrm{e}^{\mathrm{i}\lambda x}\rangle}{\langle\mathrm{e}^{\mathrm{i}\lambda x}\rangle} + \frac{\lambda^2}{2}\frac{\langle-x^2\mathrm{e}^{\lambda x}\rangle\langle\mathrm{e}^{\lambda x}\rangle - \langle\mathrm{i}x\mathrm{e}^{\lambda x}\rangle^2}{\langle\mathrm{e}^{\lambda x}\rangle^2} + \cdots \tag{20.19}$$

Setting now $\lambda = 1$ gives

$$\ln(\langle\mathrm{e}^{\mathrm{i}x}\rangle) = \langle\mathrm{i}x\rangle - \frac{1}{2}(\langle x^2\rangle - \langle x\rangle^2)$$

$$+\frac{1}{6}(\langle-\mathrm{i}x^3\rangle - 3\langle-x^2\rangle\langle\mathrm{i}x\rangle + 2\langle\mathrm{i}x\rangle^3) - \cdots \tag{20.20}$$

which for a distribution with zero mean simplifies to

$$\ln(\langle\mathrm{e}^{\mathrm{i}x}\rangle) = \frac{1}{2}\langle x^2\rangle + \frac{1}{6}\langle x^3\rangle + \cdots \tag{20.21}$$

Especially for a Gaussian distribution

$$W(x) = \frac{1}{\Delta\sqrt{2\pi}}e^{-x^2/2\Delta^2}, \tag{20.22}$$

all cumulants except the first and second order vanish. This can be seen from

$$\langle e^{ix}\rangle = \int_{-\infty}^{\infty} e^{ix}W(x)\mathrm{d}x = e^{-\Delta^2/2}. \tag{20.23}$$

If for instance the frequency fluctuations are described by diffusive motion in a harmonic potential, then the distribution of the integral $\int_0^t \delta\omega(t)\mathrm{d}t$ is Gaussian. Finally, we find

$$
\begin{aligned}
F(t) &= \exp\left(-\frac{1}{2}\left\langle\left(\int_0^t \delta\omega(t')\mathrm{d}t'\right)^2\right\rangle\right)\\
&= \exp\left(-\frac{1}{2}\left\langle\int_0^t \mathrm{d}t'\int_0^t \mathrm{d}t''\delta\omega(t')\delta\omega(t'')\right\rangle\right)\\
&= \exp\left(-\frac{1}{2}\int_0^t \mathrm{d}t'\int_0^t \mathrm{d}t''\langle\delta\omega(t'-t'')\delta\omega(0)\rangle\right).
\end{aligned}
\tag{20.24}
$$

We assume that in the simplest case, the correlation of the frequency fluctuations decays exponentially

$$\langle\delta\omega(t)\delta\omega(0)\rangle = \Delta^2 e^{-t/\tau_c}. \tag{20.25}$$

Then integration gives

$$F(t) = \exp\left(-\Delta^2\tau_c^2(e^{-|t|/\tau_c} - 1 + |t|/\tau_c)\right). \tag{20.26}$$

Let us look at the limiting cases.

20.2.1 Long Correlation Time

The limit of long correlation time corresponds to the inhomogeneous case where $\langle\delta\omega(t)\delta\omega(0)\rangle = \Delta^2$ is constant. We expand the exponential for $t \ll \tau_c$:

$$e^{-t/\tau_c} - 1 + t/\tau_c = \frac{t^2}{2\tau_c^2} + \cdots \tag{20.27}$$

$$F(t) = e^{-\Delta^2 t^2/2} \tag{20.28}$$

and the transition probability is

$$P(t) = \frac{|\mu_0|^2}{\hbar^2}\int_0^t \mathrm{d}t'\int_0^t \mathrm{d}t'' e^{i(t''-t')(\omega-\langle\omega_{21}\rangle)}e^{-\Delta^2(t''-t')^2/2}. \tag{20.29}$$

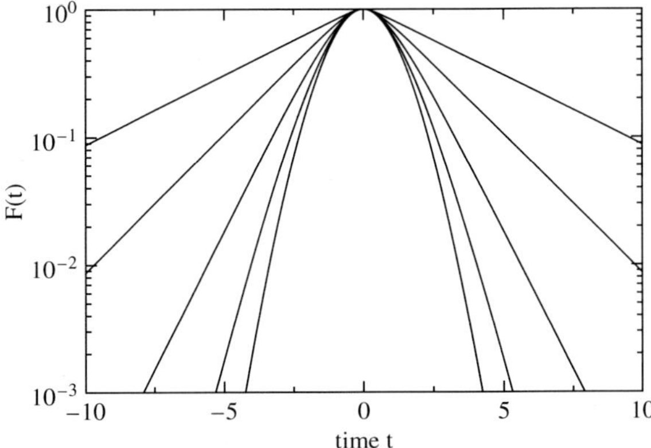

Fig. 20.1. Time correlation function of the Kubo model. $F(t)$ from (20.26) is shown for $\Delta = 1$ and several values of τ_c. For large correlation times, the Gaussian $\exp(-\Delta^2 t^2/2)$ is approximated whereas for short correlation times the correlation function becomes approximately exponential and the width increases with τ^{-1}. Correspondingly, the Fourier transform becomes sharp in this case (motional narrowing)

Since $F(t)$ is symmetric, we find

$$
\begin{aligned}
P(t) &= \frac{|\mu_0|^2}{4\hbar^2} \int_{-t}^{t} \mathrm{d}t' \int_{-t}^{t} \mathrm{d}t''\, \mathrm{e}^{\mathrm{i}(t''-t')(\omega-\langle\omega_{21}\rangle)} \mathrm{e}^{-\Delta^2(t''-t')^2/2} \\
&\approx \frac{|\mu_0|^2}{2\hbar^2}\, t \int_{-\infty}^{\infty} \mathrm{d}t\, \mathrm{e}^{\mathrm{i}t(\omega-\langle\omega_{21}\rangle)} \mathrm{e}^{-\Delta^2 t^2/2}
\end{aligned}
\tag{20.30}
$$

and the lineshape has the form of a Gaussian (Fig. 20.1)

$$
\lim \frac{P(t)}{t} \sim \exp\left(-\frac{(\omega-\langle\omega_{21}\rangle)^2}{2\Delta^2}\right).
\tag{20.31}
$$

20.2.2 Short Correlation Time

For very short correlation time $\tau_\mathrm{c} \ll t$, we approximate

$$
F(t) = \exp\left(-\Delta^2\tau_\mathrm{c}|t|\right) = \exp\left(-|t|/T_2\right)
\tag{20.32}
$$

with the dephasing time

$$
T_2 = (\Delta^2\tau_\mathrm{c})^{-1}.
\tag{20.33}
$$

The lineshape has now the form of a Lorentzian

$$
\int_{-\infty}^{\infty} \mathrm{d}t\, \mathrm{e}^{\mathrm{i}t(\omega-\langle\omega_{21}\rangle)} \mathrm{e}^{-|t|/T_2} = \frac{2T_2^{-1}}{T_2^{-2} + (\omega-\langle\omega_{21}\rangle)^2}.
\tag{20.34}
$$

This result shows the motional narrowing effect when the correlation time is short or the motion very fast.

20.3 Markovian Modulation

Another model which can be analytically investigated describes the frequency fluctuations by a Markovian random walk. We discuss the simplest case of a dichotomous process (9.2), that is, the oscillator frequency switches randomly between two values $\omega_\pm$. This is, for instance, relevant for NMR spectra of a species which undergoes a chemical reaction

$$A+B \rightleftharpoons AB, \tag{20.35}$$

where the NMR resonance frequencies depend on the chemical environment.

For a dichotomous Markovian process switching between two states $X = \pm$, we have for the conditional transition probability

$$P(X, t + \tau | X_0 t_0) = P(X, t + \tau | +, t) P(+, t | X_0 t_0)$$
$$+ P(X, t + \tau | -, t) P(-, t | X_0 t_0). \tag{20.36}$$

For small time increment τ, linearization gives

$$P(+, t + \tau | +, t) = 1 - \alpha\tau,$$
$$P(-, t + \tau | -, t) = 1 - \beta\tau,$$
$$P(-, t + \tau | +, t) = \alpha\tau,$$
$$P(+, t + \tau | -, t) = \beta\tau, \tag{20.37}$$

hence

$$P(+, t + \tau | + t_0)$$
$$= P(+, t + \tau | +, t) P(+, t | + t_0) + P(+, t + \tau | -, t) P(-, t | + t_0)$$
$$= (1 - \alpha\tau) P(+, t | +, t_0) + \beta\tau P(-, t | +, t_0) \tag{20.38}$$

and we obtain the differential equation

$$\frac{\partial}{\partial t} P(+, t | +, t_0) = -\alpha P(+, t | +, t_0) + \beta P(-, t | +, t_0). \tag{20.39}$$

Similarly, we derive

$$\frac{\partial}{\partial t} P(-, t | -, t_0) = -\beta P(-, t | -, t_0) + \alpha P(+, t | -, t_0),$$
$$\frac{\partial}{\partial t} P(-, t | +, t_0) = -\beta P(-, t | +, t_0) + \alpha P(+, t | +, t_0),$$
$$\frac{\partial}{\partial t} P(+, t | -, t_0) = -\alpha P(+, t | -, t_0) + \beta P(-, t | -, t_0). \tag{20.40}$$

In a stationary system, the probabilities depend only on $t - t_0$ and the differential equations can be written as a matrix equation

$$\frac{\partial}{\partial t} \mathbf{P}(t) = A\, \mathbf{P}(t) \tag{20.41}$$

with

$$\mathbf{P}(t) = \begin{pmatrix} P(+,t|+,0) & P(+,t|-,0) \\ P(-,t|+,0) & P(-,t|-,0) \end{pmatrix} \tag{20.42}$$

and the rate matrix

$$A = \begin{pmatrix} -\alpha & \beta \\ \alpha & -\beta \end{pmatrix}. \tag{20.43}$$

Let us define the quantity

$$Q(X,t|X',0) = \left\langle e^{-i\int_0^t \omega(t')dt'} \right\rangle_{X,X'}, \tag{20.44}$$

where the average is taken under the conditions that the system is in state X at time t and in state X' at time 0. Then we find

$$
\begin{aligned}
Q(X,t+\tau|X',0) &= \left\langle e^{-i\int_0^t \omega(t')dt'} e^{i\omega(t)\tau} \right\rangle_{X,X'} \\
&= Q(X,t+\tau|+,t)Q(+,t|X',0) \\
&\quad + Q(X,t+\tau|-,t)Q(-,t|X',0).
\end{aligned} \tag{20.45}
$$

Expansion for small τ gives

$$
\begin{aligned}
Q(+,t+\tau|+,t) &= \left\langle e^{-i\omega_+\tau} \right\rangle_{+,+} = 1 - (i\omega_+ + \alpha)\tau, \\
Q(-,t+\tau|-,t) &= \left\langle e^{-i\omega_-\tau} \right\rangle_{-,-} = 1 - (i\omega_- + \beta)\tau, \\
Q(-,t+\tau|+,t) &= \left\langle e^{-i\omega_+\tau} \right\rangle_{-,+} = \alpha\tau, \\
Q(+,t+\tau|-,t) &= \left\langle e^{-i\omega_-\tau} \right\rangle_{+,-} = \beta\tau,
\end{aligned} \tag{20.46}
$$

and for the time derivatives

$$
\begin{aligned}
\frac{\partial}{\partial t}Q(+,t|+,0) &= \lim_{\tau\to 0} \frac{1}{\tau}\left[(-i\omega_+ - \alpha)\tau Q(+,t|+,0) + \beta\tau Q(-,t|+,0)\right] \\
&= -(i\omega_+ + \alpha)Q(+,t|+,0) + \beta Q(-,t|+,0), \\
\frac{\partial}{\partial t}Q(-,t|-,0) &= -(i\omega_- - +\beta)Q(-,t|-,0) + \alpha Q(+,t|-,0), \\
\frac{\partial}{\partial t}Q(+,t|-,0) &= \lim_{\tau\to 0} \frac{1}{\tau}\left[(-i\omega_+ - \alpha)\tau Q(+,t|-,0) + \beta\tau Q(-,t|-,0)\right] \\
&= -(i\omega_+ + \alpha)Q(+,t|-,0) + \beta Q(-,t|-,0), \\
\frac{\partial}{\partial t}Q(-,t|+,0) &= -(i\omega_- + \beta)Q(-,t|+,0) + \alpha Q(+,t|+,0),
\end{aligned} \tag{20.47}
$$

or in matrix notation

$$\frac{\partial}{\partial t}Q = (-i\Omega + A)Q \tag{20.48}$$

with

$$Q = \begin{pmatrix} Q(+,t|+,0) & Q(+,t|-,0) \\ Q(-,t|+,0) & Q(-,t|-,0) \end{pmatrix} \tag{20.49}$$

and

$$\Omega = \begin{pmatrix} \omega_+ & \\ & \omega_- \end{pmatrix}. \tag{20.50}$$

Equation (20.48) is formally solved by

$$Q(t) = \exp\left\{(-\mathrm{i}\Omega + A)t\right\}. \tag{20.51}$$

For a steady-state system, we have to average over the initial state and to sum over the final states to obtain the dephasing function. This can be expressed as

$$F(t) = (1,1)Q(t)\begin{pmatrix} \frac{\beta}{\alpha+\beta} \\ \frac{\alpha}{\alpha+\beta} \end{pmatrix} = (1,1)\exp\left\{(-\mathrm{i}\Omega + A)t\right\}\begin{pmatrix} \frac{\beta}{\alpha+\beta} \\ \frac{\alpha}{\alpha+\beta} \end{pmatrix}. \tag{20.52}$$

Laplace transformation gives

$$\widetilde{F}(s) = \int_0^\infty e^{-st}F(t)\mathrm{d}t = (1,1)(s + \mathrm{i}\Omega - A)^{-1}\begin{pmatrix} \frac{\beta}{\alpha+\beta} \\ \frac{\alpha}{\alpha+\beta} \end{pmatrix}, \tag{20.53}$$

which can be easily evaluated to give

$$\begin{aligned}
\widetilde{F}(s) &= \frac{1}{(s + \mathrm{i}\omega_1)(s + \mathrm{i}\omega_2) + (\alpha + \beta)s + \mathrm{i}(\alpha\omega_2 + \beta\omega_1)} \\
&\quad \times (1,1)\begin{pmatrix} s + \mathrm{i}\omega_2 - \beta & -\beta \\ -\alpha & s + \mathrm{i}\omega_1 - \alpha \end{pmatrix}\begin{pmatrix} \frac{\beta}{\alpha+\beta} \\ \frac{\alpha}{\alpha+\beta} \end{pmatrix} \\
&= \frac{s + (\alpha + \beta) + \mathrm{i}\frac{\beta\omega_2 + \alpha\omega_1}{\alpha+\beta}}{(s + \mathrm{i}\omega_1)(s + \mathrm{i}\omega_2) + (\alpha + \beta)s + \mathrm{i}(\alpha\omega_2 + \beta\omega_1)}.
\end{aligned} \tag{20.54}$$

Since the dephasing function generally obeys the symmetry $F(-t) = F(t)^*$, the lineshape is obtained from the real part

$$\begin{aligned}
2\pi\hat{F}(\omega) &= \int_{-\infty}^\infty e^{\mathrm{i}\omega t}F(t)\mathrm{d}t = \left(\int_0^\infty e^{\mathrm{i}\omega t}F(t)\mathrm{d}t + \int_0^\infty e^{-\mathrm{i}\omega t}F^*(t)\mathrm{d}t\right) \\
&= 2\Re\left(\widetilde{F}(-\mathrm{i}\omega)\right) \\
&= 2\frac{\alpha\beta}{\alpha + \beta}\frac{(\omega_1 - \omega_2)^2}{(\omega - \omega_1)^2(\omega - \omega_2)^2 + (\alpha + \beta)^2\left(\omega - \frac{\alpha\omega_2 + \beta\omega_1}{\alpha+\beta}\right)^2}.
\end{aligned} \tag{20.55}$$

Let us introduce the average frequency

$$\bar{\omega} = \frac{\alpha\omega_2 + \beta\omega_1}{\alpha + \beta} \tag{20.56}$$

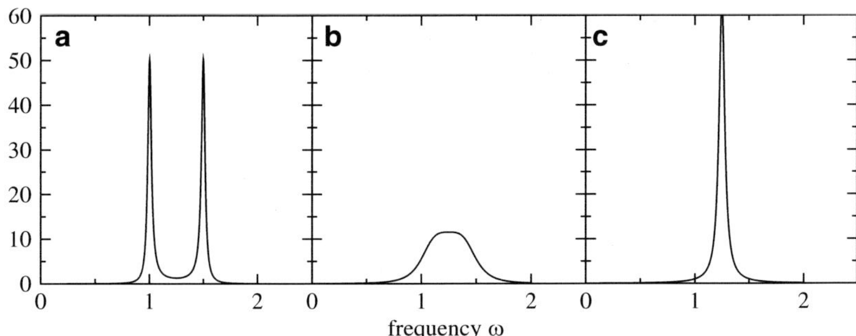

Fig. 20.2. Motional narrowing. The lineshape (20.55) is evaluated for $\omega_1 = 1.0$, $\omega_2 = 1.5$ and $\alpha = \beta =$ (**a**) 0.02, (**b**) 0.18, and (**c**) 1.0

and the correlation time of the dichotomous process

$$\tau_c = \omega_c^{-1} = (\alpha + \beta)^{-1}. \tag{20.57}$$

In the limit of slow fluctuations ($\omega_c \to 0$), two sharp resonances appear at $\omega = \omega_{1,2}$ with relative weights given by the equilibrium probabilities

$$P(+) = \beta/(\alpha + \beta), \quad P(-) = \alpha/(\alpha + \beta). \tag{20.58}$$

With increasing ω_c, the two resonances become broader and finally merge into one line. For further increasing ω_c the resonance, which is now centered at $\overline{\omega}$, becomes very narrow (this is known as the motional narrowing effect) (Fig. 20.2).

Problems

20.1. Motional Narrowing
Discuss the poles of the lineshape function and the motional narrowing effect (20.55) for the symmetrical case $\alpha = \beta = (2\tau_c)^{-1}$ and $\omega_{1,2} = \overline{\omega} \pm \Delta\omega/2$.

21

Crossing of Two Electronic States

21.1 Adiabatic and Diabatic States

We consider the avoided crossing of two electronic states along a nuclear coordinate Q. The two states are described by adiabatic wave functions

$$\psi_{1,2}^{\mathrm{ad}}(r, Q) \tag{21.1}$$

and the wave function

$$\psi_1^{\mathrm{ad}}(r, Q)\chi_1(Q) + \psi_2^{\mathrm{ad}}(r, Q)\chi_2(Q) \tag{21.2}$$

will be also written in matrix notation [57] as

$$\left(\psi_1^{\mathrm{ad}}(r, Q) \ \psi_2^{\mathrm{ad}}(r, Q) \right) \begin{pmatrix} \chi_1(Q) \\ \chi_2(Q) \end{pmatrix}. \tag{21.3}$$

We will assume that the electronic wave functions are chosen real-valued (no magnetic field). From the adiabatic energies

$$\int \mathrm{d}r \, \psi_s^{\mathrm{ad}}(T_{\mathrm{el}} + V)\psi_{s'}^{\mathrm{ad}} = \delta_{s,s'} E_s^{\mathrm{ad}}(Q) \tag{21.4}$$

and the matrix elements of the nuclear kinetic energy

$$-\frac{\hbar^2}{2m} \int \mathrm{d}r \, \psi_s^{\mathrm{ad}}(r, Q) \left(\frac{\partial^2}{\partial Q^2} \right) \psi_{s'}^{\mathrm{ad}}(r, Q)$$

$$= -\frac{\hbar^2}{2m} V_{s,s'}^{\mathrm{nad}}(Q)\frac{\partial}{\partial Q} - \frac{\hbar^2}{2m} V_{s,s'}^{\mathrm{nad},(2)}(Q) - \frac{\hbar^2}{2m}\frac{\partial^2}{\partial Q^2} \tag{21.5}$$

with the first- and second-order nonadiabatic coupling matrix elements

$$-\frac{\hbar^2}{2m} V_{s,s'}^{\mathrm{nad}}(Q) = -\frac{\hbar^2}{2m} \int \mathrm{d}r \, \psi_s^{\mathrm{ad}}(r, Q) \left(\frac{\partial}{\partial Q}\psi_{s'}^{\mathrm{ad}}(r, Q) \right), \tag{21.6}$$

$$-\frac{\hbar^2}{2m}V_{s,s'}^{\mathrm{nad},(2)}(Q) = -\frac{\hbar^2}{2m}\int \mathrm{d}r\,\psi_s^{\mathrm{ad}}(r,Q)\left(\frac{\partial^2}{\partial Q^2}\psi_{s'}^{\mathrm{ad}}(r,Q)\right), \qquad (21.7)$$

we construct the Hamiltonian for the nuclear wave function

$$H(Q) = -\frac{\hbar^2}{2m}\frac{\partial^2}{\partial Q^2} + E^{\mathrm{ad}}(Q) - \frac{\hbar^2}{2m}V^{\mathrm{nad}}(Q)\frac{\partial}{\partial Q} - \frac{\hbar^2}{2m}V_{s,s'}^{\mathrm{nad},(2)}(Q) \quad (21.8)$$

with the matrices

$$E^{\mathrm{ad}}(Q) = \begin{bmatrix} E_1^{\mathrm{ad}}(Q) & \\ & E_2^{\mathrm{ad}}(Q) \end{bmatrix}, \quad V^{\mathrm{nad}}(Q) = \begin{bmatrix} V_{11}^{\mathrm{nad}}(Q) & V_{12}^{\mathrm{nad}}(Q) \\ V_{21}^{\mathrm{nad}}(Q) & V_{22}^{\mathrm{nad}}(Q) \end{bmatrix},$$

$$V^{\mathrm{nad},(2)}(Q) = \begin{bmatrix} V_{11}^{\mathrm{nad},(2)}(Q) & V_{12}^{\mathrm{nad},(2)}(Q) \\ V_{21}^{\mathrm{nad},(2)}(Q) & V_{22}^{\mathrm{nad},(2)}(Q) \end{bmatrix}. \qquad (21.9)$$

The nuclear part of the wave function is determined by the eigenvalue equation:

$$H(Q)\begin{pmatrix} \chi_1(Q) \\ \chi_2(Q) \end{pmatrix} = E\begin{pmatrix} \chi_1(Q) \\ \chi_2(Q) \end{pmatrix}. \qquad (21.10)$$

Now for real-valued functions, we have

$$\begin{aligned}
V_{ss'}^{\mathrm{nad}}(Q) &= \int \mathrm{d}r\,\psi_s^{\mathrm{ad}}(r,Q)\frac{\partial}{\partial Q}\psi_{s'}^{\mathrm{ad}} \\
&= \frac{\partial}{\partial Q}\int \mathrm{d}r\,\psi_s^{\mathrm{ad}}(r,Q)\psi_{s'}^{\mathrm{ad}}(r,Q) - \int \mathrm{d}r\left(\frac{\partial}{\partial Q}\psi_s^{\mathrm{ad}}\right)\psi_{s'}^{\mathrm{ad}} \\
&= -\left(V_{s's}^{\mathrm{nad}}(Q)\right)
\end{aligned} \qquad (21.11)$$

and the matrix V^{nad} is antisymmetric[1]

$$V^{\mathrm{nad}} = \begin{pmatrix} 0 & V_{12}^{\mathrm{nad}} \\ -V_{12}^{\mathrm{nad}} & 0 \end{pmatrix}. \qquad (21.12)$$

We want to define a new basis of so-called diabatic electronic states for which the nonadiabatic interaction is simplified. Therefore, we introduce a transformation with a unitary matrix

$$C = \begin{pmatrix} c_{11} & c_{12} \\ c_{21} & c_{22} \end{pmatrix} \qquad (21.13)$$

and form the linear combinations

$$\psi^{\mathrm{dia}} = \left(\psi_1^{\mathrm{dia}}(r,Q)\ \psi_2^{\mathrm{dia}}(r,Q)\right) = \left(\psi_1^{\mathrm{ad}}(r,Q)\ \psi_2^{\mathrm{ad}}(r,Q)\right)C = \psi^{\mathrm{ad}}C. \quad (21.14)$$

---

[1] For complex wave functions instead $V_{s,s'}^{\mathrm{nad}} = -(V_{s',s}^{\mathrm{nad}})^*$.

The nuclear kinetic energy transforms into

$$
-\frac{\hbar^2}{2m}\int dr\,\psi^{\mathrm{dia}\,\dagger}(r,Q)\frac{\partial^2}{\partial Q^2}\psi^{\mathrm{dia}}(Q)
$$

$$
= -\frac{\hbar^2}{2m}\int dr\,C^\dagger\psi^{\mathrm{ad}\,\dagger}\left\{\frac{\partial^2\psi^{\mathrm{ad}}}{\partial Q^2}C + \psi^{\mathrm{ad}}\frac{\partial^2 C}{\partial Q^2} + \psi^{\mathrm{ad}}C\frac{\partial^2}{\partial Q^2}\right.
$$

$$
\left.+ 2\frac{\partial\psi^{\mathrm{ad}}}{\partial Q}\frac{\partial C}{\partial Q} + 2\frac{\partial\psi^{\mathrm{ad}}}{\partial Q}C\frac{\partial}{\partial Q} + 2\psi^{\mathrm{ad}}\frac{\partial C}{\partial Q}\frac{\partial}{\partial Q}\right\} \tag{21.15}
$$

$$
= -\frac{\hbar^2}{2m}\left\{C^\dagger V^{\mathrm{nad},(2)}C + C^\dagger\frac{\partial^2 C}{\partial Q^2} + \frac{\partial^2}{\partial Q^2} + 2C^\dagger V^{\mathrm{nad}}\frac{\partial C}{\partial Q}\right.
$$

$$
\left.+ 2C^\dagger V^{\mathrm{nad}}C\frac{\partial}{\partial Q} + 2C^\dagger\frac{\partial C}{\partial Q}\frac{\partial}{\partial Q}\right\}. \tag{21.16}
$$

The gradient operator vanishes if we chose C to be a solution of

$$
\frac{\partial C}{\partial Q} = -V^{\mathrm{nad}}(Q)C. \tag{21.17}
$$

This can be formally solved by

$$
C = \exp\left\{\int_Q^\infty V^{\mathrm{nad}}(Q)\mathrm{d}Q\right\}. \tag{21.18}
$$

For antisymmetric V^{nad}, the exponential function can be easily evaluated[2] to give

$$
C = \begin{pmatrix} \cos\zeta(Q) & \sin\zeta(Q) \\ -\sin\zeta(Q) & \cos\zeta(Q) \end{pmatrix}, \quad \zeta(Q) = \int_Q^\infty V_{12}^{\mathrm{nad}}(Q)\mathrm{d}Q. \tag{21.19}
$$

Due to (21.17), the last two summands in (21.16) cancel and

$$
-\frac{\hbar^2}{2m}\int dr\,\psi^{\mathrm{dia}\,\dagger}(r,Q)\frac{\partial^2}{\partial Q^2}\psi^{\mathrm{dia}}(Q)
$$

$$
= -\frac{\hbar^2}{2m}\left\{C^\dagger V^{\mathrm{nad},(2)}C + C^\dagger\frac{\partial^2 C}{\partial Q^2} + \frac{\partial^2}{\partial Q^2} + 2C^\dagger V^{\mathrm{nad}}\frac{\partial C}{\partial Q}\right\}. \tag{21.20}
$$

Applying (21.17) once more, we find that

$$
V^{\mathrm{nad},(2)}C + \frac{\partial^2 C}{\partial Q^2} + 2V^{\mathrm{nad}}\frac{\partial C}{\partial Q}
$$

$$
= V^{\mathrm{nad},(2)}C - \frac{\partial V^{\mathrm{nad}}}{\partial Q}C - V^{\mathrm{nad}}\frac{\partial C}{\partial Q} + 2V^{\mathrm{nad}}\frac{\partial C}{\partial Q}
$$

$$
= V^{\mathrm{nad},(2)}C - \frac{\partial V^{\mathrm{nad}}}{\partial Q}C - (V^{\mathrm{nad}})^2 C. \tag{21.21}
$$

---

[2] for instance by a series expansion

The derivative of the nonadiabatic coupling matrix is

$$\frac{\partial V^{\mathrm{nad}}}{\partial Q} = \int \mathrm{d}r \left(\frac{\partial \psi}{\partial Q} \frac{\partial \psi'}{\partial Q}\right) + \int \mathrm{d}r\, \psi \left(\frac{\partial^2 \psi'}{\partial Q^2}\right)$$
$$= (V^{\mathrm{nad}})^T V^{\mathrm{nad}} + V^{\mathrm{nad},(2)} = -(V^{\mathrm{nad}})^2 + V^{\mathrm{nad},(2)} \qquad (21.22)$$

and therefore (21.21) vanishes. Together, we have finally

$$-\frac{\hbar^2}{2m} \int \mathrm{d}r\, \psi^{\mathrm{dia}\,\dagger}(r,Q) \frac{\partial^2}{\partial Q^2} \psi^{\mathrm{dia}}(Q) = -\frac{\hbar^2}{2m} \frac{\partial^2}{\partial Q^2}. \qquad (21.23)$$

From

$$\int \mathrm{d}r\, \psi^{\mathrm{dia}\,\dagger}(T_{\mathrm{el}} + V)\psi^{\mathrm{dia}} = \int \mathrm{d}r\, C^\dagger \psi^{\mathrm{ad}\,\dagger}(T_{\mathrm{el}} + V)\psi^{\mathrm{ad}}C, \qquad (21.24)$$

we find the matrix elements of the electronic Hamiltonian

$$\widetilde{H} = C^\dagger E^{\mathrm{ad}}(Q)C$$
$$= \begin{pmatrix} \cos^2 \zeta E_1^{\mathrm{ad}} + \sin^2 \zeta E_2^{\mathrm{ad}} & \sin \zeta \cos \zeta (E_1^{\mathrm{ad}} - E_2^{\mathrm{ad}}) \\ \sin \zeta \cos \zeta (E_1^{\mathrm{ad}} - E_2^{\mathrm{ad}}) & \cos^2 \zeta E_2^{\mathrm{ad}} + \sin^2 \zeta E_1^{\mathrm{ad}} \end{pmatrix}. \qquad (21.25)$$

At the crossing point

$$(\cos^2 \zeta_0 - \sin^2 \zeta_0)(E_1^{\mathrm{ad}} - E_2^{\mathrm{ad}}) = 0, \qquad (21.26)$$

which implies

$$\cos^2 \zeta_0 = \sin^2 \zeta_0 = \frac{1}{2}. \qquad (21.27)$$

Expanding the sine and cosine functions around the crossing point, we have

$$\widetilde{H} \approx \begin{pmatrix} \frac{E_1+E_2}{2} + (E_2 - E_1)(\zeta - \zeta_0) + \cdots & (E_1 - E_2)\left(\frac{1}{2} + (\zeta - \zeta_0)^2 + \cdots\right) \\ (E_1 - E_2)\left(\frac{1}{2} + (\zeta - \zeta_0)^2 + \cdots\right) & \frac{E_1+E_2}{2} - (E_2 - E_1)(\zeta - \zeta_0) + \cdots \end{pmatrix}. \qquad (21.28)$$

Expanding furthermore the matrix elements

$$E_1^{\mathrm{ad}} = \bar{E} + \frac{\Delta}{2} + g_1(Q - Q_0) + \cdots, \qquad (21.29)$$

$$E_2^{\mathrm{ad}} = \overline{E} - \frac{\Delta}{2} + g_2(Q - Q_0) + \cdots, \qquad (21.30)$$

$$\zeta - \zeta_0 = -(Q - Q_0)V_{12}^{\mathrm{nad}}(Q_0), \qquad (21.31)$$

where Δ is the splitting of the adiabatic energies at the crossing point, the interaction matrix becomes

$$\widetilde{H} = \overline{E} + \frac{g_1 + g_2}{2}Q + \begin{pmatrix} -(Q - Q_0)V_{12}^{\mathrm{nad}}(Q_0)\Delta & \frac{\Delta}{2} + \frac{g_1-g_2}{2}(Q - Q_0) \\ \frac{\Delta}{2} + \frac{g_1-g_2}{2}(Q - Q_0) & (Q - Q_0)V_{12}^{\mathrm{nad}}(Q_0)\Delta \end{pmatrix} + \cdots \qquad (21.32)$$

We see that in the diabatic basis, the interaction is given by half the splitting of the adiabatic energies at the crossing point (Fig. 21.1).

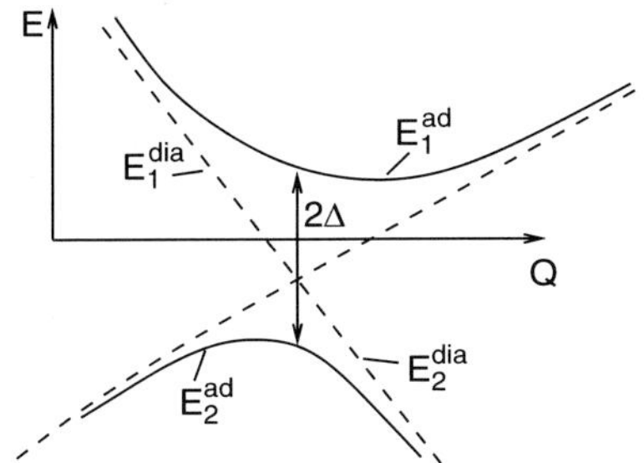

Fig. 21.1. Curve crossing

21.2 Semiclassical Treatment

Landau [58] and Zener [59] investigated the curve-crossing process treating
the nuclear motion classically by introducing a trajectory

$$Q(t) = Q_0 + vt, \tag{21.33}$$

where they assumed a constant velocity in the vicinity of the crossing point.
They investigated the time-dependent model Hamiltonian

$$H(t) = \begin{pmatrix} E_1(Q(t)) & V \\ V & E_2(Q(t)) \end{pmatrix} = \begin{pmatrix} -\frac{1}{2}\frac{\partial \Delta E}{\partial t}t & V \\ V & \frac{1}{2}\frac{\partial \Delta E}{\partial t}t \end{pmatrix} \tag{21.34}$$

and solved the time-dependent Schrödinger equation

$$i\hbar\frac{\partial}{\partial t}\begin{pmatrix} c_1 \\ c_2 \end{pmatrix} = H(t)\begin{pmatrix} c_1 \\ c_2 \end{pmatrix}, \tag{21.35}$$

which reads explicitly

$$i\hbar\frac{\partial c_1}{\partial t} = E_1(t)c_1 + Vc_2,$$
$$i\hbar\frac{\partial c_2}{\partial t} = E_2(t) + Vc_1. \tag{21.36}$$

Substituting

$$c_1 = a_1 e^{(1/i\hbar)\int E_1(t)\mathrm{d}t},$$
$$c_2 = a_2 e^{(1/i\hbar)\int E_2(t)\mathrm{d}t}, \tag{21.37}$$

the equations simplify

$$i\hbar\frac{\partial a_1}{\partial t} = V e^{(1/i\hbar)\int (E_2(t)-E_1(t))\mathrm{d}t}a_2,$$
$$i\hbar\frac{\partial a_2}{\partial t} = V e^{-(1/i\hbar)\int (E_2(t)-E_1(t))\mathrm{d}t}a_1. \tag{21.38}$$

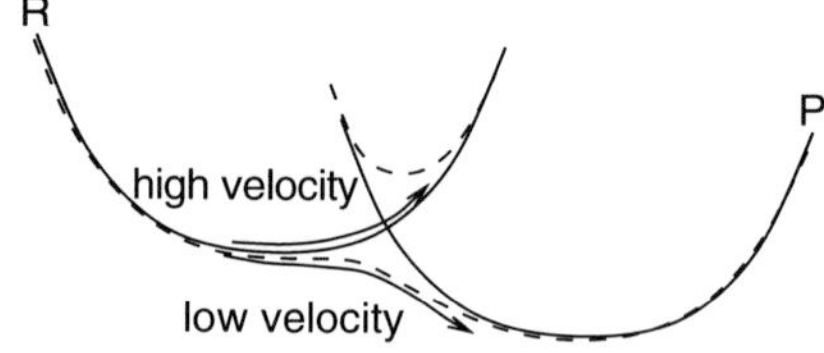

Fig. 21.2. Velocity dependence of the transition. At low velocities, the electronic wave function follows the nuclear motion adiabatically corresponding to a transition between the diabatic states. At high velocities, the probability for this transition (21.42) becomes slow. *Full*: diabatic states and *dashed*: adiabatic states

Let us consider the limit of small V and calculate the transition probability in lowest order. From the initial condition $a_1(-\infty) = 1$, $a_2(-\infty) = 0$, we get

$$\int_0^t (E_2(t') - E_1(t'))\mathrm{d}t' = \frac{\partial \Delta E}{\partial t} \frac{t^2}{2}, \tag{21.39}$$

$$a_2(\infty) \approx \frac{1}{\mathrm{i}\hbar} V \int_{-\infty}^{\infty} \exp\left\{ -\frac{1}{\mathrm{i}\hbar} \frac{\partial \Delta E}{\partial t} \frac{t^2}{2} \right\} \mathrm{d}t = \frac{V}{\mathrm{i}\hbar} \sqrt{\frac{2\pi\hbar}{-\mathrm{i}(\partial \Delta E/\partial t)}} \tag{21.40}$$

and the transition probability is

$$P_{12} = |a_2(\infty)|^2 = \frac{2\pi V^2}{\hbar |\partial \Delta E/\partial t|}. \tag{21.41}$$

Landau and Zener calculated the transition probability for arbitrary coupling strength (Fig. 21.2)

$$P_{12}^{\mathrm{LZ}} = 1 - \exp\left(-\frac{2\pi V^2}{\hbar |\partial \Delta E/\partial t|} \right). \tag{21.42}$$

21.3 Application to Diabatic ET

If we describe the diabatic potentials as displaced harmonic oscillators

$$E_1(Q) = \frac{m\omega^2}{2} Q^2, \quad E_2(Q) = \Delta G + \frac{m\omega^2}{2}(Q - Q_1)^2, \tag{21.43}$$

the energy gap is with

$$Q_1 = \sqrt{\frac{2\lambda}{m\omega^2}}, \tag{21.44}$$

$$E_2 - E_1 = \Delta G + \frac{m\omega^2}{2}(Q_1^2 - 2Q_1 Q) = \Delta G + \lambda - \omega\sqrt{2m\lambda}Q. \tag{21.45}$$

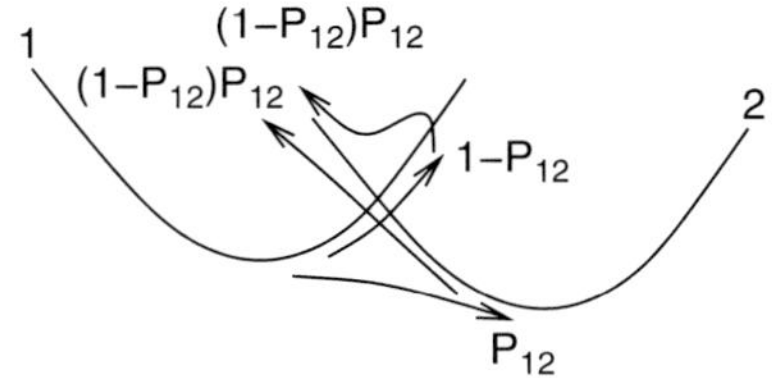

Fig. 21.3. Multiple crossing. The probability of crossing during one period of the oscillation is $2P_{12}(1 - P_{12})$ if the two crossings are independent

The time derivative of the energy gap is

$$\frac{\partial(E_2(Q) - E_1(Q))}{\partial t} = -\omega\sqrt{2m\lambda}\frac{\partial Q}{\partial t} \tag{21.46}$$

and the average velocity is

$$\left\langle\frac{\partial Q}{\partial t}\right\rangle = \frac{\int_{-\infty}^{\infty}|v|e^{-mv^2/2kT}dV}{\int_{-\infty}^{\infty}e^{-mv^2/2kT}dV} = \sqrt{\frac{2kT}{m\pi}}. \tag{21.47}$$

The probability of curve crossing during one period of the oscillation is given by $2P_{12}(1 - P_{12}) \approx 2P_{12}$ (see Fig. 21.3).

Together, this gives a transmission coefficient

$$\kappa_{\mathrm{el}} = 2\frac{2\pi V^2}{\hbar\left|\omega\sqrt{2\lambda}\sqrt{\frac{2kT}{\pi}}\right|} = \frac{2\pi V^2}{\hbar}\frac{2\pi}{\omega}\frac{1}{\sqrt{4\pi\lambda kT}} \tag{21.48}$$

and a rate of

$$k = \frac{2\pi V^2}{\hbar}\frac{1}{\sqrt{4\pi\lambda kT}}e^{-\Delta G_{\mathrm{a}}/kT}. \tag{21.49}$$

21.4 Crossing in More Dimensions

The definition of a diabatic basis is not so straightforward in the general case of more nuclear degrees of freedom [60]. Generalization of (21.17) gives

$$\frac{\partial C}{\partial Q_j} = -V_j^{\mathrm{nad}}(Q)C. \tag{21.50}$$

Now consider

$$\frac{\partial}{\partial Q_i}V_{ss',j}^{\mathrm{nad}}(Q) - \frac{\partial}{\partial Q_j}V_{ss',i}^{\mathrm{nad}}(Q)$$

$$= \int dr\left(\frac{\partial}{\partial Q_i}\psi_s^{\mathrm{ad}}(r,Q)\right)\left(\frac{\partial}{\partial Q_j}\psi_{s'}^{\mathrm{ad}}(r,Q)\right)$$

$$- \int dr\left(\frac{\partial}{\partial Q_j}\psi_s^{\mathrm{ad}}(r,Q)\right)\left(\frac{\partial}{\partial Q_i}\psi_{s'}^{\mathrm{ad}}(r,Q)\right). \tag{21.51}$$

We assume completeness of the states ψ_s^{ad} and insert

$$\sum_s \psi_s^{\mathrm{ad}}(r)\psi_s^{\mathrm{ad}}(r') = \delta(r - r') \tag{21.52}$$

to obtain

$$\sum_{s''} \int \mathrm{d}r'\mathrm{d}r \left\{ \left(\frac{\partial}{\partial Q_i}\psi_s^{\mathrm{ad}}(r,Q) \right) \psi_{s''}^{\mathrm{ad}}(r)\psi_{s''}^{\mathrm{ad}}(r') \left(\frac{\partial}{\partial Q_j}\psi_{s'}^{\mathrm{ad}}(r',Q) \right) \right.$$

$$\left. - \left(\frac{\partial}{\partial Q_j}\psi_s^{\mathrm{ad}}(r,Q) \right) \psi_{s''}^{\mathrm{ad}}(r)\psi_{s''}^{\mathrm{ad}}(r') \left(\frac{\partial}{\partial Q_i}\psi_{s'}^{\mathrm{ad}}(r,Q) \right) \right\}, \tag{21.53}$$

which can be written as

$$\sum_{s'' \neq s,s'} \left(V_{s,s'',i}^{\mathrm{nad}} V_{s'',s',j}^{\mathrm{nad}} - V_{s,s'',j}^{\mathrm{nad}} V_{s'',s',i}^{\mathrm{nad}} \right). \tag{21.54}$$

If only two states are involved, then this sum vanishes

$$\frac{\partial}{\partial Q_i} V_{1,2,j}^{\mathrm{nad}}(Q) - \frac{\partial}{\partial Q_j} V_{1,2,i}^{\mathrm{nad}}(Q) = 0 \tag{21.55}$$

and a potential function exists, which provides a solution to (21.50). For more than two states, an integrability condition like (21.55) is not fulfilled and only part of the nonadiabatic coupling can be eliminated.[3] Often a crude diabatic basis is used, which refers to a fixed configuration near the crossing point and does not depend on the nuclear coordinates at all. In a diabatic representation, the Hamiltonian of a two-state system has the general form

$$H = \begin{pmatrix} H_{11}(Q_i) & H_{12}(Q_i) \\ H_{12}(Q_i)^\dagger & H_{22}(Q_i) \end{pmatrix}. \tag{21.56}$$

In one dimension a crossing of the corresponding adiabatic curves (the eigenvalues of H) is very unlikely, since it occurs only if the two conditions

$$H_{11}(Q) - H_{22}(Q) = H_{12}(Q) = 0 \tag{21.57}$$

are fulfilled simultaneously.[4] In two dimensions, this is generally the case in a single point, which leads to a so-called conical intersection. This type of curve crossing is very important for ultrafast transitions. In more than two dimensions, crossings appear at higher-dimensional surfaces. The terminology of conical intersections is also used here (Fig. 21.4).

---

[3] This is similar to a vector field $\mathbf{v}$ in three dimensions which can be divided into a longitudinal or irrotational part with $\mathrm{rot}\,\mathbf{v} = 0$ and a rotational part with $\mathrm{div}\,\mathbf{v} = 0$.

[4] If the two states are of different symmetry, then $H_{12} = 0$ and a crossing is possible in one dimension.

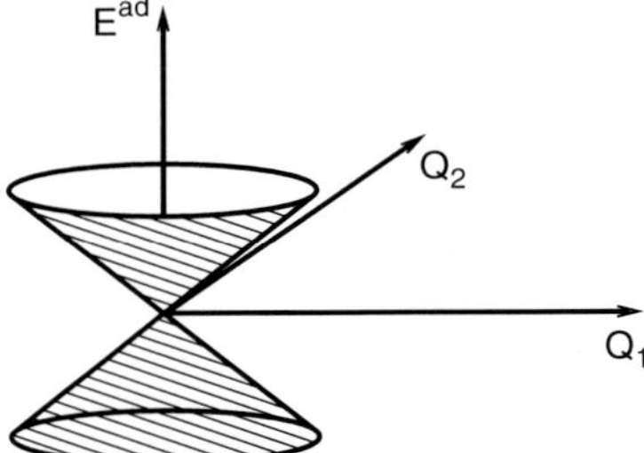

Fig. 21.4. Conical intersection

Problems

21.1. Crude Adiabatic Model

Consider the crossing of two electronic states along a coordinate Q. As basis functions, we use two coordinate-independent electronic wave functions which diagonalize the Born–Oppenheimer Hamiltonian at the crossing point Q_0:

$$(T_{el} + V(Q_0))\varphi^{1,2} = e^{1,2}\varphi^{1,2}.$$

Use the following ansatz functions

$$\Psi_1(r, Q) = (\cos\zeta(Q)\varphi^1(r) - \sin\zeta(Q)\varphi^2(r))\chi^1(Q),$$

$$\Psi_2(r, Q) = (\sin\zeta(Q)\varphi^1(r) + \cos\zeta(Q)\varphi^2(r))\chi^2(Q),$$

which can be written in more compact form

$$(\Psi_1, \Psi_2) = (\varphi^1, \varphi^2)\begin{pmatrix} c & s \\ -s & c \end{pmatrix}\begin{pmatrix} \chi_1 \\ \chi_2 \end{pmatrix}.$$

The Hamiltonian is partitioned as

$$H = T_N + T_{el} + V(r, Q_0) + \Delta V(r, Q).$$

Calculate the matrix elements of the Hamiltonian

$$\begin{pmatrix} H_{11} & H_{12} \\ H_{21} & H_{22} \end{pmatrix}$$
$$= \begin{pmatrix} c & -s \\ s & c \end{pmatrix}\begin{pmatrix} \varphi_1^\dagger \\ \varphi_2^\dagger \end{pmatrix}\left(-\frac{\hbar^2}{2m}\frac{\partial^2}{\partial Q^2} + T_{el} + V(Q_0) + \Delta V\right)(\varphi^1, \varphi^2)\begin{pmatrix} c & s \\ -s & c \end{pmatrix},$$

$$(21.58)$$

where $\chi(Q)$ and $\zeta(Q)$ depend on the coordinate Q whereas the basis functions $\varphi^{1,2}$ do not. Chose $\zeta(Q)$ such that $T_{el} + V$ is diagonalized. Evaluate the nonadiabatic interaction terms at the crossing point Q_0.

Dynamics of an Excited State

In this chapter, we would like to describe the dynamics of an excited state $|s\rangle$ which is prepared, for example, by electronic excitation, due to absorption of radiation. This state is an eigenstate of the diabatic Hamiltonian with energy E_s^0. Close in energy to $|s\rangle$ is a manifold of other states $\{|l\rangle\}$, which are not populated during the short-time excitation, since from the ground state only the transition $|0\rangle \to |s\rangle$ is optically allowed. The l-states are weakly coupled to a continuum of bath states[1] and therefore have a finite lifetime (Fig. 22.1).

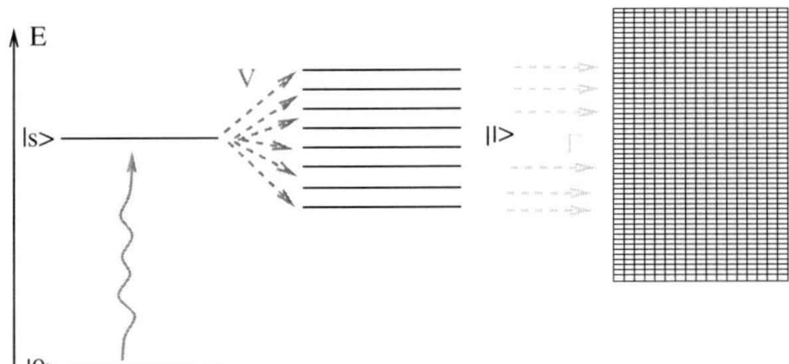

Fig. 22.1. Dynamics of an excited state. The excited state $|s\rangle$ decays into a manifold of states $|l\rangle$ which are weakly coupled to a continuum

The bath states will not be considered explicitly. Instead, we use a non-Hermitian Hamiltonian for the subspace spanned by $|s\rangle$ and $\{|l\rangle\}$. We assume that the Hamiltonian is already diagonal[2] with respect to the manifold $\{|l\rangle\}$, which has complex energies[3]

---

[1] For instance, the field of electromagnetic radiation.

[2] For non-Hermitian operators, we have to distinguish left eigenvectors and right eigenvectors.

[3] This is also known as the damping approximation.

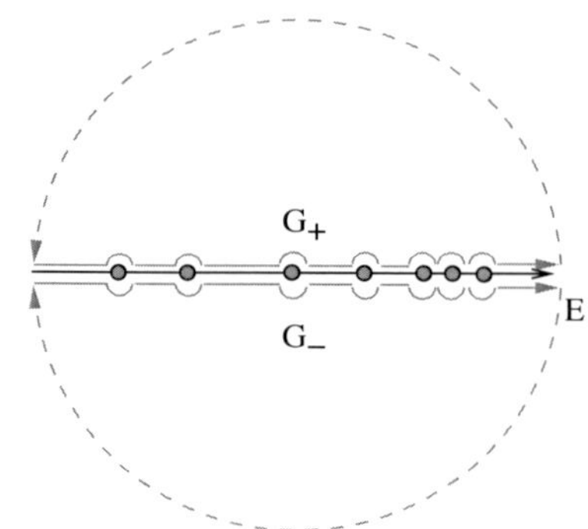

Fig. 22.2. Integration contour for $G_\pm(t)$

$$E_l^0 = \epsilon_l - \mathrm{i}\frac{\Gamma_l}{2}. \tag{22.1}$$

This describes the exponential decay $\sim \mathrm{e}^{-\Gamma_l t}$ of the l-states into the continuum states.

Thus, the model Hamiltonian takes the form

$$H = H^0 + V = \begin{pmatrix} E_s^0 & V_{s1} & \cdots & V_{sL} \\ V_{1s} & E_1^0 & & \\ \vdots & & \ddots & \\ V_{Ls} & & & E_L \end{pmatrix}. \tag{22.2}$$

22.1 Green's Formalism

The corresponding Green's operator or resolvent [61] is defined as

$$G(E) = (E - H)^{-1} = \sum_n |n\rangle \frac{1}{E - E_n} \langle n|. \tag{22.3}$$

For a Hermitian Hamiltonian, the poles of the Green's operator are on the real axis and the time evolution operator (the so-called propagator) is defined by (Fig. 22.2)

$$\widetilde{G}(t) = G_+(t) - G_-(t), \tag{22.4}$$

$$G_\pm(t) = \frac{1}{2\pi\mathrm{i}} \int_{-\infty\pm\mathrm{i}\epsilon}^{\infty\pm\mathrm{i}\epsilon} \mathrm{e}^{-(\mathrm{i}E/\hbar)t} G(E)\,\mathrm{d}E. \tag{22.5}$$

$\widetilde{G}(t)$ is given by an integral, which encloses all the poles E_n. Integration between two poles does not contribute, since the integration path can be taken to be on the real axis and the two contributions vanish. The integration

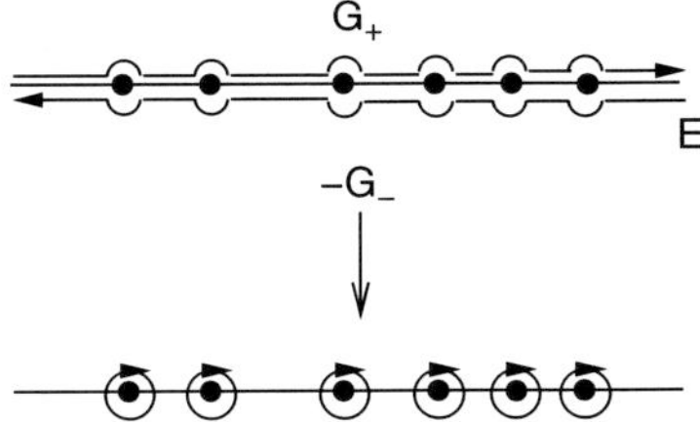

Fig. 22.3. Integration contour for $\widetilde{G}(t) = G_+(t) - G_-(t)$

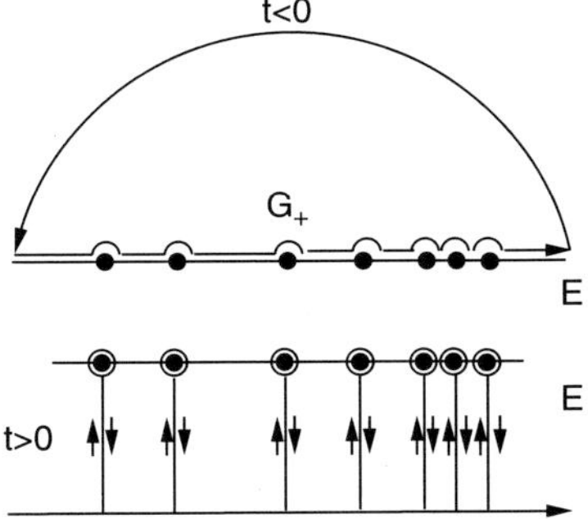

Fig. 22.4. Deformation of the integration contour for $G_+(t)$

along a small circle around a pole gives $2\pi i$ times the residual value which is given by $e^{-iE_n t/\hbar}$ (Fig. 22.3).

Hence, we find

$$\widetilde{G}(t) = \sum |n\rangle e^{-(iE/\hbar)t} \langle n| = \exp\left(\frac{t}{i\hbar}H\right). \tag{22.6}$$

For times $t < 0$, the integration path for G_+ can be closed in the upper half of the complex plane where the integrand $e^{-(iE/\hbar)t}G(E)$ vanishes exponentially for large $|E|$ (Fig. 22.4).

Hence

$$G_+(t) = 0 \quad \text{for} \quad t < 0. \tag{22.7}$$

We shift the integration path for times $t > 0$ into the lower half of the complex plane, where again the integrand vanishes exponentially and we have to sum over all residuals to find

$$G_+(t) = \widetilde{G}(t) \quad \text{for} \quad t > 0. \tag{22.8}$$

Hence, $G_+(t) = \Theta(t)\widetilde{G}(t)$ is the time evolution operator for $t > 0$.[4] There are additional interactions for times $t < 0$ which prepare the initial state

---

[4] Similarly, we find $G_-(t) = (\Theta(t-1))\widetilde{G}(t)$ is the time evolution operator for negative times.

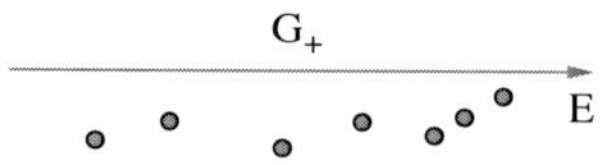

Fig. 22.5. Integration contour for a non-Hermitian Hamiltonian

$$\psi(t = 0) = |s\rangle. \tag{22.9}$$

The integration contour for a non-Hermitian Hamiltonian can be chosen as the real axis for $G_+(t)$, which now becomes the Fourier transform of the resolvent (Fig. 22.5).

Dividing the Hamiltonian in the diagonal part H^0 and the interaction V, we have

$$(G^0)^{-1} = E - H^0, \tag{22.10}$$

$$G^{-1} = E - H = (G^0)^{-1} - V. \tag{22.11}$$

Multiplication from the left with G^0 and from the right with G gives the Dyson equation:

$$G = G^0 + G^0 V G. \tag{22.12}$$

We iterate that once more

$$G = G^0 + G^0 V (G^0 + G^0 V G) = G^0 + G^0 V G^0 + G^0 V G^0 V G \tag{22.13}$$

and project on the state $|s\rangle$

$$\langle s|G|s\rangle = \langle s|G^0|s\rangle + \langle s|G^0 V G^0|s\rangle + \langle s|G^0 V G^0 V G|s\rangle. \tag{22.14}$$

Now, G^0 is diagonal and V is nondiagonal. Therefore

$$\langle s|G^0 V G^0|s\rangle = \langle s|G^0|s\rangle\langle s|V|s\rangle\langle s|G^0|s\rangle = 0 \tag{22.15}$$

and

$$\langle s|G^0 V G^0 V G|s\rangle = \langle s|G^0|s\rangle\langle s|V|l\rangle\langle l|G^0|l\rangle\langle l|V|s\rangle\langle s|G|s\rangle, \tag{22.16}$$

$$G_{ss} = G^0_{ss} + \sum_l G^0_{ss} V_{sl} G^0_{ll} V_{ls} G_{ss}, \tag{22.17}$$

$$G_{ss} = \frac{G^0_{ss}}{1 - G^0_{ss} \sum_l V_{sl} G^0_{ll} V_{ls}} = \frac{1}{E - E^0_s - R_s}, \tag{22.18}$$

with the level shift operator

$$R_s = \sum_l \frac{|V_{sl}|^2}{E - \epsilon_l + i(\Gamma_l/2)}. \tag{22.19}$$

The poles of the Green's function G_{ss} are given by the implicit equation

$$E_p = E_s^0 + \sum_l \frac{|V_{sl}|^2}{E_p - E_l^0}.$$ (22.20)

Generally, the Green's function is meromorphic and can be represented as

$$G_{ss}(E) = \sum_p \frac{A_p}{E - E_p},$$ (22.21)

where the residuals are defined by

$$A_p = \lim_{E \to E_p} G_{ss}(E)(E - E_p).$$ (22.22)

The probability of finding the system still in the state $|s\rangle$ at time $t > 0$ is

$$P_s(t) = |\langle s|\widetilde{G}(t)|s\rangle|^2 = |\widetilde{G}_{ss}(t)|^2,$$ (22.23)

where the propagator is the Fourier transform of the Green's function:

$$\widetilde{G}_{+ss}(t) = \frac{1}{2\pi i} \int_{-\infty}^{\infty} e^{(E/i\hbar)t} G_{ss}(E) \mathrm{d}E = \theta(t) \sum_p A_p e^{(E_p/i\hbar)t}.$$ (22.24)

22.2 Ladder Model

We now want to study a simplified model which can be solved analytically. The energy of $|s\rangle$ is set to zero. The manifold $\{|l\rangle\}$ consists of infinitely equally spaced states with equal width

$$E_l^0 = \alpha + l\Delta\epsilon - i\frac{\Gamma}{2}$$ (22.25)

and the interaction $V_{sl} = V$ is taken independent of l. With this simplification, the poles are solutions of

$$E_p = \sum_{l=-\infty}^{l=\infty} \frac{V^2}{E_p - \alpha - l\Delta\epsilon + i(\Gamma/2)},$$ (22.26)

which can be written using the identity[5]

$$\cot(z) = \sum_{l=-\infty}^{\infty} \frac{1}{z - l\pi}$$ (22.27)

---

[5] $\cot(z)$ has single poles at $z = l\pi$ with residues $\lim_{z \to l\pi} \cot(z)(z - l\pi) = 1$.

as

$$E_p = \frac{V^2\pi}{\Delta\epsilon} \cot\left(\frac{\pi}{\Delta\epsilon}\left(E_p - \alpha + \mathrm{i}\frac{\Gamma}{2}\right)\right). \tag{22.28}$$

For the following discussion, it is convenient to measure all energy quantities in units of $\pi/\Delta\epsilon$ and define

$$\widetilde{\alpha} = \alpha\pi/\Delta\epsilon, \quad \widetilde{\Gamma} = \Gamma\pi/\Delta\epsilon, \tag{22.29}$$

$$\widetilde{E}_p = E_p\pi/\Delta\epsilon, \quad \widetilde{V} = V\pi/\Delta\epsilon, \tag{22.30}$$

to have

$$\widetilde{E}_p = \widetilde{E}_p^r - \frac{\mathrm{i}\widetilde{\Gamma}_p}{2} = \widetilde{V}^2 \cot\left\{\widetilde{E}_p - \widetilde{\alpha} + \mathrm{i}\frac{\widetilde{\Gamma}}{2}\right\}, \tag{22.31}$$

which can be split into real and imaginary part:

$$\frac{\widetilde{E}_p^r}{\widetilde{V}^2} = \frac{\sin(2(\widetilde{E}_p^r - \widetilde{\alpha}))}{\cosh(\widetilde{\Gamma}_p - \widetilde{\Gamma}) - \cos(2(\widetilde{E}_p^r - \widetilde{\alpha}))}, \tag{22.32}$$

$$-\frac{\widetilde{\Gamma}_p}{2\widetilde{V}^2} = \frac{\sinh(\widetilde{\Gamma}_p - \widetilde{\Gamma})}{\cosh(\widetilde{\Gamma}_p - \widetilde{\Gamma}) - \cos(2(\widetilde{E}_p^r - \widetilde{\alpha}))}. \tag{22.33}$$

We now have to distinguish two limiting cases.

The Small Molecule Limit

For small molecules, the density of state $1/\Delta\epsilon$ is small. If we keep the lifetime of the l-states fixed, then $\widetilde{\Gamma}$ is small in this limit. The poles are close to the real axis.[6] We approximate the real part by solving the real equation (Fig. 22.6)

$$\widetilde{E}_p^{(0)} = \widetilde{V}^2 \cot\left(\widetilde{E}_p^{(0)} - \widetilde{\alpha}\right) \tag{22.34}$$

and consider first the solution which is closest to zero. Expanding around $\widetilde{E}_p^{(0)}$, we find

$$\widetilde{E}_p^{(0)} + \xi = \widetilde{V}^2 \cot\left(\widetilde{E}_p^{(0)} + \xi - \widetilde{\alpha} + \frac{\mathrm{i}\widetilde{\Gamma}}{2}\right)$$

$$= \widetilde{V}^2 \cot\left(\widetilde{E}_p^{(0)} - \widetilde{\alpha}\right) - \widetilde{V}^2\left(1 + \cot\left(\widetilde{E}_p^{(0)} - \widetilde{\alpha}\right)^2\right)\left(\xi + \frac{\mathrm{i}\widetilde{\Gamma}}{2}\right), \tag{22.35}$$

---

[6] We assume that $\alpha \neq 0$.

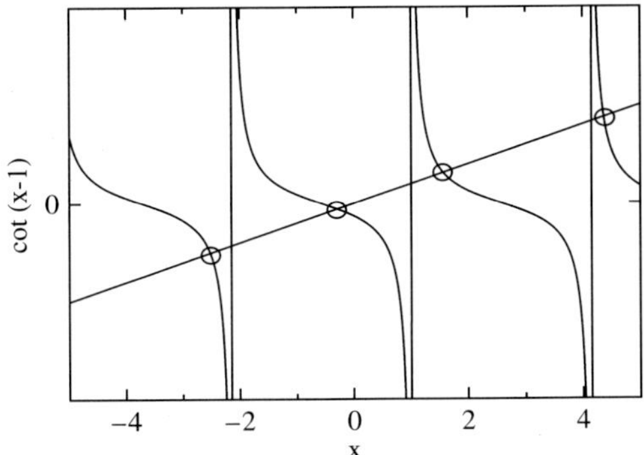

Fig. 22.6. Small molecule limit. The figure shows the graphical solution of $x = \cot(x - 1)$

which simplifies to

$$\xi = -\widetilde{V}^2 \left(1 + \cot\left(\widetilde{E}_p^{(0)} - \widetilde{\alpha}\right)^2\right)\left(\xi + \frac{\mathrm{i}\widetilde{\Gamma}}{2}\right) \tag{22.36}$$

and has the solution

$$\xi = -\frac{\mathrm{i}}{2}\widetilde{\Gamma}_p = -\frac{\mathrm{i}}{2}\widetilde{\Gamma}\frac{\widetilde{V}^2\left(1 + \frac{\widetilde{E}_p^{(0)2}}{\widetilde{V}^4}\right)}{1 + \widetilde{V}^2\left(1 + \frac{\widetilde{E}_p^{(0)2}}{\widetilde{V}^4}\right)}, \tag{22.37}$$

which in the case $\widetilde{V}$ small but $\widetilde{V} > \widetilde{E}_p^{(0)}$ gives

$$\widetilde{\Gamma}_p \approx \widetilde{V}^2\widetilde{\Gamma}, \tag{22.38}$$

$$\Gamma_p \approx \pi^2 \frac{V^2}{\Delta\epsilon^2}\Gamma. \tag{22.39}$$

The other poles are approximately undisturbed

$$\widetilde{E}_p^r = \widetilde{\alpha} + l\pi, \tag{22.40}$$

$$\widetilde{\Gamma}_p = \widetilde{\Gamma}\frac{\widetilde{V}^2 + \frac{\widetilde{E}_p^{r2}}{\widetilde{V}^2}}{1 + \widetilde{V}^2 + \frac{\widetilde{E}_p^{r2}}{\widetilde{V}^2}} \rightarrow \widetilde{\Gamma} \tag{22.41}$$

and the residuum of the Green's function follows from

$$\frac{1}{(\widetilde{E}_p + \mathrm{d}E) - \widetilde{V}^2 \cot\left(\widetilde{E}_p + \mathrm{d}E - \widetilde{\alpha} + \frac{\mathrm{i}\widetilde{\Gamma}}{2}\right)}$$

$$= \frac{1}{(\widetilde{E}_p + \mathrm{d}\widetilde{E}) - \widetilde{V}^2 \cot\left(\widetilde{E}_p - \widetilde{\alpha} + \frac{\mathrm{i}\widetilde{\Gamma}}{2}\right) + \widetilde{V}^2 \mathrm{d}\widetilde{E}(1 + \cot(\widetilde{E}_p - \widetilde{\alpha} + \frac{\mathrm{i}\widetilde{\Gamma}}{2})^2) + \cdots},$$

$$\frac{1}{\mathrm{d}\widetilde{E}\left(1 + \widetilde{V}^2\left(1 + \cot\left(\widetilde{E}_p - \widetilde{\alpha} + \frac{\mathrm{i}\widetilde{\Gamma}}{2}\right)^2\right)\right)} = \frac{1}{\mathrm{d}\widetilde{E}}\frac{1}{\widetilde{V}^2 + \widetilde{E}_p}, \qquad (22.42)$$

which shows that only contributions from poles close to zero are important.

The Statistical Limit

Consider now the limit $\widetilde{\Gamma}_p > 1$ and $\widetilde{V} > 1$. This condition is usually fulfilled in biomolecules (Fig. 22.7).

If the poles are far away enough from the real axis, then on the real axis the cotangent is very close to $-\mathrm{i}$ and

$$R_s = \frac{V^2 \pi}{\Delta\epsilon} \cot\left(\frac{\pi}{\Delta\epsilon}\left(E_p - \alpha + \mathrm{i}\frac{\Gamma}{2}\right)\right) \approx -\mathrm{i}\frac{V^2 \pi}{\Delta\epsilon}. \qquad (22.43)$$

Thus

$$G_{ss}(E) = \frac{1}{E + \mathrm{i}(V^2\pi/\Delta\epsilon)} \qquad (22.44)$$

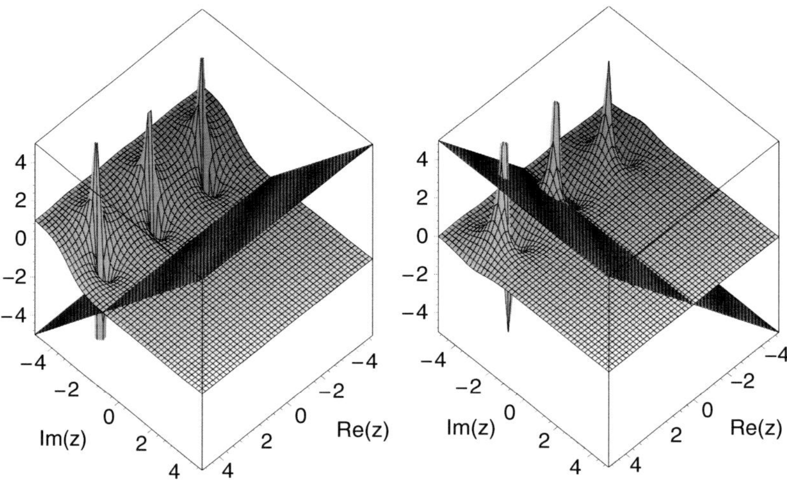

Fig. 22.7. Complex cotangent function. *Left*: imaginary part and *right*: real part of $\cot(z + 3\mathrm{i})$ are shown together with $\Im(z)$ and $\Re(z)$, respectively

and the initial state decays as

$$P(t) = |e^{(-iV^2\pi/\hbar\Delta\epsilon)t}|^2 = e^{-kt} \tag{22.45}$$

with the rate

$$k = \frac{2\pi V^2}{\hbar\Delta\epsilon} = \frac{2\pi V^2}{\hbar}\rho, \tag{22.46}$$

where the density of states is

$$\rho = \frac{1}{\Delta\epsilon}. \tag{22.47}$$

22.3 A More General Ladder Model

We now consider a more general model where the spacing of the states is not constant and the matrix elements are different. The Hamiltonian has the form

$$H = \begin{pmatrix} E_s & V_{s1} & \cdots & V_{sL} \\ V_{1s} & E_1 & & \\ \vdots & & \ddots & \\ V_{Ls} & & & E_L \end{pmatrix}. \tag{22.48}$$

We take the energies relative to the lowest l-state (Fig. 22.8)

$$E_s = E_0 + \Delta E, \quad E_l = E_0 + \epsilon_l. \tag{22.49}$$

We start from the golden rule expression

$$k = \sum_l \frac{2\pi V_{sl}^2}{\hbar}\delta(E_s - E_l) = \sum_l \frac{2\pi V_{sl}^2}{\hbar}\delta(\Delta E - \epsilon_l), \tag{22.50}$$

where the density of final states is given by

$$\rho(E_s) = \sum_l \delta(E_s - E_l) = \sum_l \delta(\Delta E - \epsilon_l). \tag{22.51}$$

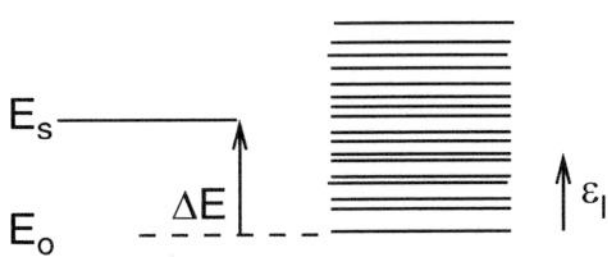

Fig. 22.8. Generalized ladder model

We represent the δ-function by a Fourier integral:

$$\delta(\omega) = \frac{1}{2\pi} \int_{-\infty}^{\infty} e^{-i\omega t} dt, \qquad (22.52)$$

$$\omega = \frac{E}{\hbar} \rightarrow \delta(E) = \frac{1}{2\pi\hbar} \int_{-\infty}^{\infty} e^{(E/i\hbar)t} dt, \qquad (22.53)$$

$$k = \sum_l \frac{V_{sl}^2}{\hbar^2} \int_{-\infty}^{\infty} e^{(\Delta E - \epsilon_l/i\hbar)t} dt. \qquad (22.54)$$

Interchanging integration and summation, we find

$$k = \int_{-\infty}^{\infty} dt \sum_l \frac{V_{sl}^2}{\hbar^2} \exp\left\{ \frac{\Delta E - \epsilon_l}{i\hbar} t \right\}. \qquad (22.55)$$

We introduce

$$z = \sum_l \frac{V_{sl}^2}{\hbar^2} e^{(i\epsilon_l/\hbar)t} \qquad (22.56)$$

and write the rate as

$$k = \int_{\infty}^{\infty} dt\, e^{-(i/\hbar)\Delta E\, t + \ln(z)}. \qquad (22.57)$$

This integral will now be evaluated approximately using the saddle point (p. 339) method.[7] The saddle point equation for the ladder model becomes

$$-\frac{i}{\hbar}\Delta E + \frac{d\ln z(t_s)}{dt} = 0. \qquad (22.58)$$

From the derivative

$$\frac{d\ln z}{dt} = \frac{1}{z} \sum_l \frac{V_{sl}^2}{\hbar^2} \frac{i\epsilon_l}{\hbar} e^{it\epsilon_l/\hbar}, \qquad (22.59)$$

we find

$$\Delta E = \frac{1}{z} \sum_l \frac{V_{sl}^2}{\hbar^2} \epsilon_l e^{it\epsilon_l/\hbar}. \qquad (22.60)$$

Since ΔE is real, we look for a saddle point on the imaginary axis and write

$$\frac{it_s}{\hbar} = -\beta, \qquad (22.61)$$

where the new variable β is real. The saddle point equation now has a quasithermodynamic meaning:

$$\Delta E = \frac{1}{z} \sum_l \frac{V_{sl}^2}{\hbar^2} \epsilon_l e^{-\beta\epsilon_l} = \langle \epsilon_l \rangle, \qquad (22.62)$$

---

[7] Also known as method of steepest descent or Laplace method.

where β plays the role of $1/k_{\mathrm{B}}T$ and $g_l = V_{sl}^2/\hbar^2$ that of a degeneracy factor. The second derivative relates to the width of the energy distribution:

$$\frac{\mathrm{d}^2 \ln z}{\mathrm{d}t^2} = \frac{\mathrm{d}}{\mathrm{d}t}\left(\frac{\frac{\mathrm{d}z}{\mathrm{d}t}}{z}\right) = \frac{\frac{\mathrm{d}^2 z}{\mathrm{d}t^2}}{z} - \left(\frac{\frac{\mathrm{d}z}{\mathrm{d}t}}{z}\right)^2$$

$$= \frac{1}{z}\sum_l g_l \frac{-\epsilon_l^2}{\hbar^2}\mathrm{e}^{-\beta\epsilon_l} - \frac{1}{z^2}\left(\sum_l g_l \frac{\mathrm{i}\epsilon_l}{\hbar}\mathrm{e}^{-\beta\epsilon}\right)^2$$

$$= -\left\langle\frac{\epsilon^2}{\hbar^2}\right\rangle + \left\langle\frac{\epsilon}{\hbar}\right\rangle^2 . \tag{22.63}$$

We approximate now the integral by the contribution around the saddle point:

$$k = \int_\infty^\infty \mathrm{d}t\, \mathrm{e}^{-(\mathrm{i}/\hbar)\Delta E\, t + \ln(z)} = z(t_s)\mathrm{e}^{\beta\Delta E}\sqrt{\frac{2\pi\hbar^2}{\langle\epsilon^2\rangle - \langle\epsilon\rangle^2}}$$

$$= \sum_l \frac{V_{sl}^2}{\hbar}\mathrm{e}^{-\beta(\epsilon_l - \Delta E)}\sqrt{\frac{2\pi}{\langle\epsilon^2\rangle - \langle\epsilon\rangle^2}}. \tag{22.64}$$

For comparison, let us consider the simplified model with $V_{sl} = V$ and $\epsilon_l = l\Delta\epsilon$. Then

$$z = \frac{V^2}{\hbar^2}\sum_{l=0}^{\infty}\mathrm{e}^{-\beta l\Delta\epsilon} = \frac{V^2}{\hbar^2}\frac{1}{1 - \mathrm{e}^{-\beta\Delta\epsilon}}. \tag{22.65}$$

The saddle point equation is

$$\Delta E = -\frac{\partial}{\partial\beta}\ln z = \frac{\Delta\epsilon}{\mathrm{e}^{\beta\Delta\epsilon} - 1} \tag{22.66}$$

which determines the "temperature"

$$\beta = \frac{1}{\Delta\epsilon}\ln\left(1 + \frac{\Delta\epsilon}{\Delta E}\right) \tag{22.67}$$

$$\approx \frac{1}{\Delta E} \quad \text{if} \quad \Delta\epsilon \ll \Delta E. \tag{22.68}$$

Then

$$\langle\epsilon_l\rangle = \frac{\Delta\epsilon}{\mathrm{e}^{\Delta\epsilon/\Delta E} - 1} \approx \Delta E, \tag{22.69}$$

$$\langle\epsilon_l^2\rangle - \langle\epsilon\rangle^2 = \frac{\partial^2}{\partial\beta^2}\ln z = \frac{\Delta\epsilon^2 \mathrm{e}^{\Delta\epsilon/\Delta E}}{\left(\mathrm{e}^{\Delta\epsilon/\Delta E} - 1\right)^2} \approx \Delta e^2, \tag{22.70}$$

$$z(t_s) = \frac{V^2}{\hbar^2}\frac{1}{1 - \mathrm{e}^{-\Delta\epsilon/\Delta E}} \approx \frac{V^2}{\hbar^2}\frac{\Delta E}{\Delta\epsilon}, \tag{22.71}$$

and the rate becomes

$$k = \frac{V^2}{\hbar^2}\frac{\Delta E}{\Delta\epsilon}\mathrm{e}^1\sqrt{\frac{2\pi\hbar^2}{\Delta e^2}} = \frac{2\pi V^2}{\hbar\Delta\epsilon}\mathrm{e}^1\sqrt{\frac{1}{2\pi}}, \tag{22.72}$$

where the numerical value of $e/\sqrt{2\pi} = 1.08$

22.4 Application to the Displaced Oscillator Model

We now want to discuss a displaced oscillator model (p. 205) for the transition between the vibrational manifolds of two electronic states.

$$k = \frac{2\pi}{\hbar} V^2 FCD(\Delta E)$$

$$= \frac{V^2}{\hbar^2} \int_{-\infty}^{\infty} dt\, e^{(-it/\hbar)\Delta E} \langle e^{(it/\hbar)H_{\mathrm{f}}} e^{-(it/\hbar)H_{\mathrm{i}}} \rangle = \frac{V^2}{\hbar^2} \int_{-\infty}^{\infty} dt\, e^{(-it/\hbar)\Delta E} F(t),$$

$$(22.73)$$

where H_{f} is the Hamiltonian of the nuclear vibrations in the final state and the correlation function $F(t) = \exp(g(t))$ for independent displaced oscillators is taken from (31.12). The rate becomes

$$k = \frac{V^2}{\hbar^2} \int_{-\infty}^{\infty} dt\, e^{(-it/\hbar)\Delta E}$$

$$\times \exp\left\{ \sum_r g_r^2 \left[(\overline{n}_r + 1)(e^{i\omega_r t} - 1) + \overline{n}_r (e^{-i\omega_r t} - 1) \right] \right\}.$$

$$(22.74)$$

The short-time expansion gives

$$k = \frac{V^2}{\hbar^2} \sqrt{\frac{2\pi\hbar^2}{\Delta^2}} \exp\left\{ -\frac{(\Delta E - E_r)^2}{2\Delta^2} \right\}. \qquad (22.75)$$

If all modes can be treated classically $\hbar\omega_r \ll k_{\mathrm{B}}T$, the phonon number is $\overline{n}_r = k_{\mathrm{B}}T/\hbar\omega_r$ and

$$\Delta^2 \approx 2k_{\mathrm{B}}T \sum_r g_r^2 \hbar\omega_r = 2k_{\mathrm{B}}T E_r \qquad (22.76)$$

which gives for the rate in the classical limit the Marcus expression

$$k = \frac{2\pi V^2}{\hbar} \sqrt{\frac{1}{4\pi k_{\mathrm{B}}T E_r}} \exp\left\{ -\frac{(\Delta E - E_r)^2}{4 E_r k_{\mathrm{B}}T} \right\}. \qquad (22.77)$$

The saddle point equation is

$$\frac{i}{\hbar}\Delta E = \sum_r g_r^2 \left[i\omega_r (\overline{n}_r + 1)e^{i\omega_r t} - i\omega_r \overline{n}_r e^{-i\omega_r t} \right]$$

$$= i \sum_r g_r^2 \omega_r \left[(\overline{n}_r + 1)e^{-i\omega_r t} - \overline{n}_r e^{i\omega_r t} \right], \qquad (22.78)$$

which we solve approximately by linearization

$$i\Delta E = i\hbar \sum_r g_r^2 \omega_r \left[1 - (2\bar{n}_r + 1)i\omega_r t + \cdots\right]$$

$$= iE_r - \Delta^2 \frac{t}{\hbar} + \cdots , \tag{22.79}$$

$$t_s = -i\hbar \frac{\Delta E - E_r}{\Delta^2}. \tag{22.80}$$

At the saddle point, we have

$$-\frac{it_s}{\hbar}\Delta E + g(t_s) = -\frac{it_s}{\hbar}\Delta E + \sum_r g_r^2 \left[(\bar{n}_r + 1)(i\omega_r t_s) - \bar{n}_r i\omega_r t_s + \cdots\right]$$

$$= -\frac{it_s}{\hbar}\Delta E + i\sum_r g_r^2 \omega_r t_s - \frac{\Delta^2}{\hbar^2}\frac{t_s^2}{2}$$

$$= -\frac{it_s}{\hbar}(\Delta E - E_r) - \Delta^2 \frac{-(E_r - \Delta E)^2}{2\Delta^4}$$

$$= -\frac{(\Delta E - E_r)^2}{2\Delta^2}. \tag{22.81}$$

The second derivative is

$$-\sum_r g_r^2 \omega_r^2 \left[(\bar{n}_r + 1)e^{-i\omega_r t} + \bar{n}_r e^{i\omega_r t}\right] = -\frac{\Delta^2}{\hbar^2} + \cdots \tag{22.82}$$

and finally the rate again is

$$k = \frac{V^2}{\hbar^2}\sqrt{\frac{2\pi\hbar^2}{\Delta^2}}\exp\left\{-\frac{(\Delta E - E_r)^2}{2\Delta^2}\right\}. \tag{22.83}$$

At low temperatures $k_{\mathrm{B}}T \ll \hbar\omega$, the saddle point equation simplifies to

$$\Delta E = \sum_r g_r^2 \hbar\omega_r e^{i\omega_r t}. \tag{22.84}$$

To solve this equation, we introduce an average frequency (major accepting modes) $\widetilde{\omega}$ with

$$\sum_r g_r^2 \hbar\omega_r = \sum_r g_r^2 \hbar\widetilde{\omega} = S\hbar\widetilde{\omega}, \tag{22.85}$$

$$\widetilde{\omega} = \frac{\sum_r g_r^2 \omega_r}{S}, \quad S = \sum_r g_r^2, \tag{22.86}$$

which, after insertion into the saddle point equation, gives the approximation

$$\Delta E = S\hbar\widetilde{\omega}e^{it_s\widetilde{\omega}} \tag{22.87}$$

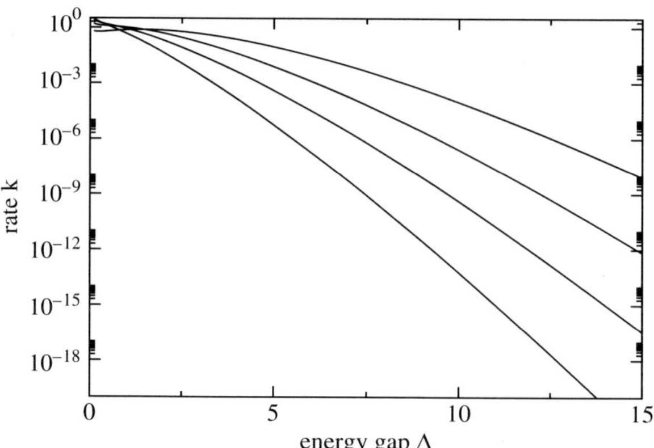

Fig. 22.9. Energy gap law. The rate (22.90) is shown as a function of the energy gap in units of the average frequency for $S = 0.2, 0.5, 1, 2$

with the solution

$$\mathrm{i}t_s = \frac{1}{\widetilde{\omega}} \ln \frac{\Delta E}{S\hbar\widetilde{\omega}}. \tag{22.88}$$

The second derivative is

$$-\sum_r g_r^2 \omega_r^2 \mathrm{e}^{\mathrm{i}\omega_r t_s} \approx -S\widetilde{\omega}^2 \mathrm{e}^{\ln(\Delta E/S\hbar\widetilde{\omega})} = -\frac{1}{\hbar}\Delta E\widetilde{\omega} \tag{22.89}$$

and the rate is

$$
\begin{aligned}
k &= \frac{V^2}{\hbar^2}\sqrt{\frac{2\pi\hbar}{\Delta E\widetilde{\omega}}} \exp\left\{-\frac{\Delta E}{\hbar}\frac{1}{\widetilde{\omega}}\ln\frac{\Delta E}{S\hbar\widetilde{\omega}} + S\frac{\Delta E}{S\hbar\widetilde{\omega}} - S\right\} \\
&= \frac{V^2}{\hbar}\sqrt{\frac{2\pi}{\Delta E\hbar\widetilde{\omega}}} \exp\left\{-S - \frac{\Delta E}{\hbar\widetilde{\omega}}\left[\ln\frac{\Delta E}{S\hbar\widetilde{\omega}} - 1\right]\right\}.
\end{aligned}
\tag{22.90}
$$

The dependence on ΔE is known as the energy gap law (Fig. 22.9)

Problems

22.1. Ladder Model

Solve the time evolution for the ladder model approximately

$$H = \begin{pmatrix} 0 & V & \cdots & V \\ V & E_1 & & \\ \vdots & & \ddots & \\ V & & & E_n \end{pmatrix}, \qquad E_j = \alpha + (j-1)\Delta\epsilon.$$

First derive an integral equation for $C_0(t)$ only by substitution. Then replace the sum by integration over $\omega = j(\Delta\epsilon/\hbar)$ and extend the integration over the whole real axis. Replace the integral by a δ-function and show that the initial state decays exponentially with a rate

$$k = \frac{2\pi V^2}{\hbar \Delta\epsilon}.$$

Part VII

Elementary Photoinduced Processes

Photophysics of Chlorophylls and Carotenoids

Chlorophylls and carotenoids are very important light receptors. Both classes of molecules have a large π-electron system which is delocalized over many conjugated bonds and is responsible for strong absorption bands in the visible region (Figs. 23.1 and 23.2).

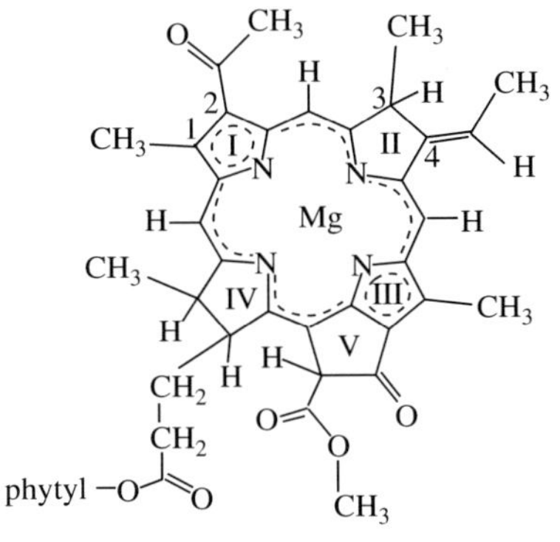

Fig. 23.1. Structure of bacteriochlorophyll-b [62]. This is the principal green pigment of the photosynthetic bacterium *Rhodopseudomonas viridis*. *Dashed curves* indicate the delocalized π-electron system. Chlorophylls have an additional double bond between positions 3 and 4. Variants have different side chains at positions 2 and 3

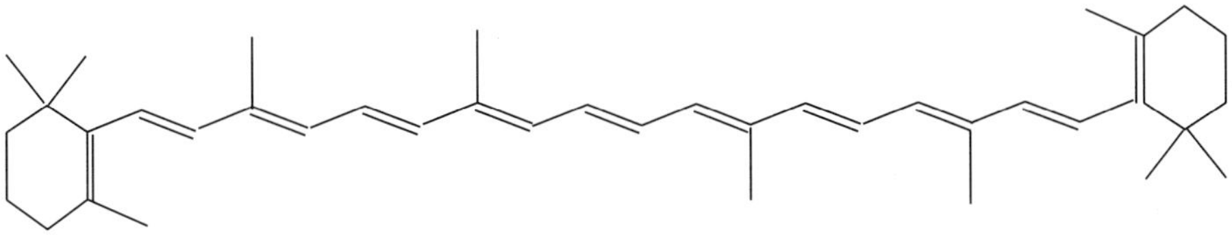

Fig. 23.2. Structure of β-carotene. The basic structure of the carotenoids is a conjugated chain made up of isoprene units. Variants have different end groups

23.1 MO Model for the Electronic States

The electronic wave functions of larger molecules are usually described by introducing one-electron wave functions $\phi(r)$ and expanding the wave function in terms of one or more Slater determinants. Singlet ground states can be in most cases described quite sufficiently by one determinant representing a set of doubly occupied orbitals:

$$|S_0\rangle = |\phi_{1\uparrow}\phi_{1,\downarrow}\cdots\phi_{\text{nocc},\uparrow}\phi_{\text{nocc},\downarrow}|$$

$$= \frac{1}{\sqrt{(2N_{\text{occ}})!}}
\begin{vmatrix}
\phi_{1\uparrow}(r_1) & \phi_{1\downarrow}(r_1) & \cdots & \phi_{\text{nocc},\downarrow}(r_1) \\
\phi_{1\uparrow}(r_2) & \phi_{1\downarrow}(r_2) & \cdots & \phi_{\text{nocc},\downarrow}(r_2) \\
\vdots & \vdots & & \vdots \\
\phi_{1\uparrow}(r_{2N_{\text{occ}}}) & \phi_{1\downarrow}(r_{2N_{\text{occ}}}) & & \phi_{\text{nocc},\downarrow}(r_{2N_{\text{occ}}})
\end{vmatrix}. \quad (23.1)$$

Excited states can be described as a linear combination of excited electronic configurations. The lowest excited state can often be reasonably approximated as the transition of one electron from the highest occupied orbital ϕ_{nocc} (HOMO) to the lowest unoccupied orbital $\phi_{\text{nocc}+1}$ (LUMO). The singlet excitation is given by

$$|S_1\rangle = \frac{1}{\sqrt{2}}(|\phi_{1\uparrow}\phi_{1,\downarrow}\cdots\phi_{\text{nocc},\uparrow}\phi_{\text{nocc}+1,\downarrow}| - |\phi_{1\uparrow}\phi_{1,\downarrow}\cdots\phi_{\text{nocc},\downarrow}\phi_{\text{nocc}+1,\uparrow}|).$$

$$(23.2)$$

The molecular orbitals can be determined with the help of more or less sophisticated methods (Fig. 23.3).

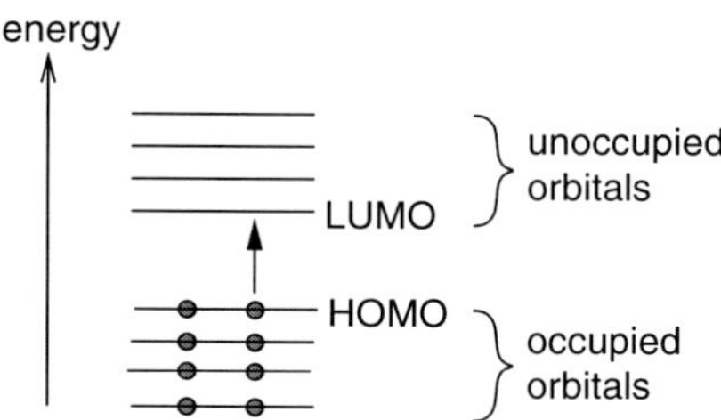

Fig. 23.3. HOMO–LUMO transition. The singlet ground state is approximated by a Slater determinant of doubly occupied molecular orbitals. The lowest singlet excitation can be described by promotion of an electron from the highest occupied to the lowest unoccupied orbital for most chromophores

23.2 The Free Electron Model for Polyenes

We approximate a polyene with N double bonds by a one-dimensional box of length $L = (2N + 1)d$.[1] The orbitals of the free electron model

$$H = -\frac{\hbar^2}{2m_e}\frac{\partial^2}{\partial x^2} + V(r) \tag{23.3}$$

have to fulfill the boundary condition $\phi(0) = \phi(L) = 0$. Therefore, they are given by

$$\phi_s(x) = \sqrt{\frac{2}{L}}\sin\frac{\pi s x}{L} \tag{23.4}$$

with energies (Fig. 23.4)

$$E_s = \frac{2}{L}\int_0^L \frac{\hbar^2}{2m_e}\left(\frac{\pi s}{L}\right)^2\left(\sin\frac{\pi s x}{L}\right)^2 \mathrm{d}x = \frac{1}{2m_e}\left(\frac{\pi\hbar s}{L}\right)^2. \tag{23.5}$$

Since there are $2N$ π-electrons, the energy of the lowest excitation is estimated as

$$\begin{aligned}
\Delta E &= E_{n+1} - E_n \\
&= \frac{\pi^2\hbar^2}{2m_e d^2(2N+1)^2}((N+1)^2 - N^2) = \frac{\pi^2\hbar^2}{2m_e d^2(2N+1)}.
\end{aligned} \tag{23.6}$$

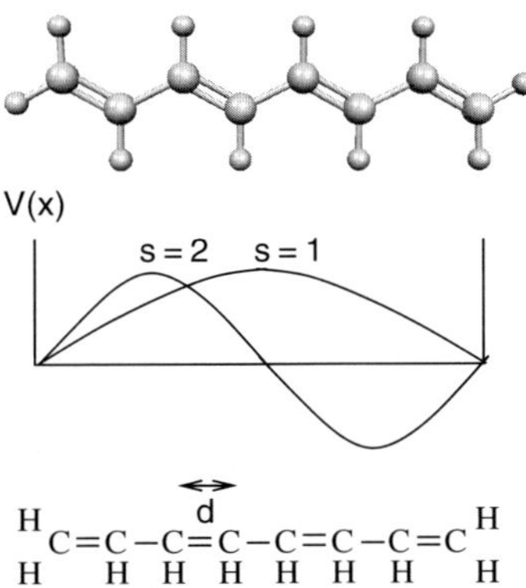

Fig. 23.4. Octatetraene. *Top*: geometry-optimized structure. *Middle*: potential energy and lowest two eigenfunctions of the free electron model. *Bottom*: linear model with equal bond lengths d

---

[1] The end of the box is one bond length behind the last C-atom.

The transition dipole matrix element for the singlet–singlet transition is

$$\boldsymbol{\mu} = \left\langle |\phi_{\mathrm{nocc}\uparrow}\phi_{\mathrm{nocc}\downarrow}| \sum (-e\mathbf{r})| \frac{1}{\sqrt{2}} (|\phi_{\mathrm{nocc}\uparrow}\phi_{\mathrm{nocc}+1\downarrow}| - |\phi_{\mathrm{nocc}\downarrow}\phi_{\mathrm{nocc}+1\uparrow}|) \right\rangle$$

$$= -\sqrt{2}e \int dr\, \phi_{\mathrm{nocc}}(r)\mathbf{r}\phi_{\mathrm{nocc}+1}(r)$$

$$= -\sqrt{2}e \frac{2}{L} \int_0^L \sin\frac{(N+1)\pi x}{L}\, x\, \sin\frac{N\pi x}{L} dx$$

$$= \frac{2\sqrt{2}e}{L} \frac{4L^2 N(N+1)}{\pi^2(2N+1)^2} = \frac{8\sqrt{2}ed}{\pi^2} \frac{n(N+1)}{2N+1}, \tag{23.7}$$

which grows with increasing length of the polyene. The absorption coefficient is proportional to

$$\alpha \sim \mu^2 \Delta E \sim \frac{N^2(N+1)^2}{(2N+1)^3}, \tag{23.8}$$

which depends close to linear on the number of double bonds N.

23.3 The LCAO Approximation

The molecular orbitals are usually expanded in a basis of atomic orbitals

$$\phi(r) = \sum_s C_s \varphi_s(r), \tag{23.9}$$

where the atomic orbitals are centered on the nuclei and the coefficients are determined from diagonalization of a certain one-electron Hamiltonian (for instance, Hartree–Fock, Kohn–Sham, semiempirical approximations such as AM1)

$$H\psi = E\psi. \tag{23.10}$$

Inserting the LCAO wave function gives

$$\sum_s C_s H\varphi_s(r) = E\sum_s C_s \varphi_s(r) \tag{23.11}$$

and projection on one of the atomic orbitals $\varphi_{s'}$ gives a generalized eigenvalue problem:

$$0 = \int d^3 r\, \varphi_{s'}(r) \sum_s C_s(H - E)\varphi_s(r)$$

$$= \sum_s C_s(H_{s's} - ES_{s's}). \tag{23.12}$$

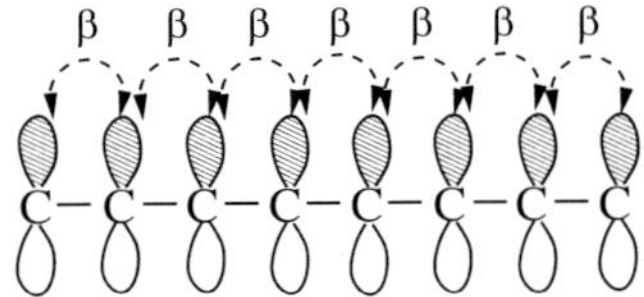

Fig. 23.5. Hückel model for polyenes

23.4 Hückel Approximation

The Hückel approximation [63] is a very simple LCAO model for the π-electrons. It makes the following approximations (Fig. 23.5):

- The diagonal matrix elements $H_{ss} = \alpha$ have the same value (Coulomb integral) for all carbon atoms.
- The overlap of different p_z-orbitals is neglected $S_{ss'} = \delta_{ss'}$.
- The interaction between bonded atoms is $H_{ss'} = \beta$ (resonance integral) and has the same value for all bonds.
- The interaction between nonbonded atoms is zero $H_{ss'} = 0$.

The Hückel matrix for a linear polyene has the form of a tridiagonal matrix

$$H = \begin{pmatrix} \alpha & \beta & & & \\ \beta & \alpha & \beta & & \\ & \ddots & \ddots & \ddots & \\ & & \beta & \alpha & \beta \\ & & & \beta & \alpha \end{pmatrix} \tag{23.13}$$

and can be easily diagonalized with the eigenvectors (Fig. 23.6)

$$\phi_r = \sqrt{\frac{2}{2N+1}} \sum_{s=1}^{2N} \sin \frac{rs\pi}{2N+1} \, \varphi_s, \quad r = 1, 2, \ldots, 2N. \tag{23.14}$$

The eigenvalues are

$$E_r = \alpha + 2\beta \cos \frac{r\pi}{2N+1}. \tag{23.15}$$

The symmetry of the molecule is C_{2h} (Fig. 23.7). All π-orbitals are antisymmetric with respect to the vertical reflection σ_h. With respect to the C_2 rotation, they have alternating a- or b-symmetry. Since $\sigma_h \times C_2 = i$, the orbitals are of alternating a_u- and b_g-symmetry. The lowest transition energy is in the Hückel model

$$\Delta E = E_{N+1} - E_N = 2\beta \left(\cos \frac{(N+1)\pi}{2N+1} - \cos \frac{N\pi}{2N+1} \right), \tag{23.16}$$

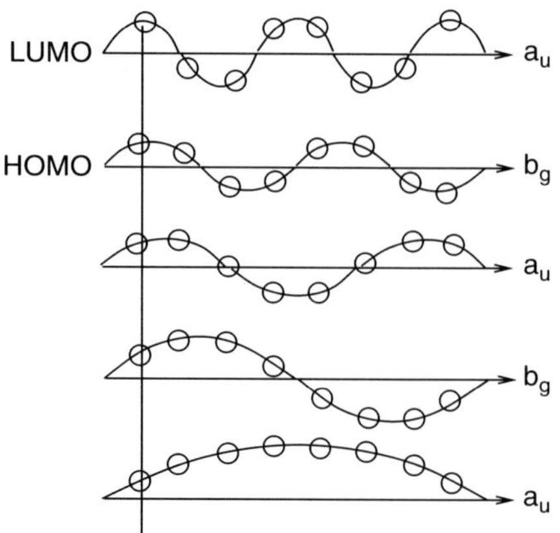

Fig. 23.6. Hückel orbitals for octatetraene

C_{2h}	E	C_2	σ	i
A_g	1	1	1	1
A_u	1	1	−1	−1
B_g	1	−1	−1	1
B_u	1	−1	1	−1

Fig. 23.7. Character table of the symmetry group C_{2h}. All π-orbitals are antisymmetric with respect to the reflection σ and are therefore of a_u- or b_g-symmetry

which can be simplified with the help of $(x = \pi/(2N+1))$

$$\cos(N+1)x - \cos Nx$$
$$= \frac{1}{2}\left(e^{i(N+1)x} + e^{-i(N+1)x} - e^{iNx} - e^{-iNx}\right)$$
$$= \frac{1}{2}e^{i(N+1/2)x}\left(e^{ix/2} - e^{-ix/2}\right) + \frac{1}{2}e^{-i(N+1/2)x}\left(e^{-ix/2} - e^{ix/2}\right)$$
$$= -2\sin\left(\left(N+\frac{1}{2}\right)x\right)\sin\frac{x}{2} \tag{23.17}$$

to give[2]

$$\Delta E = -4\beta\sin\frac{\pi}{4N+2} \sim \frac{1}{N} \quad \text{for large } n. \tag{23.18}$$

This approximation can be improved if the bond length alternation is taken into account, by using two alternating β-values [64]. The resulting energies are

$$E_k = \alpha \pm \sqrt{\beta^2 + \beta'^2 + 2\beta\beta'\cos k}, \tag{23.19}$$

---

[2] β is a negative quantity.

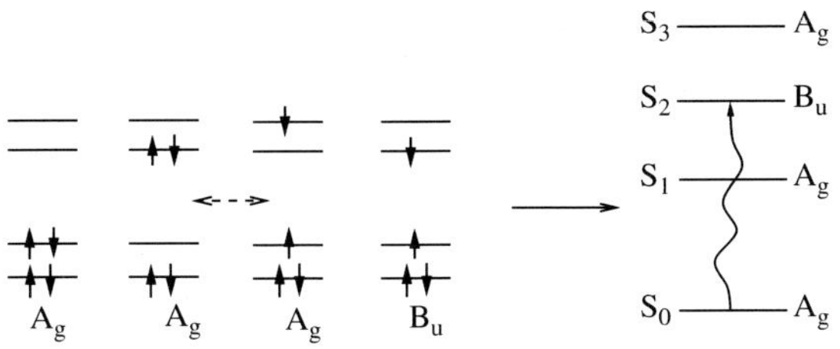

Fig. 23.8. Simplified CI model for linear polyenes

where the k-values are solutions of

$$\beta \sin(N+1)k + \beta' \sin Nk = 0. \tag{23.20}$$

In the Hückel model, the lowest excited state has the symmetry $a_u \times b_g = B_u$ and is strongly allowed. However, it has been well established that for longer polyenes, the lowest excited singlet is an optically forbidden totally symmetric A_g state [65,66]. Such a low-lying dark state has been observed experimentally also for carotenoids [67] and can be only understood if correlation effects are taken into account [68].

23.5 Simplified CI Model for Polyenes

In a very simple model, Kohler [69] treats only transitions from the b_g-HOMO into the a_u-LUMO and b_g-LUMO+1 orbitals. The HOMO–LUMO+1 transition as well as the double HOMO–LUMO transition are both of A_g-symmetry and can therefore interact. If the interaction is strong enough, then the lowest excited state will be of A_g-symmetry and therefore optically forbidden (Fig. 23.8).

23.6 Cyclic Polyene as a Model for Porphyrins

For a cyclic polyene with N carbon atoms the Hückel matrix has the form

$$H = \begin{pmatrix} \alpha & \beta & & & \beta \\ \beta & \alpha & \beta & & \\ & \ddots & \ddots & \ddots & \\ & & \beta & \alpha & \beta \\ \beta & & & \beta & \alpha \end{pmatrix} \tag{23.21}$$

with eigenvectors

$$\phi_k = \frac{1}{\sqrt{N}} \sum_{s=1}^{N} e^{iks} \varphi_s, \quad k = 0, \frac{2\pi}{N}, 2\frac{2\pi}{N}, \ldots, (N-1)\frac{2\pi}{N} \tag{23.22}$$

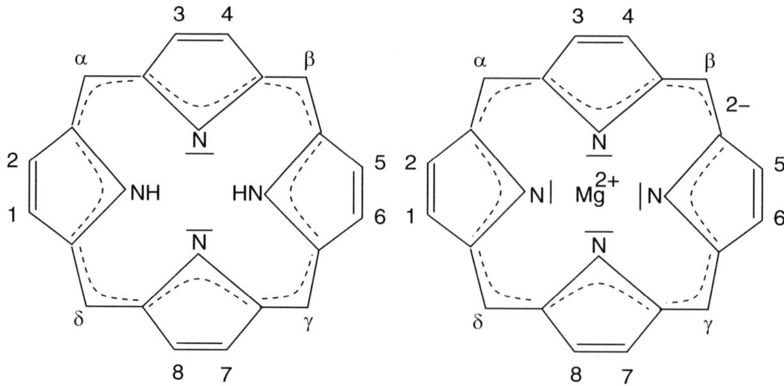

Fig. 23.9. Free base porphin and Mg-porphin. The bond character was assigned on the basis of the bond lengths from HF-optimized structures

and eigenvalues

$$E_k = \alpha + 2\beta \cos k. \tag{23.23}$$

This can be used as a model for the class of porphyrin molecules [70, 71] (Fig. 23.9).

For the metal–porphyrin, there are in principle two possibilities for the assignment of the essential π-system, an inner ring with 16 atoms or an outer ring with 20 atoms, both reflecting the D_{4h}-symmetry of the molecule. Since it is not possible to draw a unique chemical structure, we have to count the π-electrons. The free base porphin is composed of 14 H-atoms (of which the peripheral ones are not shown), 20 C-atoms and 4 N-atoms which provide a total of $14 + 4 \times 20 + 5 \times 4 = 114$ valence electrons. There are 42 σ-bonds (including those to the peripheral H-atoms) and two lonely electron pairs involving a total of 88 electrons. Therefore, the number of π-electrons is $114 - 88 = 26$. Usually, it is assumed that the outer double bonds are not so strongly coupled to the π-system and a model with 18 π-electrons distributed over the $N = 16$-ring is used. For the metal porphin, the number of H-atoms is only 12 but there are four lonely electron pairs instead of two. Therefore, the number of non-π valence electrons is the same as for the free base porphin. The total number of valence electrons is again 114 since the Mg atom formally donates two electrons.

23.7 The Four Orbital Model for Porphyrins

Gouterman [72–74] introduced the four orbital model which considers only the doubly degenerate HOMO ($k = \pm 4 \times 2\pi/N$) and LUMO ($k = \pm 5 \times 2\pi/N$) orbitals. There are four HOMO–LUMO transitions (Fig. 23.10). Their intensities are given by

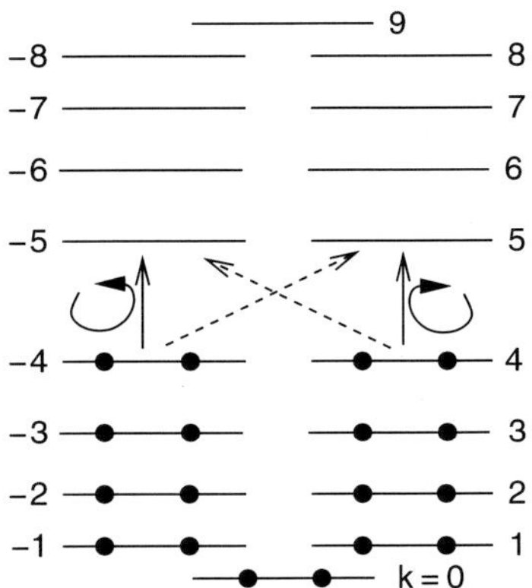

Fig. 23.10. Porphin orbitals

$$\boldsymbol{\mu} = \frac{1}{N}\sum_{ss'} e^{-\mathrm{i}k's'}\, e^{\mathrm{i}ks} \int \varphi_{s'}^{*}(r)(-e\mathbf{r})\varphi_s(r)\mathrm{d}^3 r \qquad (23.24)$$

which is approximately[3]

$$\boldsymbol{\mu} = \frac{1}{N}\sum_{ss'} e^{-\mathrm{i}k's'}\, e^{\mathrm{i}ks}\,\delta_{ss'}(-eR) \begin{pmatrix} \cos\left(s\times\frac{2\pi}{N}\right) \\ \sin\left(s\times\frac{\pi}{N}\right) \\ 0 \end{pmatrix}, \qquad (23.25)$$

where the z-component vanishes and the perpendicular components have circular polarization[4]

$$\boldsymbol{\mu}\left(\frac{1}{\sqrt{2}},\pm\frac{\mathrm{i}}{\sqrt{2}},0\right) = \frac{-eR}{\sqrt{2}N}\sum_{s} e^{\mathrm{i}(k-k'\pm 2\pi/N)s} = \frac{-eR}{\sqrt{2}}\delta_{k-k',\pm 2\pi/N}. \qquad (23.26)$$

The selection rule is

$$k - k' = \pm 2\pi/N. \qquad (23.27)$$

Hence, two of the HOMO–LUMO transitions are allowed and circularly polarized:

$$4\frac{2\pi}{N} \to 5\frac{2\pi}{N},$$

$$-4\frac{2\pi}{N} \to -5\frac{2\pi}{N}. \qquad (23.28)$$

If configuration interaction is taken into account, these four transitions split into two degenerate excited states of E-symmetry. The higher one carries most of the intensity and corresponds to the Soret band in the UV. The

---

[3] Neglecting differential overlaps and assuming a perfect circular arrangement.

[4] For an average $R = 3\,\text{Å}$, this gives a total intensity of $207\,\text{Debye}^2$ which is comparable to the $290\,\text{Debye}^2$ from a HF/CI calculation.

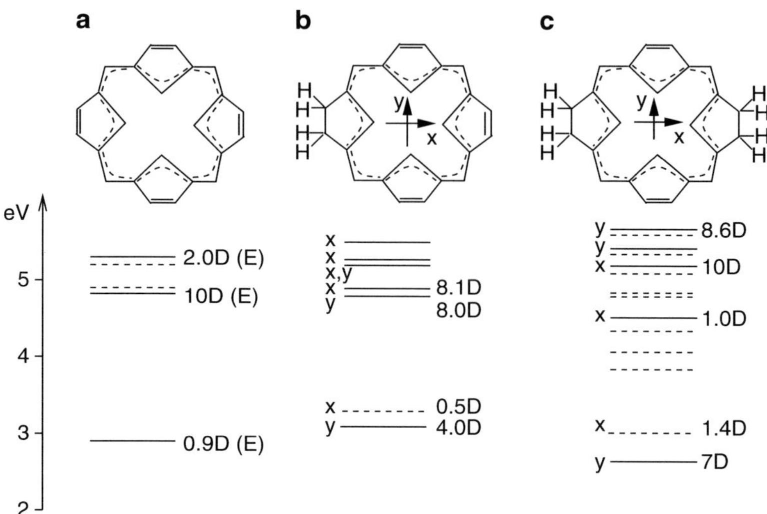

Fig. 23.11. Electronic structure of porphins and porphin derivatives. 631G-HF/CI calculations were performed with GAMESS [132]. *Numbers* give transition dipoles in Debyes. *Dashed lines* indicate very weak or dipole-forbidden transitions. (**a**) Mg-porphin has two allowed transitions of *E*-symmetry, corresponding to the weak *Q*-band at lower and the strong *B*-band at higher energy. (**b**) In Mg-chlorin (dihydroporphin), the double bond 1–2 is saturated similar to chlorophylls. The *Q*-band splits into the lower and stronger Q_y-band and the weaker Q_x-band. (**c**) In Mg-tetrahydroporphin, double bonds 1–2 and 5–6 are saturated similar to bacteriochlorophyll. The Q_y-band is shifted to lower energies

lower one is very weak and corresponds to the *Q*-band in the visible [75]. As a result of ab initio calculations, the four orbitals of the Gouterman model stay approximately separated from the rest of the MO orbitals [76]. If the symmetry is disturbed as for chlorin, the *Q*-band splits into the lower Q_y-band which gains large intensity and the higher weak Q_x-band. Figure 23.11 shows calculated spectra for Mg-porphin, Mg-chlorin, and Mg-tetrahydroporphin.

23.8 Energy Transfer Processes

Carotenoids and chlorophylls are very important for photosynthetic systems [77]. Chlorophyll molecules with different absorption maxima are used to harvest light energy and to direct it to the reaction center where the special pair dimer has the lowest absorption band of the chlorophylls. Carotenoids can act as additional light-harvesting pigments in the blue region of the spectrum [78, 79] and transfer energy to chlorophylls which absorb at longer wavelengths. Carotenoids are also important as triplet quenchers to prevent

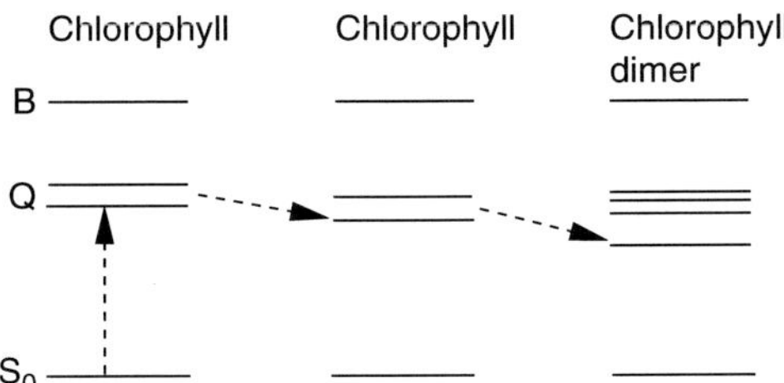

Fig. 23.12. Chlorophyll–chlorophyll energy transfer. Energy is transferred via a sequence of chromophores with decreasing absorption maxima

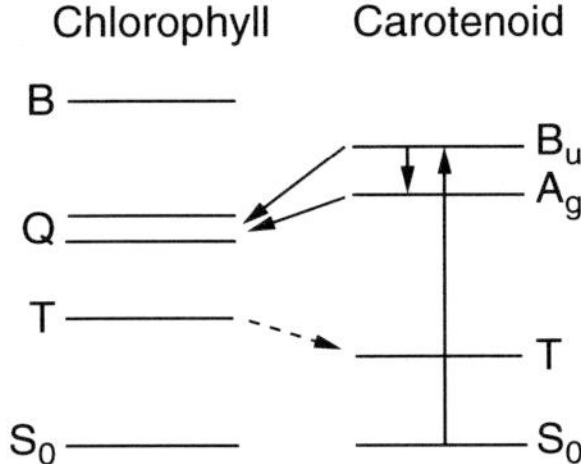

Fig. 23.13. Carotenoid–chlorophyll energy transfer. Carotenoids absorb light in the blue region of the spectrum and transfer energy to chlorophylls which absorb at longer wavelengths (*solid arrows*). Carotenoid triplet states are below the lowest chlorophyll triplet and therefore important triplet quenchers (*dashed arrow*)

the formation of triplet oxygen and for dissipation of excess energy [80] (Figs. 23.12 and 23.13).

Problems

23.1. Polyene with Bond Length Alternation

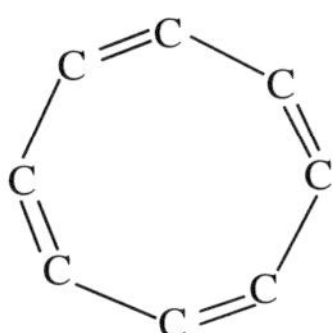

Consider a cyclic polyene with $2N$-carbon atoms with alternating bond lengths.

The Hückel matrix has the form

$$H = \begin{pmatrix} \alpha & \beta & & & & \beta' \\ \beta & \alpha & \beta' & & & \\ & \beta' & \ddots & \ddots & & \\ & & & \ddots & \alpha & \beta \\ \beta' & & & & \beta & \alpha \end{pmatrix}.$$

(a) Show that the eigenvectors can be written as

$$c_{2n} = e^{ikn}, \quad c_{2n-1} = e^{i(kn+\chi)}.$$

(b) Determine the phase angle χ and the eigenvalues for $\beta \neq \beta'$

(c) We want now to find the eigenvectors of a linear polyene. Therefore, we use the real-valued functions

$$c_{2n} = \sin kn, \quad c_{2n-1} = \sin(kn + \chi), \quad n = 1, \ldots, N.$$

with the phase angle as in (b).

We now add two further carbon atoms with indices 0 and $2N + 1$. The first of these two obviously has no effect since $c_0 = \sin(0 \times k) = 0$. For the atom $2N + 1$, we demand that the wave function again vanishes which restricts the possible k-values:

$$0 = c_{2N+1} = \sin((N + 1)k + \chi) = \Im(e^{i\chi + i(N+1)k}).$$

For these k-values, the cyclic polyene with $2N + 2$ atoms is equivalent to the linear polyene with $2N$-atoms as the off-diagonal interaction becomes irrelevant. Show that the k-values obey the equation

$$\beta \sin((N + 1)k) + \beta' \sin(Nk) = 0.$$

(d) Find a similar treatment for a linear polyene with odd number of C-atoms.

Incoherent Energy Transfer

We consider the transfer of energy from an excited donor molecule D^* to an acceptor molecule A

$$D^* + A \rightarrow D + A^*. \tag{24.1}$$

24.1 Excited States

We assume that the optical transitions of both molecules can be described by the transition between the highest occupied (HOMO) and lowest unoccupied (LUMO) molecular orbitals. The Hartree–Fock ground state of one molecule can be written as a Slater determinant of doubly occupied molecular orbitals:

$$|\mathrm{HF}\rangle = |\phi_{1\uparrow}\phi_{1\downarrow}\cdots\phi_{\mathrm{HO}\uparrow}\phi_{\mathrm{HO}\downarrow}|. \tag{24.2}$$

Promotion of an electron from the HOMO to the LUMO creates a singlet or a triplet state both of which are linear combinations of the four determinants:

$$\begin{aligned}
|\uparrow\downarrow\rangle &= |\phi_{1\uparrow}\phi_{1\downarrow}\cdots\phi_{\mathrm{HO}\uparrow}\phi_{\mathrm{LU}\downarrow}|, \\
|\downarrow\uparrow\rangle &= |\phi_{1\uparrow}\phi_{1\downarrow}\cdots\phi_{\mathrm{HO}\downarrow}\phi_{\mathrm{LU}\uparrow}|, \\
|\uparrow\uparrow\rangle &= |\phi_{1\uparrow}\phi_{1\downarrow}\cdots\phi_{\mathrm{HO}\uparrow}\phi_{\mathrm{LU}\uparrow}|, \\
|\downarrow\downarrow\rangle &= |\phi_{1\uparrow}\phi_{1\downarrow}\cdots\phi_{\mathrm{HO}\downarrow}\phi_{\mathrm{LU}\downarrow}|.
\end{aligned} \tag{24.3}$$

Obviously, the last two determinants are the components of the triplet state with $S_z = \pm 1$:

$$\begin{aligned}
|1, +1\rangle &= |\uparrow\uparrow\rangle, \\
|1, -1\rangle &= |\downarrow\downarrow\rangle.
\end{aligned} \tag{24.4}$$

Linear combination of the first two determinants gives the triplet state with $S_z = 0$:

$$|1, 0\rangle = \frac{1}{\sqrt{2}}(|\uparrow\downarrow\rangle + |\downarrow\uparrow\rangle) \tag{24.5}$$

and the singlet state

$$|0,0\rangle = \frac{1}{\sqrt{2}}(|\uparrow\downarrow\rangle - |\downarrow\uparrow\rangle). \tag{24.6}$$

Let us now consider the states of the molecule pair DA. The singlet ground state is

$$|\phi_{1\uparrow}\phi_{1\downarrow}\cdots\phi_{\mathrm{HO,D},\uparrow}\phi_{\mathrm{HO,D},\downarrow}\phi_{\mathrm{HO,A},\uparrow}\phi_{\mathrm{HO,A},\downarrow}|, \tag{24.7}$$

which will simply be denoted as

$$|^1\mathrm{DA}\rangle = |\mathrm{D}_\uparrow \mathrm{D}_\downarrow \mathrm{A}_\uparrow \mathrm{A}_\downarrow|. \tag{24.8}$$

The excited singlet state of the donor is

$$|^1\mathrm{D}^*\mathrm{A}\rangle = \frac{1}{\sqrt{2}}(|\mathrm{D}_\uparrow^* \mathrm{D}_\downarrow \mathrm{A}_\uparrow \mathrm{A}_\downarrow| - |\mathrm{D}_\downarrow^* \mathrm{D}_\uparrow \mathrm{A}_\uparrow \mathrm{A}_\downarrow|) \tag{24.9}$$

and the excited state of the acceptor is

$$|^1\mathrm{DA}^*\rangle = \frac{1}{\sqrt{2}}(|\mathrm{D}_\uparrow \mathrm{D}_\downarrow \mathrm{A}_\uparrow^* \mathrm{A}_\downarrow| - |\mathrm{D}_\uparrow \mathrm{D}_\downarrow \mathrm{A}_\downarrow^* \mathrm{A}_\uparrow|). \tag{24.10}$$

24.2 Interaction Matrix Element

The interaction responsible for energy transfer is the Coulombic electron–electron interaction:

$$V_{\mathrm{ee}} = \frac{e^2}{4\pi\varepsilon|r_1 - r_2|}. \tag{24.11}$$

With respect to the basis of molecular orbitals, its matrix elements are denoted as

$$V(\phi_1\phi_{1'}\phi_2\phi_{2'})$$
$$= \int \mathrm{d}^3 r_1 \mathrm{d}^3 r_2 \phi_{1\sigma}^*(r_1)\phi_{2,\sigma'}^*(r_2)\frac{e^2}{4\pi\varepsilon|r_1 - r_2|}\phi_{1'\sigma}(r_1)\phi_{2',\sigma'}(r_2). \tag{24.12}$$

The transfer interaction

$$\begin{aligned}
V_{\mathrm{if}} &= \langle ^1\mathrm{D}^*\mathrm{A}|V_{\mathrm{ee}}|^1\mathrm{DA}^*\rangle \\
&= \frac{1}{2}\langle |\mathrm{D}_\uparrow^* \mathrm{D}_\downarrow \mathrm{A}_\uparrow \mathrm{A}_\downarrow |V_{\mathrm{ee}}|\mathrm{D}_\uparrow \mathrm{D}_\downarrow \mathrm{A}_\uparrow^* \mathrm{A}_\downarrow|\rangle \\
&\quad - \frac{1}{2}\langle |\mathrm{D}_\uparrow^* \mathrm{D}_\downarrow \mathrm{A}_\uparrow \mathrm{A}_\downarrow |V_{\mathrm{ee}}|\mathrm{D}_\uparrow \mathrm{D}_\downarrow \mathrm{A}_\downarrow^* \mathrm{A}_\uparrow\rangle \\
&\quad - \frac{1}{2}\langle |\mathrm{D}_\downarrow^* \mathrm{D}_\uparrow \mathrm{A}_\uparrow \mathrm{A}_\downarrow |V_{\mathrm{ee}}||\mathrm{D}_\uparrow \mathrm{D}_\downarrow \mathrm{A}_\uparrow^* \mathrm{A}_\downarrow|\rangle \\
&\quad + \frac{1}{2}\langle |\mathrm{D}_\downarrow^* \mathrm{D}_\uparrow \mathrm{A}_\uparrow \mathrm{A}_\downarrow |V_{\mathrm{ee}}|\mathrm{D}_\uparrow \mathrm{D}_\downarrow \mathrm{A}_\downarrow^* \mathrm{A}_\uparrow|\rangle
\end{aligned} \tag{24.13}$$

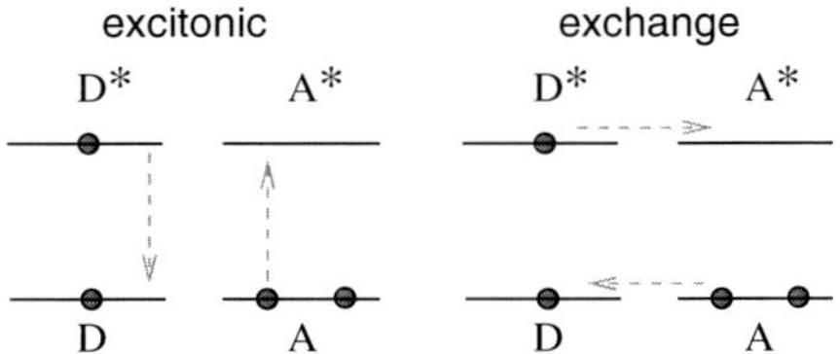

Fig. 24.1. Intramolecular energy transfer. *Left*: the excitonic or Förster mechanism [81, 82] depends on transition intensities and spectral overlap. *Right*: the exchange or Dexter mechanism [83] depends on the overlap of the wave functions

consists of four summands. The first one has two contributions

$$\frac{1}{2}\langle|D_\uparrow^* D_\downarrow A_\uparrow A_\downarrow|V_{ee}|D_\uparrow D_\downarrow A_\uparrow^* A_\downarrow|\rangle = \frac{1}{2}V(D^*DA\,A^*) - \frac{1}{2}V(D^*A^*AD), \quad (24.14)$$

where the first part is the excitonic interaction and the second part is the exchange interaction (Fig. 24.1).

The second summand

$$-\frac{1}{2}\langle|D_\uparrow^* D_\downarrow A_\uparrow A_\downarrow|V_{ee}|D_\uparrow D_\downarrow A_\downarrow^* A_\uparrow|\rangle = \frac{1}{2}V(D^*DA\,A^*) \quad (24.15)$$

has no exchange contribution due to the spin orientations. The two remaining summands are just mirror images of the first two. Altogether, the interaction for singlet energy transfer is

$$\langle{}^1D^*A|V_{ee}|{}^1DA^*\rangle = 2V(D^*DAA^*) - V(D^*A^*AD). \quad (24.16)$$

In the triplet case,

$$\begin{aligned}
V_{\mathrm{if}} &= \langle{}^3D^*A|V_{ee}|{}^3DA^*\rangle \\
&= \frac{1}{2}\langle|D_\uparrow^* D_\downarrow A_\uparrow A_\downarrow|V_{ee}|D_\uparrow D_\downarrow A_\uparrow^* A_\downarrow|\rangle \\
&\quad +\frac{1}{2}\langle|D_\uparrow^* D_\downarrow A_\uparrow A_\downarrow|V_{ee}|D_\uparrow D_\downarrow A_\downarrow^* A_\uparrow\rangle \\
&\quad +\frac{1}{2}\langle|D_\downarrow^* D_\uparrow A_\uparrow A_\downarrow|V_{ee}||D_\uparrow D_\downarrow A_\uparrow^* A_\downarrow|\rangle \\
&\quad +\frac{1}{2}|D_\downarrow^* D_\uparrow A_\uparrow A_\downarrow|V_{ee}|D_\uparrow D_\downarrow A_\downarrow^* A_\uparrow|\rangle \\
&= -V(D^*A^*AD). \quad (24.17)
\end{aligned}$$

Here, energy can only be transferred by the exchange coupling [83]. Since this involves the overlap of electronic wave functions, it is important at small distances. In the singlet state, the excitonic interaction [81, 82] allows for energy transfer also at larger distances.

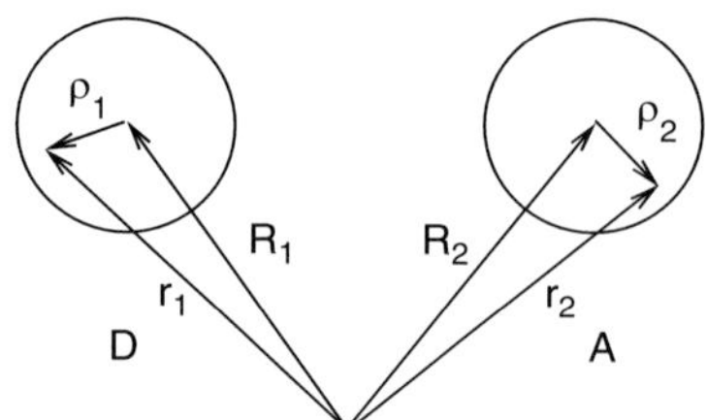

Fig. 24.2. Multipole expansion. For large intermolecular distance $R = |\mathbf{R}_1 - \mathbf{R}_2|$, the distance of the electrons $\mathbf{r}_1 - \mathbf{r}_2$ can be expanded as a Taylor series with respect to $(\mathbf{r}_1 - \mathbf{R}_1)/R$ and $(\mathbf{r}_2 - \mathbf{R}_2)/R$

24.3 Multipole Expansion of the Excitonic Interaction

We will now apply a multipole expansion to the excitonic matrix element (Fig. 24.2)

$$V_{\text{exc}} = 2V(\text{D}^*\text{DAA}^*)$$

$$= 2\int d^3r_1 d^3r_2 \phi_{\text{D}^*}^*(r_1)\phi_{\text{A}'}^*(r_2)\frac{e^2}{4\pi\varepsilon|r_1 - r_2|}\phi_{\text{D}}(r_1)\phi_{\text{A}^*}(r_2). \quad (24.18)$$

We take the position of the electrons relative to the centers of the molecules

$$r_{1,2} = R_{1,2} + \varrho_{1,2} \quad (24.19)$$

and expand the Coulombic interaction

$$\frac{1}{|r_1 - r_2|} = \frac{1}{|R_1 - R_2 + \varrho_1 - \varrho_2|} \quad (24.20)$$

using the Taylor series

$$\frac{1}{|R + \varrho|} = \frac{1}{|R|} - \frac{1}{|R|^2}\frac{\varrho R}{|R|} + \frac{1}{2}\frac{3|R|(R\varrho)^2 - |R|^3\varrho^2}{|R|^6} + \cdots \quad (24.21)$$

for

$$R = R_1 - R_2, \quad \varrho = \varrho_1 - \varrho_2. \quad (24.22)$$

The zero-order term vanishes due to the orthogonality of the orbitals. The first-order term gives

$$-2\frac{e^2}{4\pi\varepsilon|R|^3}R\int d^3\varrho_1 d^3\varrho_2 \phi_{\text{D}^*}^*(r_1)\phi_{\text{D}}(r_1)(\varrho_1 - \varrho_2)\phi_{\text{A}}^*(r_2)\phi_{\text{A}^*}(r_2) \quad (24.23)$$

and also vanishes due to the orthogonality. The second-order term gives the leading contribution:

$$\frac{e^2}{4\pi\varepsilon|R|^6} \int \mathrm{d}^3\varrho_1 \mathrm{d}^3\varrho_2 \phi_{\mathrm{D}^*}^*(r_1)\phi_{\mathrm{D}}(r_1)$$
$$\times (3|R|(R\varrho_1 - R\varrho_2)^2 - |R|^3(\varrho_1 - \varrho_2)^2)\phi_{\mathrm{A}}^*(r_2)\phi_{\mathrm{A}^*}(r_2). \quad (24.24)$$

Only the integrals over mixed products of ϱ_1 and ϱ_2 are not zero. They can be expressed with the help of the transition dipoles:

$$\mu_{\mathrm{D}} = \langle {}^1\mathrm{D}|\mathbf{er}|{}^1\mathrm{D}^*\rangle = \sqrt{2} \int \mathrm{d}^3\varrho_1 \phi_{\mathrm{D}^*}^*(\varrho_1)e\varrho_1\phi_{\mathrm{D}}(\varrho_1),$$

$$\mu_{\mathrm{A}} = \langle {}^1\mathrm{A}|\mathbf{er}|{}^1\mathrm{A}^*\rangle = \sqrt{2} \int \mathrm{d}^3\varrho_2 \phi_{\mathrm{A}}^*(\varrho_2)e\varrho_2\phi_{\mathrm{A}^*}(\varrho_2). \quad (24.25)$$

Finally[1] the leading term of the excitonic interaction is the dipole–dipole term:

$$V_{\mathrm{exc}} = \frac{e^2}{4\pi\varepsilon|R|^5}\left(|R|^2\mu_{\mathrm{D}}\mu_{\mathrm{A}} - 3(R\mu_{\mathrm{D}})(R\mu_{\mathrm{A}})\right). \quad (24.26)$$

This is often written with an orientation-dependent factor K as

$$V_{\mathrm{exc}} = \frac{K}{|R|^3}\,|\mu_{\mathrm{D}}||\mu_{\mathrm{A}}|. \quad (24.27)$$

24.4 Energy Transfer Rate

We consider excitonic interaction of two molecules. The Hamilton operator is divided into the zero-order Hamiltonian

$$H_0 = |\mathrm{D}^*\mathrm{A}\rangle H_{\mathrm{D}^*\mathrm{A}}\langle \mathrm{D}^*\mathrm{A}| + |\mathrm{DA}^*\rangle H_{\mathrm{DA}^*}\langle \mathrm{DA}^*| \quad (24.28)$$

and the interaction operator

$$H' = |\mathrm{D}^*\mathrm{A}\rangle V_{\mathrm{exc}}\langle \mathrm{DA}^*| + |\mathrm{DA}^*\rangle V_{\mathrm{exc}}\langle \mathrm{D}^*\mathrm{A}|. \quad (24.29)$$

We neglect electron exchange between the two molecules and apply the Condon approximation. Then taking energies relative to the electronic ground state $|\mathrm{DA}\rangle$, we have

$$H_{\mathrm{D}^*\mathrm{A}} = \hbar\omega_{\mathrm{D}^*} + H_{\mathrm{D}^*} + H_{\mathrm{A}},$$
$$H_{\mathrm{DA}^*} = \hbar\omega_{\mathrm{A}^*} + H_{\mathrm{D}} + H_{\mathrm{A}^*}, \quad (24.30)$$

with electronic excitation energies $\hbar\omega_{\mathrm{D}^*(\mathrm{A}^*)}$ and nuclear Hamiltonians H_{D}, $H_{\mathrm{D}^*}, H_{\mathrm{A}}, H_{\mathrm{A}^*}$. Now, consider once more Fermi's golden rule for the transition between vibronic states $|i\rangle \to |f\rangle$:

---

[1] Local dielectric constant and local field correction [84, 85] will be taken into account implicitly by considering the transition dipoles as effective values.

$$
\begin{aligned}
k &= \frac{2\pi}{\hbar} \sum_{i,f} P_i |\langle i|H'|f\rangle|^2 \delta(\omega_f - \omega_i) \\
&= \frac{1}{\hbar^2} \int \sum_{i,f} P_i \langle i|H'|f\rangle \langle f|H'|i\rangle e^{i(\omega_f - \omega_i)t} \mathrm{d}t \\
&= \frac{1}{\hbar^2} \int \sum_{i,f} P_i \langle i|H'|f\rangle e^{i\omega_f t} \langle f|H'|e^{i\omega_i t} i\rangle \mathrm{d}t \qquad (24.31) \\
&= \frac{1}{\hbar^2} \int \sum_{i} \langle i|Q^{-1} e^{-H_0/k_{\mathrm{B}}T} H' e^{itH_0/\hbar} H' e^{itH_0/\hbar}|i\rangle \mathrm{d}t \\
&= \frac{1}{\hbar^2} \int \mathrm{d}t \, \langle H'(0)H'(t)\rangle . \qquad (24.32)
\end{aligned}
$$

Initially, only the donor is excited. Then the average is restricted to the vibrational states of $\mathrm{D^*A}$:

$$
k = \frac{1}{\hbar^2} \int \mathrm{d}t \, \langle H'(0)H'(t)\rangle_{\mathrm{D^*A}} . \qquad (24.33)
$$

With the transition dipole operators

$$
\begin{aligned}
\hat{\mu}_{\mathrm{D}} &= |\mathrm{D}\rangle \mu_{\mathrm{D}} \langle \mathrm{D^*}| + |\mathrm{D^*}\rangle \mu_{\mathrm{D}} \langle \mathrm{D}|, \\
\hat{\mu}_{\mathrm{A}} &= |\mathrm{A}\rangle \mu_{\mathrm{A}} \langle \mathrm{A^*}| + |\mathrm{A^*}\rangle \mu_{\mathrm{A}} \langle \mathrm{A}|, \qquad (24.34)
\end{aligned}
$$

the rate becomes

$$
k = \frac{K^2}{\hbar^2 R^6} \int \mathrm{d}t \, \langle \hat{\mu}_{\mathrm{D}}(0)\hat{\mu}_{\mathrm{A}}(0)\hat{\mu}_{\mathrm{D}}(t)\hat{\mu}_{\mathrm{A}}(t)\rangle_{\mathrm{D^*A}} .
$$

Here, we assumed that the orientation does not change on the relevant time scale. Since each of the dipole operators acts only on one of the molecules, we have

$$
\begin{aligned}
k &= \frac{1}{\hbar^2} \frac{K^2}{|R|^6} \int \mathrm{d}t \, \langle \hat{\mu}_{\mathrm{D}}(0)\hat{\mu}_{\mathrm{D}}(t)\hat{\mu}_{\mathrm{A}}(0)\hat{\mu}_{\mathrm{A}}(t)\rangle_{\mathrm{D^*A}} \\
&= \frac{1}{\hbar^2} \frac{K^2}{|R|^6} \int \mathrm{d}t \, \langle \hat{\mu}_{\mathrm{D}}(0)\hat{\mu}_{\mathrm{D}}(t)\rangle_{\mathrm{D^*A}} \, \langle \hat{\mu}_{\mathrm{A}}(0)\hat{\mu}_{\mathrm{A}}(t)\rangle_{\mathrm{D^*A}} \\
&= \frac{1}{\hbar^2} \frac{K^2}{|R|^6} \int \mathrm{d}t \, \langle \hat{\mu}_{\mathrm{D}}(0)\hat{\mu}_{\mathrm{D}}(t)\rangle_{\mathrm{D^*}} \, \langle \hat{\mu}_{\mathrm{A}}(0)\hat{\mu}_{\mathrm{A}}(t)\rangle_{\mathrm{A}} . \qquad (24.35)
\end{aligned}
$$

24.5 Spectral Overlap

The two factors are related to the acceptor absorption and donor fluorescence spectra. Consider optical transitions between the singlet states $|^1\mathrm{D^*}\rangle \rightarrow |^1\mathrm{D}\rangle$ and $|^1\mathrm{A} \rightarrow\, ^1\mathrm{A^*}\rangle$. The number of fluorescence photons per time is given by the

Einstein coefficient for spontaneous emission[2]

$$A_{D^*D} = \frac{2\omega_{D^*D}^3}{3\varepsilon_0 hc^3}|\mu_D|^2 \tag{24.36}$$

with the donor transition dipole moment (24.25)

$$\mu_D = \sqrt{2}\int d^3r\,\phi_{D^*}(er)\phi_D. \tag{24.37}$$

The frequency-resolved total fluorescence is then

$$I_e(\omega) = A_{D^*D}g_e(\omega) = \frac{2\omega^3}{3\varepsilon_0 hc^3}|\mu_D|^2 g_e(\omega) \tag{24.38}$$

with the normalized lineshape function $g_e(\omega)$.

The absorption cross section can be expressed with the help of the Einstein coefficient for absorption B_{12}^ω as [86]

$$\sigma_a(\omega) = B_{AA^*}^\omega\,\hbar\omega g_a(\omega)/c. \tag{24.39}$$

The Einstein coefficients for absorption and emission are related by

$$B_{AA^*}^\omega = \frac{\pi^2 c^3}{\hbar\omega_{A^*A}^3}A_{A^*A} = \frac{1}{3}\frac{\pi}{\hbar^2\varepsilon_0}|\mu_A|^2$$

and therefore (Fig. 24.3)

$$\sigma_a(\omega) = \frac{1}{3}\frac{\pi}{\hbar c\varepsilon_0}|\mu_A|^2\omega g_a(\omega).$$

In the following, we discuss absorption and emission in terms of the modified spectra:

$$\begin{aligned}
\alpha(\omega) &= \frac{3\hbar c\varepsilon_0}{\pi}\frac{\sigma_a(\omega)}{\omega} = |\hat{\mu}_A|^2 g_a(\omega)\\
&= \sum_{i,f}P_i|\langle i|\hat{\mu}_A|f\rangle|^2\delta(\omega_f - \omega_i - \omega)\\
&= \frac{1}{2\pi}\int dt\,e^{-i\omega t}\sum_i\left\langle i\left|\frac{e^{-H/k_BT}}{Q}\hat{\mu}_A\right|f\right\rangle e^{i\omega_f t}\langle f|\hat{\mu}_A e^{-i\omega_i t}|i\rangle\\
&= \frac{1}{2\pi}\int dt\,e^{-i\omega t}\langle\hat{\mu}_A(0)\hat{\mu}_A(t)\rangle_A
\end{aligned}$$

---

[2] A detailed discussion of the relationships between absorption cross section and Einstein coefficients is found in [86].

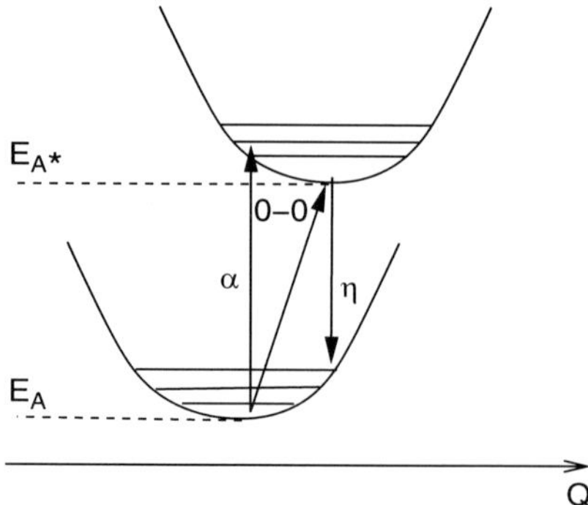

Fig. 24.3. Absorption and emission. Within the displaced oscillator model, the $0 \rightarrow 0$ transition energy is located halfway between the absorption (α) and emission maximum (η)

and[3]

$$\eta(\omega) = \frac{3\varepsilon_0 hc^3 I_{\mathrm{e}}(\omega)}{2\omega^3} = |\mu_{\mathrm{D}}|^2 g_{\mathrm{e}}(\omega)$$

$$= \sum_{i,f} P_f |\langle f|\hat{\mu}_{\mathrm{D}}|i\rangle|^2 \delta(\omega_f - \omega_i - \omega)$$

$$= \frac{1}{2\pi} \int \mathrm{d}t\, \mathrm{e}^{-\mathrm{i}\omega t} \sum_f \left\langle f\left| \frac{\mathrm{e}^{-H/k_{\mathrm{B}}T}}{Q} \mathrm{e}^{\mathrm{i}\omega_f t} \hat{\mu}_{\mathrm{D}}\right| i\right\rangle \mathrm{e}^{-\mathrm{i}\omega_i t} \langle i|\hat{\mu}_{\mathrm{D}}|f\rangle$$

$$= \frac{1}{2\pi} \int \mathrm{d}t\, \mathrm{e}^{-\mathrm{i}\omega t} \langle \hat{\mu}_{\mathrm{D}}(t)\hat{\mu}_{\mathrm{D}}(0)\rangle_{\mathrm{D}*}$$

$$= \frac{1}{2\pi} \int \mathrm{d}t\, \mathrm{e}^{\mathrm{i}\omega t} \langle \hat{\mu}_{\mathrm{D}}(0)\hat{\mu}_{\mathrm{D}}(t)\rangle_{\mathrm{D}*}. \tag{24.40}$$

Within the Condon approximation

$$\langle \hat{\mu}_{\mathrm{A}}(0)\hat{\mu}_{\mathrm{A}}(t)\rangle_{\mathrm{A}} = \mathrm{e}^{\mathrm{i}\omega_{\mathrm{A}*}t}|\mu_{\mathrm{A}}|^2 \left\langle \mathrm{e}^{\mathrm{i}tH_{\mathrm{A}*}/\hbar}\mathrm{e}^{-\mathrm{i}tH_{\mathrm{A}}/\hbar}\right\rangle_{\mathrm{A}}$$

$$= \mathrm{e}^{\mathrm{i}\omega_{\mathrm{A}*}t}|\mu_{\mathrm{A}}|^2 F_{\mathrm{A}}(t) \tag{24.41}$$

and therefore the lineshape function is the Fourier transform of the correlation function:

$$g_{\mathrm{a}}(\omega) = \frac{1}{2\pi} \int \mathrm{d}t\, \mathrm{e}^{\mathrm{i}(\omega - \omega_{\mathrm{A}*})t} F_{\mathrm{A}}(t). \tag{24.42}$$

Similarly,

$$\langle \hat{\mu}_{\mathrm{D}}(0)\hat{\mu}_{\mathrm{D}}(t)\rangle_{\mathrm{D}*} = \mathrm{e}^{-\mathrm{i}\omega_{\mathrm{D}*}t}|\mu_{\mathrm{D}}|^2 \left\langle \mathrm{e}^{\mathrm{i}tH_{\mathrm{D}}/\hbar}\mathrm{e}^{-\mathrm{i}tH_{\mathrm{D}*}/\hbar}\right\rangle_{\mathrm{D}*}$$

$$= \mathrm{e}^{-\mathrm{i}\omega_{\mathrm{D}*}t}|\mu_{\mathrm{D}}|^2 F_{\mathrm{D}*}(t) \tag{24.43}$$

---

[3] Note the sign change which appears since we now have to average over the excited state vibrations.

and

$$g_e(\omega) = \frac{1}{2\pi} \int dt\, e^{i(\omega - \omega_{D^*})t} F_{D^*}(t). \tag{24.44}$$

With the inverse Fourier integrals

$$\langle \hat{\mu}_A(0)\hat{\mu}_A(t)\rangle_A = \int d\omega e^{i\omega t} \alpha(\omega),$$

$$\langle \hat{\mu}_D(0)\hat{\mu}_D(t)\rangle_{D^*} = \int d\omega e^{-i\omega t} \eta(\omega), \tag{24.45}$$

(24.35) becomes

$$\begin{aligned}
k &= \frac{1}{\hbar^2}\frac{K^2}{|R|^6} \int dt \int d\omega e^{i\omega t}\eta(\omega) \int d\omega' e^{-i\omega' t}\alpha(\omega') \\
&= \frac{1}{\hbar^2}\frac{K^2}{|R|^6} \int d\omega d\omega'\eta(\omega)\alpha(\omega')\, 2\pi\delta(\omega - \omega') \\
&= \frac{2\pi}{\hbar^2}\frac{K^2}{|R|^6} \int d\omega \eta(\omega)\alpha(\omega) \\
&= \frac{2\pi}{\hbar^2}\frac{K^2}{|R|^6}|\mu_A|^2|\mu_D|^2 \int d\omega g_e(\omega)g_a(\omega). \tag{24.46}
\end{aligned}$$

This is the famous rate expression for Förster [81] energy transfer which involves the spectral overlap of donor emission and acceptor absorption [87, 88]. For optimum efficiency of energy transfer, the maximum of the acceptor absorption should be at longer wavelength than the maximum of the donor emission (Fig. 24.4).

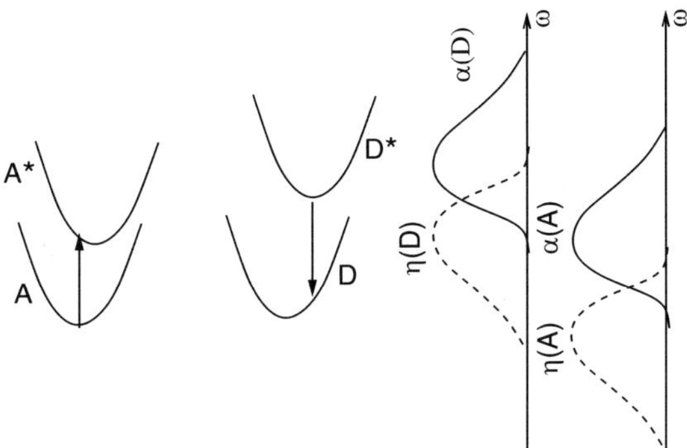

Fig. 24.4. Energy transfer and spectral overlap. Efficient energy transfer requires good spectral overlap of donor emission and acceptor absorption. This is normally only the case for transfer between unlike molecules

24.6 Energy Transfer in the Triplet State

Consider now energy transfer in the triplet state. Here, the transitions are optically not allowed. We consider a more general interaction operator which changes the electronic state of both molecules simultaneously and can be written as[4]

$$H' = |D^*\rangle|A\rangle V_x \langle A^*|\langle D| + \text{h.c.} \tag{24.47}$$

We start from the rate expression (24.32)

$$k = \frac{1}{\hbar^2} \int dt \, \langle H'(0)H'(t)\rangle. \tag{24.48}$$

The thermal average has to be taken over the initially populated state $|D^*A\rangle$

$$k = \frac{1}{\hbar^2} \int dt \, \langle H'(0)H'(t)\rangle_{D^*A}.$$

In the static case ($V_x = \text{const}$),

$$k = \frac{1}{\hbar^2} \int dt \, \left\langle H' e^{itH_{DA^*}/\hbar} H' e^{-itH_{D^*A}/\hbar} \right\rangle_{D^*A}$$

$$= \frac{1}{\hbar^2} \int dt \, e^{i(\omega_{A^*}-\omega_{D^*})t} \left\langle H' e^{it(H_D+H_{A^*})/\hbar} H' e^{-it(H_{D^*}+H_A)/\hbar} \right\rangle_{D^*A}$$

$$= \frac{V_x^2}{\hbar^2} \int dt \, e^{i(\omega_{A^*}-\omega_{D^*})t} \left\langle e^{itH_D/\hbar} e^{-itH_{D^*}/\hbar} \right\rangle_{D^*} \left\langle e^{itH_{A^*}/\hbar} e^{-itH_A/\hbar} \right\rangle_A$$

$$= \frac{V_x^2}{\hbar^2} \int dt \, e^{it(\omega_{A^*}-\omega_{D^*})} F_{D^*}(t) F_A(t) \tag{24.49}$$

with the correlation functions

$$F_A(t) = \left\langle e^{itH_{A^*}/\hbar} e^{-itH_A/\hbar} \right\rangle_A, \tag{24.50}$$

$$F_{D^*}(t) = \left\langle e^{itH_D/\hbar} e^{-itH_{D^*}/\hbar} \right\rangle_{D^*}. \tag{24.51}$$

Introducing lineshape functions (24.42) and (24.44) similar to the excitonic case, the rate becomes

$$k = \frac{V_x^2}{\hbar^2} \int dt \, e^{it(\omega_{A^*}-\omega_{D^*})} \int d\omega' e^{-i(\omega'-\omega_{D^*})t} g_e(\omega') \int d\omega e^{i(\omega-\omega_{A^*})t} g_a(\omega)$$

$$= \frac{V_x^2}{\hbar^2} \int d\omega d\omega' g_e(\omega') g_a(\omega) 2\pi\delta(\omega - \omega')$$

$$= \frac{2\pi V_x^2}{\hbar} \int d\omega g_a(\omega) g_e(\omega), \tag{24.52}$$

which is very similar to the Förster expression (24.46). The excitonic interaction is replaced by the exchange coupling matrix element and the overlap of the optical spectra is replaced by the overlap of the Franck–Condon-weighted densities of states.

---

[4] We assume that the wave function of the pair can be factorized approximately.

Coherent Excitations in Photosynthetic Systems

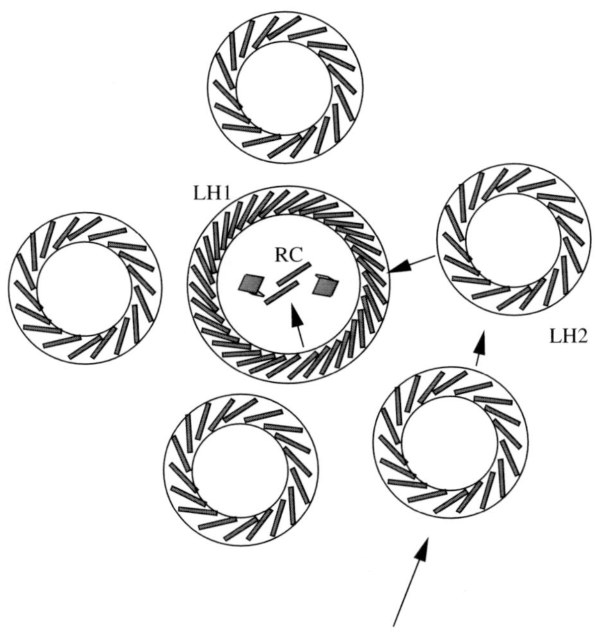

Fig. 25.1. Energy transfer in bacterial photosynthesis

Photosynthetic units of plants and bacteria consist of antenna complexes and reaction centers. Rings of closely coupled chlorophyll chromophores form the light-harvesting complexes which transfer the incoming photons very efficiently and rapidly to the reaction center where the photon energy is used to create an ion pair. In this chapter, we concentrate on the properties of strongly coupled chromophore aggregates (Fig. 25.1).

25.1 Coherent Excitations

If the excitonic coupling is large compared to fluctuations of the excitation energies, a coherent excitation of two or more molecules can be generated.

25.1.1 Strongly Coupled Dimers

Let us consider a dimer consisting of two strongly coupled molecules A and B as in the reaction center of photosynthesis. The two excited states

$$|A^*B\rangle, \quad |AB^*\rangle \tag{25.1}$$

are mixed due to the excitonic interaction. The eigenstates are given by the eigenvectors of the matrix

$$\begin{pmatrix} E_{A^*B} & V \\ V & E_{AB^*} \end{pmatrix}. \tag{25.2}$$

For a symmetric dimer, the diagonal energies have the same value and the eigenvectors can be characterized as symmetric or antisymmetric

$$\begin{pmatrix} \frac{1}{\sqrt{2}} & -\frac{1}{\sqrt{2}} \\ \frac{1}{\sqrt{2}} & \frac{1}{\sqrt{2}} \end{pmatrix} \begin{pmatrix} E_{A^*B} & V \\ V & E_{A^*B} \end{pmatrix} \begin{pmatrix} \frac{1}{\sqrt{2}} & \frac{1}{\sqrt{2}} \\ -\frac{1}{\sqrt{2}} & \frac{1}{\sqrt{2}} \end{pmatrix} = \begin{pmatrix} E_{A^*B} - V & \\ & E_{A^*B} + V \end{pmatrix}. \tag{25.3}$$

The two excitonic states are split by $2V$.[1] The transition dipoles of the two dimer bands are given by

$$\mu_\pm = \frac{1}{\sqrt{2}}(\mu_A \pm \mu_B) \tag{25.4}$$

and the intensities are given by

$$|\mu_\pm|^2 = \frac{1}{2}(\mu_A^2 + \mu_B^2 \pm 2\mu_A\mu_B). \tag{25.5}$$

For a symmetric dimer $\mu_A^2 = \mu_B^2 = \mu^2$ and

$$|\mu_\pm|^2 = \mu^2(1 \pm \cos\alpha), \tag{25.6}$$

where α denotes the angle between μ_A and μ_B. In case of an approximately C_2-symmetric structure, the components of μ_A and μ_B are furthermore related by symmetry operations (Fig. 25.2).

If we choose the C_2-axis along the z-axis, we have

$$\begin{pmatrix} \mu_{Bx} \\ \mu_{By} \\ \mu_{Bz} \end{pmatrix} = \begin{pmatrix} -\mu_{Ax} \\ -\mu_{Ay} \\ \mu_{Az} \end{pmatrix} \tag{25.7}$$

---

[1] This is known as Davidov splitting.

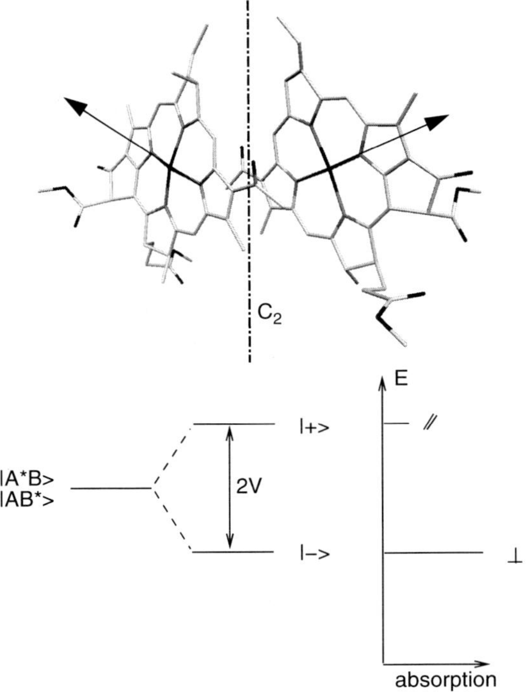

Fig. 25.2. The "special pair" dimer. *Top*: the nearly C_2-symmetric molecular arrangement for the reaction center of *Rhodopseudomonas viridis* (molekel graphics [89]). The transition dipoles of the two molecules (*arrows*) are essentially antiparallel. *Bottom*: the lower exciton component carries most of the oscillator strength

and therefore

$$\frac{1}{\sqrt{2}}(\mu_A + \mu_B) = \frac{1}{\sqrt{2}} \begin{pmatrix} 0 \\ 0 \\ 2\mu_{Az} \end{pmatrix},$$

$$\frac{1}{\sqrt{2}}(\mu_A - \mu_B) = \frac{1}{\sqrt{2}} \begin{pmatrix} 2\mu_{Ax} \\ 2\mu_{Ay} \\ 0 \end{pmatrix}, \tag{25.8}$$

which shows that the transition to the state $|+\rangle$ is polarized along the symmetry axis whereas the transition to $|-\rangle$ is polarized perpendicularly. For the special pair dimer, interaction with internal charge transfer states $|A^+B^-\rangle$ and $|A^-B^+\rangle$ has to be considered. In the simplest model, the following interaction matrix elements are important (Fig. 25.3).

The local excitation A^*B is coupled to the CT state A^+B^- by transferring an electron between the two LUMOs

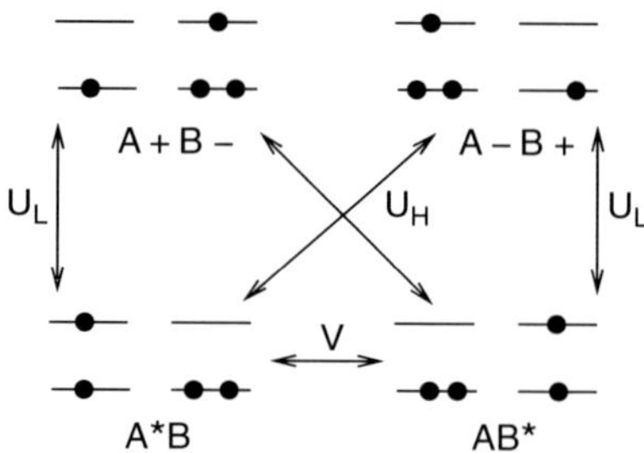

Fig. 25.3. Extended dimer model. The optical excitations A^*B and AB^* are coupled to the ionic states A^+B^- and A^-B^+

$$\langle A^*B|H|A^+B^-\rangle$$
$$= \frac{1}{2}\langle(A_\uparrow^*A_\downarrow - A_\downarrow^*A_\uparrow)B_\uparrow B_\downarrow H(B_\uparrow^*A_\downarrow - B_\downarrow^*A_\uparrow)B_\uparrow B_\downarrow\rangle$$
$$= H_{A^*,B^*} = U_L \tag{25.9}$$

and to the CT state A^-B^+ by transferring an electron between the HOMOs

$$\langle A^*B|H|A^-B^+\rangle$$
$$= \frac{1}{2}\langle(A_\uparrow^*A_\downarrow - A_\downarrow^*A_\uparrow)B_\uparrow B_\downarrow H(A_\uparrow^*B_\downarrow - A_\downarrow^*B_\uparrow)A_\uparrow A_\downarrow\rangle$$
$$= -H_{A,B} = U_H. \tag{25.10}$$

Similarly, the second local excitation couples to the CT states by

$$\langle AB^*|H|A^+B^-\rangle$$
$$= \frac{1}{2}\langle(B_\uparrow^*B_\downarrow - B_\downarrow^*B_\uparrow)A_\uparrow A_\downarrow H(B_\uparrow^*A_\downarrow - B_\downarrow^*A_\uparrow)B_\uparrow B_\downarrow\rangle$$
$$= -H_{A,B} = U_H, \tag{25.11}$$

$$\langle AB^*|H|A^-B^+\rangle$$
$$= \frac{1}{2}\langle(B_\uparrow^*B_\downarrow - B_\downarrow^*B_\uparrow)A_\uparrow A_\downarrow H(A_\uparrow^*B_\downarrow - A_\downarrow^*B_\uparrow)A_\uparrow A_\downarrow\rangle$$
$$= H_{A^*,B^*} = U_L. \tag{25.12}$$

The interaction of the four states is summarized by the matrix

$$H = \begin{pmatrix} E_{A^*B} & V & U_L & U_H \\ V & E_{B^*A} & U_H & U_L \\ U_L & U_H & E_{A^+B^-} & \\ U_H & U_L & & E_{A^-B^+} \end{pmatrix}. \tag{25.13}$$

Again for a symmetric dimer, $E_{B^*A} = E_{AB^*}$ and $E_{A^+B^-} = E_{A^-B^+}$ and the interaction matrix can be simplified by transforming to symmetrized basis functions with the transformation matrix

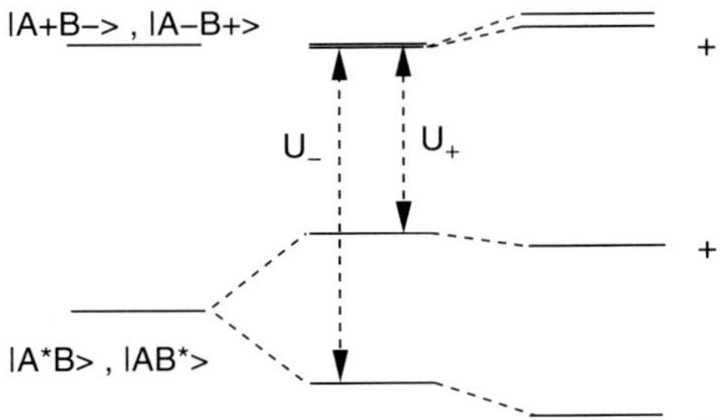

Fig. 25.4. Dimer states

$$S = \begin{pmatrix} \frac{1}{\sqrt{2}} & \frac{1}{\sqrt{2}} & & \\ -\frac{1}{\sqrt{2}} & \frac{1}{\sqrt{2}} & & \\ & & \frac{1}{\sqrt{2}} & \frac{1}{\sqrt{2}} \\ & & -\frac{1}{\sqrt{2}} & \frac{1}{\sqrt{2}} \end{pmatrix}. \tag{25.14}$$

The transformation gives

$$S^{-1} H S = \begin{pmatrix} E_* - V & & U_{\mathrm{L}} - U_{\mathrm{H}} & \\ & E_* + V & & U_{\mathrm{L}} + U_{\mathrm{H}} \\ U_{\mathrm{L}} - U_{\mathrm{H}} & & E_{\mathrm{CT}} & \\ & U_{\mathrm{L}} + U_{\mathrm{H}} & & E_{\mathrm{CT}} \end{pmatrix}, \tag{25.15}$$

where the states of different symmetry are decoupled (Fig. 25.4)

$$H_+ = \begin{pmatrix} E_* + V & U_{\mathrm{L}} + U_{\mathrm{H}} \\ U_{\mathrm{L}} + U_{\mathrm{H}} & E_{\mathrm{CT}} \end{pmatrix}, \quad H_- = \begin{pmatrix} E_* - V & U_{\mathrm{L}} - U_{\mathrm{H}} \\ U_{\mathrm{L}} - U_{\mathrm{H}} & E_{\mathrm{CT}} \end{pmatrix}. \tag{25.16}$$

25.1.2 Excitonic Structure of the Reaction Center

The reaction center of bacterial photosynthesis consists of six chromophores. Two bacteriochlorophyll molecules $(P_{\mathrm{L}}, P_{\mathrm{M}})$ form the special pair dimer. Another bacteriochlorophyll (B_{L}) and a bacteriopheophytine (H_{L}) act as the acceptors during the first electron transfer steps. Both have symmetry-related counterparts $(B_{\mathrm{M}}, H_{\mathrm{M}})$ which are not directly involved in the charge separation process. Due to the short neighbor distances (11–13 Å), delocalization of the optical excitation has to be considered to understand the optical spectra. Whereas the dipole–dipole approximation is not applicable to the strong coupling of the dimer chromophores, it has been used to estimate the remaining excitonic interactions in the reaction center. The strongest couplings are expected for the pairs $P_{\mathrm{L}} B_{\mathrm{L}}, B_{\mathrm{L}} H_{\mathrm{L}}, P_{\mathrm{M}} B_{\mathrm{M}}, B_{\mathrm{M}} H_{\mathrm{M}}$ due to favorable distances and orientational factors (Fig. 25.5; Table 25.1).

Starting from a system with full C_2-symmetry the excitations again can be classified as symmetric or antisymmetric. The lowest dimer band interacts

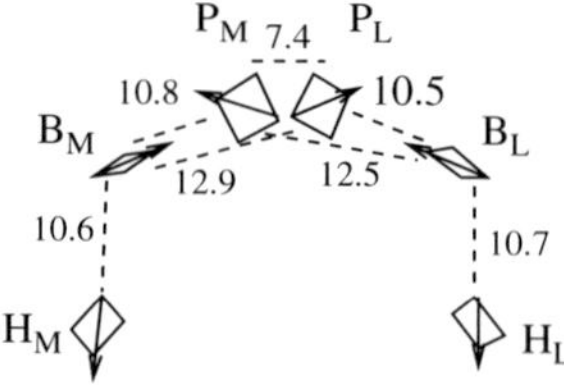

Fig. 25.5. Transition dipoles of the reaction center *Rps. viridis* [90–93]. *Arrows* show the transition dipoles of the isolated chromophores. Center–center distances (as defined by the nitrogen atoms) are given in Å

Table 25.1. Excitonic couplings for the reaction center of *Rps. viridis*

	H_M	B_M	P_M	P_L	B_L	H_L
H_M		160	33	−9	−11	6
B_M	160		−168	−37	29	−11
P_M	33	−168		770	−42	−7
P_L	−9	−37	770		−189	35
B_L	−11	29	−42	−189		167
H_L	6	−11	−7	35	167	

The matrix elements were calculated with the dipole–dipole approximation for $\mu^2 = 45\,\text{Debye}^2$ for bacteriochlorophyll and $30\,\text{Debye}^2$ for bacteriopheophytine [94]. All values in cm^{-1}

with the antisymmetric combinations of B and H excitations. Due to the differences of excitation energies, this leads only to a small amount of state mixing. For the symmetric states, the situation is quite different as the upper dimer band is close to the B excitation (Fig. 25.6).

If the symmetry is disturbed by structural differences or interactions with the protein, excitations of different symmetry character interact. Qualitatively, we expect that the lowest excitation is essentially the lower dimer band, the highest band[2] reflects absorption from the pheophytines and in the region of the B absorption, we expect mixtures of the B* excitations and the upper dimer band.

25.1.3 Circular Molecular Aggregates

We consider now a circular aggregate of N-chromophores as it is found in the light-harvesting complexes of photosynthesis [96–98] (Figs. 25.7 and 25.8).

We align the C_N-symmetry axis along the z-axis. The position of the nth molecule is

$$\mathbf{R}_n = R \begin{pmatrix} \cos(2\pi\,n/N) \\ \sin(2\pi\,n/N) \\ 0 \end{pmatrix}, \quad n = 0, 1, \ldots, N-1, \qquad (25.17)$$

---

[2] in the Q_y region

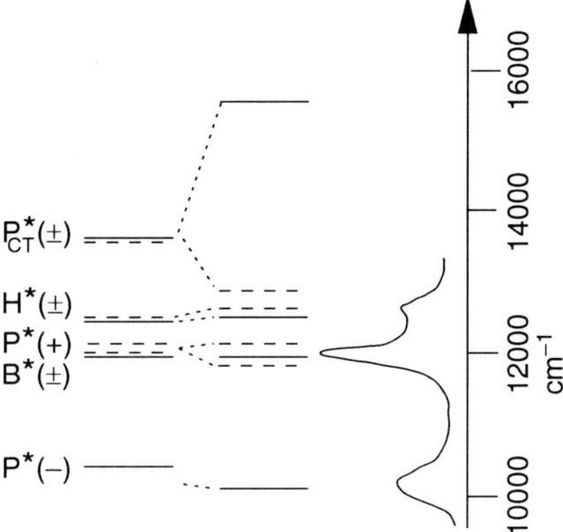

Fig. 25.6. Excitations of the reaction center. The absorption spectrum of *Rps. viridis* in the Q_y region is assigned on the basis of symmetric excitonic excitations [95]. The charge transfer states P^*_{CT} are very broad and cannot be observed directly

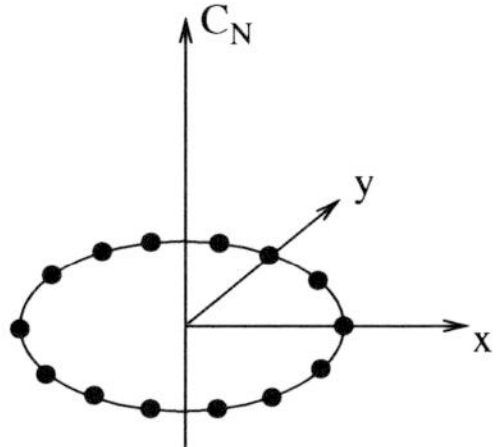

Fig. 25.7. Circular aggregate

which can also be written with the help of a rotation matrix

$$S_N = \begin{pmatrix} \cos(2\pi/N) & -\sin(2\pi/N) & \\ \sin(2\pi/N) & \cos(2\pi/N) & \\ & & 1 \end{pmatrix} \tag{25.18}$$

as

$$\mathbf{R}_n = S_N^n \mathbf{R}_0, \quad \mathbf{R}_0 = \begin{pmatrix} 1 \\ 0 \\ 0 \end{pmatrix}. \tag{25.19}$$

Similarly, the transition dipoles are given by

$$\boldsymbol{\mu}_n = S_N^n \boldsymbol{\mu}_0. \tag{25.20}$$

The component parallel to the symmetry axis is the same for all monomers

$$\mu_{n,z} = \mu_\| \tag{25.21}$$

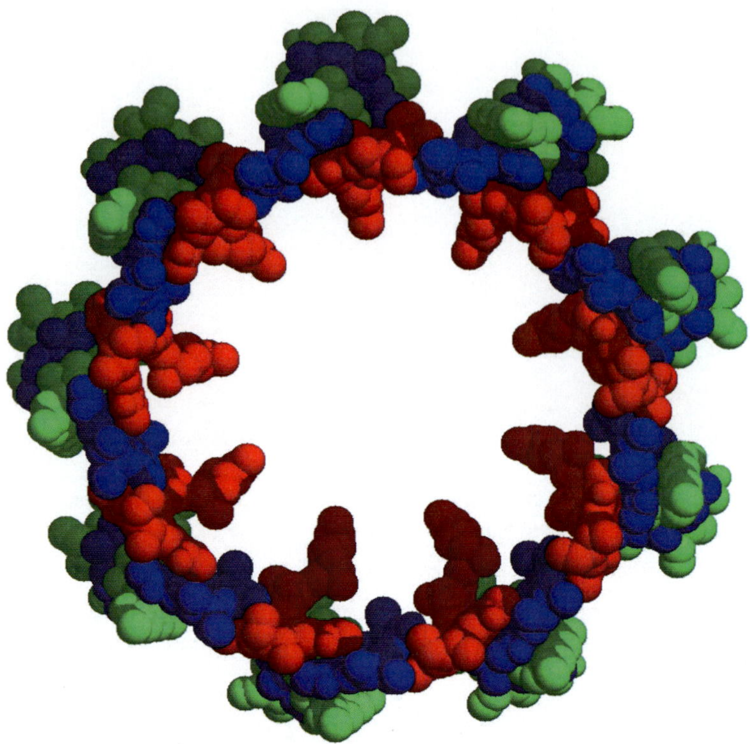

Fig. 25.8. Light-harvesting complex. The light-harvesting complex from *Rhodopseu-domonas acidophila* (structure 1kzu [99–103] from the protein data bank [104, 105]) consists of a ring of 18 closely coupled (shown in *red and blue*) and another ring of 9 less strongly coupled bacteriochlorophyll molecules (*green*). Rasmol graphics [106]

whereas for the component in the perpendicular plane

$$\begin{pmatrix} \mu_{n,x} \\ \mu_{n,y} \end{pmatrix} = \begin{pmatrix} \cos(n2\pi/N) & -\sin(n2\pi/N) \\ \sin(n2\pi/N) & \cos(n2\pi/N) \end{pmatrix} \begin{pmatrix} \mu_{0,x} \\ \mu_{0,y} \end{pmatrix}. \tag{25.22}$$

We describe the orientation of $\boldsymbol{\mu}_0$ in the $x - y$ plane by the angle ϕ:

$$\begin{pmatrix} \mu_{0,x} \\ \mu_{0,y} \end{pmatrix} = \mu_\perp \begin{pmatrix} \cos(\phi) \\ \sin(\phi) \end{pmatrix}. \tag{25.23}$$

Then we have (Fig. 25.9)

$$\begin{pmatrix} \mu_{n,x} \\ \mu_{n,y} \\ \mu_{n,z} \end{pmatrix} = \begin{pmatrix} \mu_\perp \cos(\phi + n2\pi/N) \\ \mu_\perp \cos(\phi + n2\pi/N) \\ \mu_\| \end{pmatrix}. \tag{25.24}$$

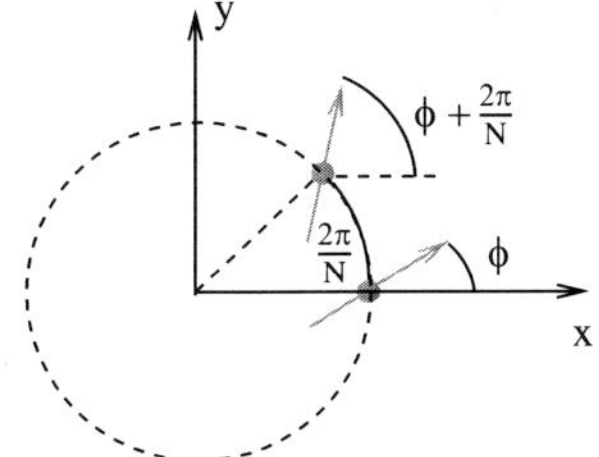

Fig. 25.9. Orientation of the transition dipoles

We denote the local excitation of the nth molecule by

$$|n\rangle = |A_0 A_2 \cdots A_n^* \cdots A_{N-1}\rangle. \tag{25.25}$$

Due to the symmetry of the system, the excitonic interaction is invariant against the S_N rotation and therefore

$$\langle m|V|n\rangle = \langle m-n|V|0\rangle = \langle 0|V|n-m\rangle. \tag{25.26}$$

Without a magnetic field, the coupling matrix elements can be chosen real and depend only on $|m-n|$. Within the dipole–dipole approximation, we have furthermore[3]

$$V_{|m-n|} = \langle m|V|n\rangle \tag{25.27}$$

$$= \frac{e^2}{4\pi\varepsilon |R_{mn}|^5}\left(|R_{mn}|^2 \boldsymbol{\mu}_m \boldsymbol{\mu}_n - 3(\mathbf{R}_{mn}\boldsymbol{\mu}_m)(\mathbf{R}_{mn}\boldsymbol{\mu}_n)\right). \tag{25.28}$$

The interaction matrix has the form (Table 25.2)

$$H = \begin{pmatrix} E_0 & V_1 & V_2 & \cdots & V_2 & V_1 \\ V_1 & E_1 & V_1 & \cdots & V_3 & V_2 \\ V_2 & V_1 & \ddots & & & \vdots \\ \vdots & \vdots & & & & V_2 \\ V_2 & V_3 & & & & V_1 \\ V_1 & V_2 & \cdots & V_2 & V_1 & E_{N-1} \end{pmatrix}. \tag{25.29}$$

The excitonic wave functions are easily constructed as

$$|k\rangle = \frac{1}{\sqrt{N}} \sum_{n=0}^{N-1} e^{ikn}|n\rangle \tag{25.30}$$

---

[3] A more realistic description based on a semiempirical INDO/S method is given by [107].

Table 25.2. Excitonic couplings

$|m - n|$	$V_{|m-n|}\ (\mathrm{cm}^{-1})$
1	-352.5
2	-43.4
3	-12.6
4	-5.1
5	-2.6
6	-1.5
7	-0.9
8	-0.7
9	-0.6

The matrix elements are calculated with the dipole–dipole approximation (25.27) for $R = 26.5\,\text{Å}$, $\phi_0 = 66°$, and $\mu^2 = 37\,\mathrm{Debye}^2$

with

$$k = \frac{2\pi}{N}l, \quad l = 0, 1, \ldots, N - 1, \tag{25.31}$$

$$\langle k'|H|k\rangle = \frac{1}{N} \sum_{n'=0}^{N-1} \sum_{n=0}^{N-1} \mathrm{e}^{\mathrm{i}kn} \mathrm{e}^{-\mathrm{i}k'n'} \langle n'|H|n\rangle$$

$$= \frac{1}{N} \sum_{n'=0}^{N-1} \sum_{n=0}^{N-1} \mathrm{e}^{\mathrm{i}kn} \mathrm{e}^{-\mathrm{i}k'n'} H_{|n-n'|}. \tag{25.32}$$

We substitute

$$m = n - n' \tag{25.33}$$

to get

$$\langle k'|H|k\rangle = \frac{1}{N} \sum_{n=0}^{N-1} \sum_{m=n}^{n-N+1} \mathrm{e}^{\mathrm{i}(k-k')n + \mathrm{i}k'm} H_{|m|}$$

$$= \delta_{k,k'} \sum_{m=0}^{N-1} \mathrm{e}^{\mathrm{i}k'm} H_{|m|}$$

$$= \delta_{k,k'} \left(E_0 + 2V_1 \cos k + 2V_2 \cos 2k + \cdots\right). \tag{25.34}$$

For even N, the lowest and highest states are not degenerate whereas for all of the other states (Fig. 25.10)

$$E_k = E_{N-k} = E_{-k}. \tag{25.35}$$

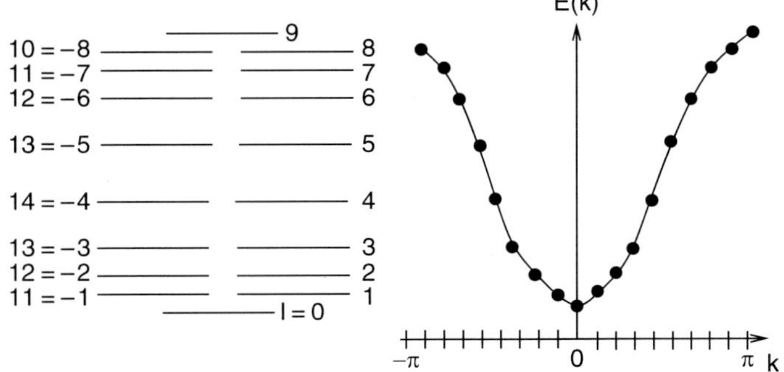

Fig. 25.10. Exciton dispersion relation. The figure shows the case of negative excitonic interaction for which the $k = 0$ state is the lowest in energy

The transition dipoles of the k-states are given by

$$
\boldsymbol{\mu}_k = \frac{1}{\sqrt{N}} \sum_{n=0}^{N-1} \mathrm{e}^{\mathrm{i}kn} \boldsymbol{\mu}_n = \frac{1}{\sqrt{N}} \sum_{n=0}^{N-1} \mathrm{e}^{\mathrm{i}kn} S_N^n \boldsymbol{\mu}_0
$$

$$
= \frac{1}{\sqrt{N}} \sum_{n=0}^{N-1} \mathrm{e}^{\mathrm{i}kn}
\begin{pmatrix}
\mu_x \cos(2\pi\, n/N) + \mu_y \sin(2\pi\, n/N) \\
\mu_y \cos(2\pi\, n/N) - \mu_x \sin(2\pi\, n/N) \\
\mu_z
\end{pmatrix}. \quad (25.36)
$$

For the z-component, we have (Fig. 25.11)

$$
\frac{1}{\sqrt{N}} \mu_z \sum_{n=0}^{N-1} \mathrm{e}^{\mathrm{i}kn} = \sqrt{N} \mu_z \delta_{k,0}. \quad (25.37)
$$

For the component in the x, y-plane, we introduce complex polarization vectors corresponding to circular polarized light:

$$
\mu_{k,\pm} = \begin{pmatrix} \frac{1}{\sqrt{2}} & \frac{\pm\mathrm{i}}{\sqrt{2}} & 0 \end{pmatrix} \boldsymbol{\mu}_k
$$

$$
= \frac{1}{\sqrt{N}} \sum_{n=0}^{N-1} \mathrm{e}^{\mathrm{i}kn} \begin{pmatrix} \frac{1}{\sqrt{2}} & \pm\frac{\mathrm{i}}{\sqrt{2}} & 0 \end{pmatrix}
\begin{pmatrix}
\cos(\frac{2\pi n}{N}) & -\sin(\frac{2\pi n}{N}) & \\
\sin(\frac{2\pi n}{N}) & \cos(\frac{2\pi n}{N}) & \\
& & 1
\end{pmatrix}
\begin{pmatrix} \mu_x \\ \mu_y \\ \mu_z \end{pmatrix}
$$

$$
= \frac{1}{\sqrt{N}} \sum_{n=0}^{N-1} \mathrm{e}^{\mathrm{i}kn} \begin{pmatrix} \frac{1}{\sqrt{2}}\mathrm{e}^{\pm\mathrm{i}2n\pi/N} & \pm\frac{\mathrm{i}}{\sqrt{2}}\mathrm{e}^{\pm\mathrm{i}2n\pi/N} & 0 \end{pmatrix}
\begin{pmatrix} \mu_x \\ \mu_y \\ \mu_z \end{pmatrix}
$$

$$
= \frac{1}{\sqrt{N}} \sum_{n=0}^{N-1} \mathrm{e}^{\mathrm{i}(k\pm 2\pi/N)} \begin{pmatrix} \frac{1}{\sqrt{2}} & \frac{\pm\mathrm{i}}{\sqrt{2}} \end{pmatrix}
\begin{pmatrix} \mu_x \\ \mu_y \end{pmatrix}
$$

$$
= \sqrt{N} \delta_{k,\mp 2\pi/N} \mu_{\pm} \quad \text{with} \quad \mu_{\pm} = \frac{1}{\sqrt{2}}(\mu_x \pm \mathrm{i}\mu_y) = \frac{1}{\sqrt{2}}\mu_\perp \mathrm{e}^{\pm\mathrm{i}\phi}. \quad (25.38)
$$

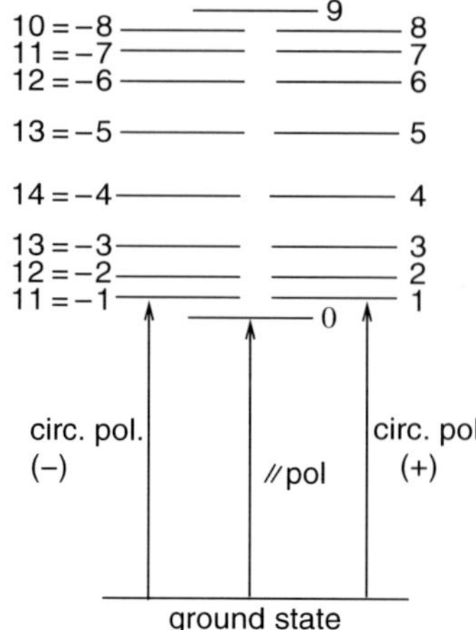

Fig. 25.11. Allowed optical transitions. The figure shows the case of negative excitonic interaction where the lowest three exciton states carry intensity

Linear polarized light with a polarization vector

$$\left(\cos\Theta \; \sin\Theta \; 0\right) = \frac{1}{\sqrt{2}}e^{i\Theta}\left(\tfrac{1}{\sqrt{2}} \; \tfrac{-i}{\sqrt{2}} \; 0\right) + \frac{1}{\sqrt{2}}e^{-i\Theta}\left(\tfrac{1}{\sqrt{2}} \; \tfrac{i}{\sqrt{2}} \; 0\right) \qquad (25.39)$$

couples to a linear combination of the $k = \pm 2\pi/N$ states since

$$\mu_{k,\Theta} = \sqrt{\frac{N}{2}}\mu_{\perp}\left(e^{i(\Phi-\Theta)}\delta_{k,-2\pi/N} + e^{-i(\Phi-\Theta)}\delta_{k,2\pi/N}\right). \qquad (25.40)$$

The absorption strength does not depend on the polarization angle Θ

$$|\mu_{-2\pi/N,\Theta}|^2 + |\mu_{2\pi/N,\Theta}|^2 = N\mu_{\perp}^2. \qquad (25.41)$$

25.1.4 Dimerized Systems of LH2

The light-harvesting complex LH2 of *Rps. acidophila* contains a ring of nine weakly coupled chlorophylls and another ring of nine stronger coupled chlorophyll dimers. The two units forming a dimer will be denoted as α, β and the number of dimers as N (Fig. 25.12).
The transition dipole moments are

$$\boldsymbol{\mu}_{n,\alpha} = \mu\begin{pmatrix}\sin\theta_\alpha\cos\left(n\frac{2\pi}{N}-\nu+\phi_\alpha\right)\\ \sin\theta_\alpha\sin\left(n\frac{2\pi}{N}-\nu+\phi_\alpha\right)\\ \cos\theta_\alpha\end{pmatrix} = \mu S_N^n R_z(-\nu+\phi_\alpha)\begin{pmatrix}\sin\theta_\alpha\\ 0\\ \cos\theta_\alpha\end{pmatrix},$$

$$\boldsymbol{\mu}_{n,\beta} = \mu\begin{pmatrix}\sin\theta_\beta\cos\left(n\frac{2\pi}{N}+\nu+\phi_\beta\right)\\ \sin\theta_\beta\sin\left(n\frac{2\pi}{N}+\nu+\phi_\beta\right)\\ \cos\theta_\beta\end{pmatrix} = \mu S_N^n R_z(+\nu+\phi_\beta)\begin{pmatrix}\sin\theta_\beta\\ 0\\ \cos\theta_\beta\end{pmatrix}.$$

$$(25.42)$$

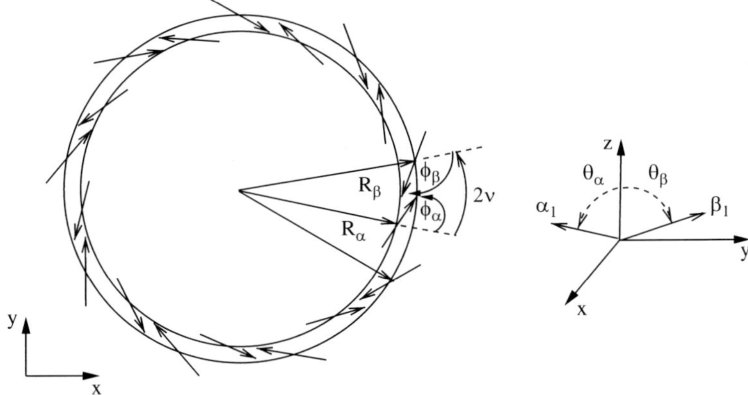

Fig. 25.12. BChl arrangement for the LH2 complex of *Rps. acidophila* [99–103]. *Arrows* represent the transition dipole moments of the BChl molecules as calculated on the 631G/HF-CI level. Distances are with respect to the center of the four nitrogen atoms. $R_\alpha = 26.1$ Å, $R_\beta = 26.9$ Å, $2\nu = 20.7°$, $\phi_\alpha = 70.5°$, $\phi_\beta = -117.7°$, $\theta_\alpha = 83.7°$, and $\theta_\beta = 80.3°$

For a circle with C_{2N}-symmetrical positions ($2\nu = \pi/N, \theta_\alpha = \theta_\beta$) and alternating transition dipole directions ($\phi_\beta = \pi + \phi_\alpha$), we find the relation

$$\boldsymbol{\mu}_{n,\beta} = R_z(2\nu + \phi_\beta - \phi_\alpha)\boldsymbol{\mu}_{n,\alpha} = R_z\left(\frac{\pi}{N} + \pi\right)\boldsymbol{\mu}_{n,\alpha}. \tag{25.43}$$

The experimental value of

$$2\nu + \phi_\beta - \phi_\alpha = 208.9° \tag{25.44}$$

is in fact very close to the value of $180° + 20°$ (or $\pi + (\pi/N)$).

We have to distinguish the following interaction matrix elements between two monomers in different unit cells (Figs. 25.13; Table 25.3):

$$V_{\alpha,\alpha,|m|}, \ V_{\alpha,\beta,m} = V_{\beta,\alpha,-m}, \ V_{\alpha,\beta,-m} = V_{\beta,\alpha,m}, \ V_{\beta,\beta,|m|}. \tag{25.45}$$

The interaction matrix of one dimer is

$$\begin{pmatrix} E_\alpha & V_{\mathrm{dim}} \\ V_{\mathrm{dim}} & E_\beta \end{pmatrix}. \tag{25.46}$$

The wave function has to be generalized as

$$|k, s\rangle = \frac{1}{\sqrt{N}} \sum_{n=0}^{N-1} e^{ikn}(C_{s\alpha}|n\alpha\rangle + C_{s\beta}|n\beta\rangle), \tag{25.47}$$

$$\langle k's'|H|ks\rangle = \frac{1}{N} \sum_{n=0}^{N-1} \sum_{n'=0}^{N-1} e^{i(kn-k'n')} \left(C_{s'\alpha}C_{s\alpha}H_{\alpha\alpha|n-n'|} \right.$$

$$\left. + C_{s'\beta}C_{s\beta}H_{\beta\beta|n-n'|} + C_{s'\alpha}C_{s\beta}H_{\alpha\beta(n-n')} + C_{s'\beta}C_{s\alpha}H_{\beta\alpha(n-n')} \right),$$

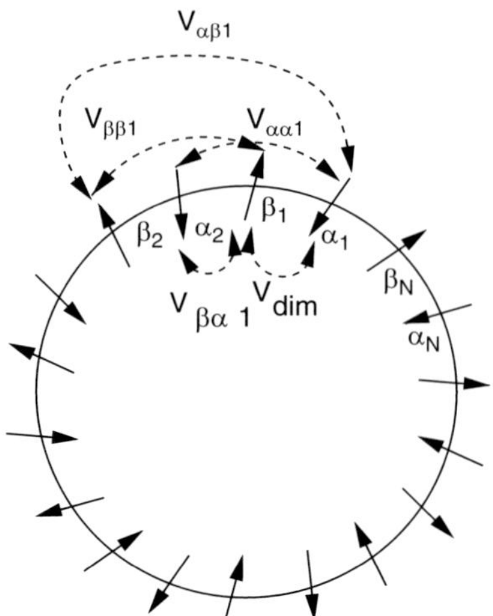

Fig. 25.13. Couplings in the dimerized ring

Table 25.3. Excitonic couplings for the LH2 complex of *Rps. acidophila*

n	$V_{\alpha\beta n}$	$V_{\beta\alpha n}$	$V_{\alpha\alpha n}$	$V_{\beta\beta n}$
0	356.1	356.1		
1	13.3	324.5	-49.8	-35.0
2	2.8	11.8	-6.1	-4.0
3	1.0	2.4	-1.8	-1.0
4	0.7	0.9	-0.9	-0.4

The matrix elements were calculated within the dipole–dipole approximation for the structure shown in Fig. 25.12 for $\mu^2 = 37\,\text{Debye}^2$. All values in cm^{-1}

$$= \frac{1}{N} \left(C_{s'\alpha}\ C_{s'\beta} \right) \sum_{n=0}^{N-1} \sum_{m=0}^{N-1} e^{i(k-k')n+ik'm} \begin{pmatrix} H_{\alpha\alpha|m|} & H_{\alpha\beta m} \\ H_{\beta\alpha m} & H_{\beta\beta|m|} \end{pmatrix} \begin{pmatrix} C_{s\alpha} \\ C_{s\beta} \end{pmatrix},$$

$$= \delta_{k,k'} \left(C_{s'\alpha}\ C_{s'\beta} \right) \begin{pmatrix} \sum_{m=0}^{N-1} e^{ikm} H_{\alpha\alpha|m|} & \sum_{m=0}^{N-1} e^{ikm} H_{\alpha\beta m} \\ \sum_{m=0}^{N-1} e^{ikm} H_{\beta\alpha m} & \sum_{m=0}^{N-1} e^{ikm} H_{\beta\beta m} \end{pmatrix} \begin{pmatrix} C_{s\alpha} \\ C_{s\beta} \end{pmatrix}.$$

$$(25.48)$$

The coefficients $C_{s\alpha}$ and $C_{s\beta}$ are determined by diagonalization of the matrix:

$$H_k = \begin{pmatrix} E_\alpha + 2V_{\alpha\alpha,1}\cos k + \cdots & V_{\text{dim}} + e^{ik}V_{\alpha\beta,1} + e^{-ik}V_{\alpha\beta,-1} + \cdots \\ V_{\text{dim}} + e^{-ik}V_{\alpha\beta,1} + e^{ik}V_{\alpha\beta,-1} + \cdots & E_\alpha + 2V_{\beta\beta,1}\cos k + \cdots \end{pmatrix}.$$

$$(25.49)$$

If we consider only interactions between nearest neighbors, this simplifies to

$$H_k = \begin{pmatrix} E_\alpha & V_{\text{dim}} + e^{-ik}W \\ V_{\text{dim}} + e^{ik}W & E_\beta \end{pmatrix} \quad \text{with} \quad W = V_{\alpha\beta,-1}. \tag{25.50}$$

In the following, we discuss the limit $E_\alpha = E_\beta$. The general case is discussed in the "Problems section." The eigenvectors of

$$H_k = \begin{pmatrix} E_\alpha & V_{\text{dim}} + e^{-ik}W \\ V_{\text{dim}} + e^{ik}W & E_\alpha \end{pmatrix}$$

are given by

$$\frac{1}{\sqrt{2}} \begin{pmatrix} 1 \\ \pm e^{i\chi} \end{pmatrix}, \tag{25.51}$$

where the angle χ is chosen such that

$$V + e^{ik}W = U(k)e^{i\chi} \tag{25.52}$$

with

$$U(k) = \text{sign}(V)|V + e^{ik}W|$$

and the eigenvalues are

$$E_{k,\pm} = E_\alpha \pm U(k) = E_\alpha \pm \text{sign}(V)\sqrt{V_{\text{dim}}^2 + W^2 + 2V_{\text{dim}}W\cos k}. \tag{25.53}$$

The transition dipoles follow from

$$\begin{aligned}
\boldsymbol{\mu}_{k,\pm} &= \frac{1}{\sqrt{2N}} \sum_{n=0}^{N-1} e^{ikn}(\boldsymbol{\mu}_{n,\alpha} \pm e^{i\chi}\boldsymbol{\mu}_{n,\beta}) \\
&= \frac{\mu}{\sqrt{2N}} \sum_{n=0}^{N-1} e^{ikn} S_N^n \\
&\quad \times \left(\begin{pmatrix} \sin\theta_\alpha \cos(-\nu + \phi_\alpha) \\ \sin\theta_\alpha \sin(-\nu + \phi_\alpha) \\ \cos\theta_\alpha \end{pmatrix} \pm e^{i\chi} \begin{pmatrix} \sin\theta_\beta \cos(\nu + \phi_\beta) \\ \sin\theta_\beta \sin(\nu + \phi_\beta) \\ \cos\theta_\beta \end{pmatrix} \right).
\end{aligned} \tag{25.54}$$

Similar selection rules as for the simple ring system follow for the first factor

$$\begin{aligned}
(0\ 0\ 1)\,\boldsymbol{\mu}_{k,\pm} &= \frac{\mu}{\sqrt{2N}} \sum_{n=0}^{N-1} e^{ikn}\left(\cos\theta_\alpha \pm e^{i\chi}\cos\theta_\beta\right) \\
&= \delta_{k,0}\,\mu\sqrt{\frac{N}{2}}\left(\cos\theta_\alpha \pm \cos\theta_\beta\right), \tag{25.55}
\end{aligned}$$

$$\left(\tfrac{1}{\sqrt{2}}\ \tfrac{i}{\sqrt{2}}\ 0\right)\boldsymbol{\mu}_{k,\pm} = \frac{1}{\sqrt{2N}} \sum_{n=0}^{N-1} e^{i(k\pm 2\pi/N)n}\left(1\ i\ 0\right)\left(\boldsymbol{\mu}_{\alpha,1} \pm e^{i\chi}\boldsymbol{\mu}_{\beta,1}\right)$$

$$= \sqrt{\frac{N}{2}}\delta_{k,\mp 2\pi/N}\mu\left(\sin\theta_\alpha e^{i(\phi_\alpha-\nu)} \pm e^{i\chi}\sin\theta_\beta e^{i(\phi,M_\beta+\nu)}\right),$$

$$(25.56)$$

$$\left(\tfrac{1}{\sqrt{2}}\ \tfrac{-i}{\sqrt{2}}\ 0\right)\boldsymbol{\mu}_{k,\pm} = \frac{1}{\sqrt{2N}}\sum_{n=0}^{N-1} e^{i(k\pm 2\pi/N)n}\left(1\ -i\ 0\right)\left(\boldsymbol{\mu}_{\alpha,1} \pm e^{i\chi}\boldsymbol{\mu}_{\beta,1}\right)$$

$$0 = \sqrt{\frac{N}{2}}\delta_{k,\mp 2\pi/N}\mu\left(\sin\theta_\alpha e^{i(\phi_\alpha-\nu)} \mp e^{i\chi}\sin\theta_\beta e^{i(\phi_\beta+\nu)}\right).$$

$$(25.57)$$

The second factor determines the distribution of intensity among the $+$ and $-$ states.

In the limit of $2N$-equivalent molecules, $V = W$, $2\nu + \phi_\beta - \phi_\alpha = \pi + (\pi/N)$, and $\theta_\alpha = \theta_\beta$ and we have

$$V + e^{ik}W = (1 + e^{ik})V = e^{ik/2}(e^{-ik/2} + e^{+ik/2})V = e^{ik/2}2V\cos k/2$$

$$(25.58)$$

hence

$$\chi = k/2, \quad U(k) = 2V\cos k/2, \tag{25.59}$$

and (25.56) becomes

$$\left(\frac{1}{\sqrt{2}}\ \frac{i}{\sqrt{2}}\ 0\right)\mu_{k,\pm} = \sqrt{\frac{N}{2}}\delta_{k,\mp 2\pi/N}\mu\sin\theta e^{i(\phi_\alpha-\nu)}\left(1\mp e^{i(k/2+\pi/N)}\right),$$

which is zero for the upper case ($+$ states) and

$$\sqrt{2N}\delta_{k,2\pi/N}\mu\sin\theta e^{i(\phi_\alpha-\nu)} \tag{25.60}$$

for the ($-$) states. Similarly (25.57) becomes

$$\frac{1}{\sqrt{2}}(\mu_{k,x} - i\mu_{k,y}) = \sqrt{2N}\delta_{k,+2\pi/N}\mu\sin\theta e^{-i(\phi_\alpha-\nu)} \tag{25.61}$$

for the $-$ states and zero for the $+$ states. (For positive V, the $+$ states are higher in energy than the $-$ states.) For the z-component, the selection rule of $k = 0$ implies (Fig. 25.14)

$$\mu_{z,+} = \sqrt{2N}\delta_{k,0}\mu\cos\theta, \quad \mu_{z,-} = 0. \tag{25.62}$$

In the LH2 complex, the transition dipoles of the two dimer halves are nearly antiparallel. The oscillator strengths of the N molecules are concentrated in the degenerate next to lowest transitions $k = \pm 2\pi/N$. This means that the lifetime of the optically allowed states will be reduced compared to the radiative lifetime of a monomer. In the LH2 complex, the transition dipoles have only a very small z-component. Therefore in a perfectly

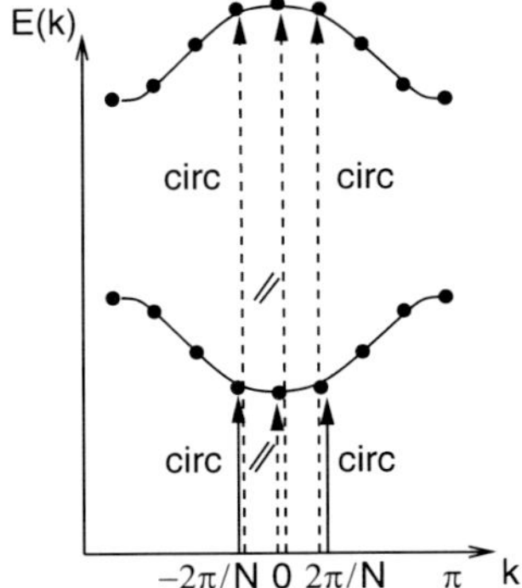

Fig. 25.14. Dispersion relation of the dimer ring

symmetric structure, the lowest $(k = 0)$ state is almost forbidden and has a longer lifetime than the optically allowed $k = \pm 2\pi/N$ states. Due to the degeneracy of the $k = \pm 2\pi/N$ states, the absorption of photons coming along the symmetry axis does not depend on the polarization.

25.2 Influence of Disorder

So far, we considered a perfectly symmetrical arrangement of the chromophores. In reality, there exist deviations due to the protein environment and to low-frequency nuclear motion which leads to variations of the site energies, the coupling matrix elements, and the transition dipoles. Experimentally, a pair of mutually perpendicular states has been observed which were assigned to the strong $k = \pm 2\pi/N$ states [96, 108]. This can be only understood if the degeneracy is lifted by some symmetry-breaking perturbation which gives as zero-order eigenstates linear combinations of the $k = \pm 2\pi/N$ states (25.40). In the following, we neglect the effects of dimerization and study a ring of N-equivalent chromophores.

25.2.1 Symmetry-Breaking Local Perturbation

We consider perturbation of the symmetry by a specific local interaction, for instance due to an additional hydrogen bond at one site. We assume that the excitation energy at the site $n_0 = 0$ is modified by a small amount δE. The Hamiltonian

$$H = \sum_{n=0}^{N-1} |n\rangle E_0 \langle n| + |0\rangle \delta E \langle 0| + \sum_{n=0}^{N-1} \sum_{n'=0, n'\neq n}^{N-1} |n\rangle V_{nn'} \langle n'| \tag{25.63}$$

is transformed to the basis of k-states (25.30) to give

$$\langle k'|H|k\rangle = \frac{1}{N}\sum_{n,n'} e^{-ik'n'+ikn}\langle n'|H|n\rangle$$

$$= \frac{1}{N}\sum_{n} e^{i(k-k')n} E_0 + \frac{\delta E}{N} + \frac{1}{N}\sum_{n}\sum_{n'\neq n} e^{-ik'n'+ikn} V_{nn'}$$

$$= \delta_{k,k'} E_k + \frac{\delta E}{N}. \tag{25.64}$$

Obviously, the energy of all k-states is shifted by the same amount:

$$\langle k|H|k\rangle = E_k + \frac{\delta E}{N}. \tag{25.65}$$

The degeneracy of the pairs $|\pm k\rangle$ is removed due to the interaction

$$\langle k|H|-k\rangle = \frac{\delta E}{N}. \tag{25.66}$$

The zero-order eigenstates are the linear combinations

$$|k_\pm\rangle = \frac{1}{\sqrt{2}}(|k\rangle \pm |-k\rangle) \tag{25.67}$$

with zero-order energies (Fig. 25.15)

$$E(k) = E_k + \frac{\delta E}{N}\quad \text{(nondegenerate states)},$$

$$E(k_-) = E_k, \quad E(k_+) = E_k + 2\frac{\delta E}{N}\quad \text{(degenerate states)}. \tag{25.68}$$

Only the symmetric states are affected by the perturbation. The interaction between degenerate pairs is given by the matrix elements

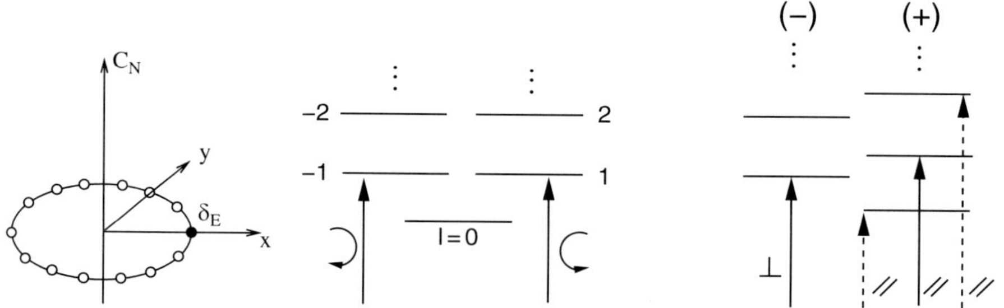

Fig. 25.15. Local perturbation of the symmetry. This figure shows the case of $\delta E > 0$

$$\langle k'_+|H|k_+\rangle = \delta_{k,k'}\,E_k + 2\frac{\delta E}{N},$$

$$\langle k'_-|H|k_-\rangle = \delta_{k,k'}\,E_k,$$

$$\langle k'_+|H|k_-\rangle = 0, \tag{25.69}$$

and the coupling to the nondegenerate $k = 0$ state is

$$\langle 0|H|k_+\rangle = \frac{\sqrt{2}\delta E}{N},$$

$$\langle 0|H|k_-\rangle = 0. \tag{25.70}$$

Optical transitions to the $|1_\pm\rangle$ states are linearly polarized with respect to perpendicular axes. First-order intensity is transferred from the $|1_+\rangle$ state to the $|0\rangle$ and the other $|k_+\rangle$ states and hence the $|1_-\rangle$ state absorbs stronger than the $|1_+\rangle$ state which is approximately

$$|1_+\rangle + \frac{\sqrt{2}\delta E}{N(E_0 - E_1)}|0\rangle + \frac{2\delta E}{N(E_2 - E_1)}|2_+\rangle + \cdots \tag{25.71}$$

Its intensity is reduced by a factor of

$$\frac{1}{1 + \frac{\delta E^2}{N^2}\left(\frac{2}{(E_0 - E_1)^2} + \frac{4}{(E_2 - E_1)^2} + \cdots\right)}. \tag{25.72}$$

25.2.2 Periodic Modulation

The local perturbation (25.64) has Fourier components

$$H'(\Delta k) = \langle k|H'|k + \Delta k\rangle = \frac{1}{N}\sum_n e^{in\Delta k}\delta E_n \tag{25.73}$$

for all possible values of Δk. In this section, we discuss the most important Fourier components separately. These are the $\Delta k = \pm 4\pi/N$ components which mix the $k = \pm 2\pi/N$ states and the $\Delta k = \pm 2\pi/N$ components which couple the $k = \pm 2\pi/N$ to the $k = 0, \pm 4\pi/N$ states and thus redistribute the intensity of the allowed states.

A general modulation of the diagonal energies with Fourier components $\Delta k = \pm 2\pi/N$ is given by

$$\delta E_n = \delta E \cos(\chi_0 + 2n\pi/N) \tag{25.74}$$

and its matrix elements are

$$H'(\Delta k) = \frac{e^{i\chi_0}\delta E}{2}\delta_{\Delta k,-2\pi/N} + \frac{e^{-i\chi_0}\delta E}{2}\delta_{\Delta k,2\pi/N}. \tag{25.75}$$

Transformation to linear combinations of the degenerate pairs

$$|k, +\rangle = \frac{1}{\sqrt{2}} \left(e^{i\chi_0}| - k\rangle + e^{-i\chi_0}|k\rangle \right) \quad k > 0,$$

$$|k, -\rangle = \frac{1}{\sqrt{2}} \left(-e^{-i\chi_0}| - k\rangle + e^{i\chi_0}|k\rangle \right) \quad k > 0 \tag{25.76}$$

leads to two decoupled sets of states with

$$\langle 0|H'|k, +\rangle = \frac{\delta E}{\sqrt{2}} \cos(2\chi_0)\delta_{k, 2\pi/N},$$

$$\langle 0|H'|k, -\rangle = 0,$$

$$\langle k, -|H'|k', +\rangle = 0,$$

$$\langle k, -|H'|k', -\rangle = \frac{\delta E}{2} \cos\chi_0(\delta_{k'-k, 2\pi/N} + \delta_{k'-k, -2\pi/N}),$$

$$\langle k, +|H'|k', +\rangle = \frac{\delta E}{2} \cos(3\chi_0)(\delta_{k'-k, 2\pi/N} + \delta_{k'-k, -2\pi/N}).$$

The $|k = 2\pi/N, \pm\rangle$ states are linearly polarized (Fig. 25.16)

$$\mu_{2\pi/N, +, \Theta} = \frac{e^{i\chi_0}}{\sqrt{2}}\mu_{-2\pi/N, \Theta} + \frac{e^{i\chi_0}}{\sqrt{2}}\mu_{2\pi/N, \Theta} = \sqrt{N}\mu_\perp \cos(\Phi + \chi_0 - \Theta),$$

$$\mu_{2\pi/N, -, \Theta} = -\frac{e^{-i\chi_0}}{\sqrt{2}}\mu_{-2\pi/N, \Theta} + \frac{e^{i\chi_0}}{\sqrt{2}}\mu_{2\pi/N, \Theta} = i\sqrt{N}\mu_\perp \sin(\Phi + \chi_0 - \Theta).$$

$$\tag{25.77}$$

Let us now consider a C_2-symmetric modulation [109]

$$\delta E_n = \delta E \cos(\chi_0 + 4\pi n/N) \tag{25.78}$$

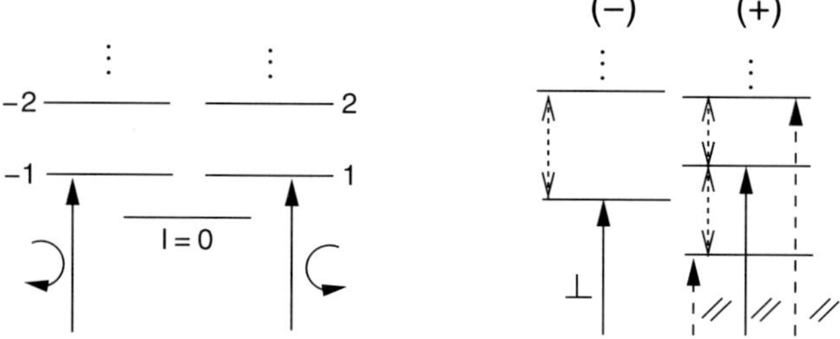

Fig. 25.16. C_1-symmetric perturbation. Modulation of the diagonal energies by the perturbation $\delta E_n = \delta E \cos(\chi_0 + 2\pi n/N)$ does not mix the $|k, +\rangle$ and $|k, -\rangle$ states. In first order, the allowed $k = \pm 2\pi/N$ states split into two states $|k = 2\pi/N, \pm\rangle$ with mutually perpendicular polarization and intensity is transferred to the $|k = 0\rangle$ and $|k = 4\pi/N, \pm\rangle$ states

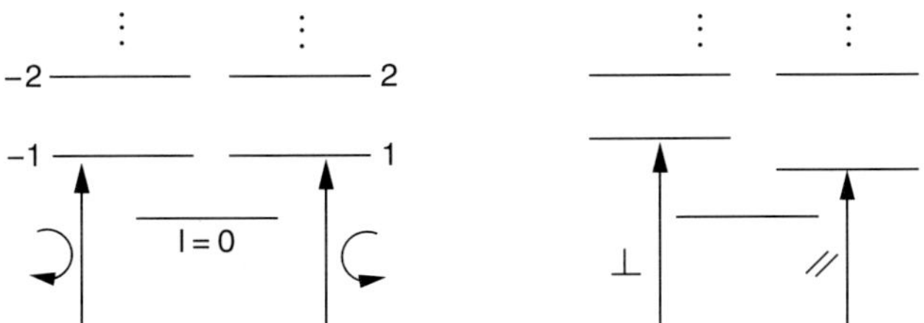

Fig. 25.17. C_2-symmetric perturbation. Modulation of the diagonal energies by the perturbation $\delta E_n = \delta E \cos(\chi_0 + 4\pi n/N)$ splits the $|k = \pm 2\pi/N\rangle$ states in zero order into a pair with equal intensity and mutually perpendicular polarization

with matrix elements

$$\langle k|H'|k'\rangle = \frac{\delta E}{2} \left(e^{i\chi_0}\delta_{k'-k,-4\pi/N} + e^{-i\chi_0}\delta_{k'-k,4\pi/N}\right). \tag{25.79}$$

The $|k = \pm 2\pi/N\rangle$ states are mixed to give the zero-order states

$$|2\pi/N, +\rangle = \frac{1}{\sqrt{2}} \left(|-2\pi/N\rangle + e^{i\chi_0}|2\pi/N\rangle\right),$$

$$|2\pi/N, -\rangle = \frac{1}{\sqrt{2}} \left(|-2\pi/N\rangle - e^{i\chi_0}|2\pi/N\rangle\right), \tag{25.80}$$

which are again linearly polarized (Fig. 25.17).

A perturbation of this kind could be due to an elliptical deformation of the ring [109], but also to interaction with a static electric field, which is given up to second order by

$$\delta E_n = -\mathbf{p}_n \mathbf{E} + \frac{1}{2}\mathbf{E}^t \alpha_n \mathbf{E}, \tag{25.81}$$

where the permanent dipole moments are given by an expression similar[4] to (25.24)

$$\mathbf{p}_n = S_N^n \mathbf{p}_0 = \begin{pmatrix} p_\perp \cos(\varPhi + n2\pi/N) \\ p_\perp \sin(\varPhi + n2\pi/N) \\ p_\parallel \end{pmatrix} \tag{25.82}$$

and the polarizabilities transform as

$$\alpha_n = S_N^n \alpha_0 S_N^{-n}. \tag{25.83}$$

With the field vector

$$\mathbf{E} = \begin{pmatrix} E_\perp \cos\xi \\ E_\perp \sin\xi \\ E_\parallel \end{pmatrix}, \tag{25.84}$$

---

[4] The angle $\varPhi$ is different for permanent and transition dipoles.

we find

$$- \mathbf{p}_n \mathbf{E} = -p_\perp E_\perp \cos(\Phi + n2\pi/N - \xi) \tag{25.85}$$

and

$$\begin{aligned}
\frac{1}{2}\mathbf{E}^t \alpha_n \mathbf{E} &= \frac{1}{2}(\mathbf{E}^t S_N^n)\alpha_0(S_N^{-n}\mathbf{E}) \\
&= \frac{E_\perp^2}{2}\left\{\alpha_{xx}\cos^2(n2\pi/N - \xi) + \alpha_{yy}\sin^2(n2\pi/N - \xi) \right. \\
&\quad + 2\alpha_{xy}\sin(n2\pi/N - \xi)\cos(n2\pi/N - \xi)\} \\
&\quad + E_| E_\perp\left\{\alpha_{xz}\cos(n2\pi/N - \xi) + \alpha_{yz}\sin(n2\pi/N - \xi)\right\} \\
&\quad + \frac{E_\|^2}{2}\alpha_{zz}.
\end{aligned} \tag{25.86}$$

The quadratic term has Fourier components $\Delta k = \pm 4\pi/N$ and can therefore act as an elliptic perturbation.

Let us assume that the polarizability is induced by the coupling to charge transfer states like in the special pair dimer. Due to the molecular arrangement, the polarizability will be zero along the C_N-symmetry axis and will have a maximum along a direction in the xy-plane. We denote the angle of this direction by η. The polarizability α_0 simplifies to

$$\alpha_0 = \begin{pmatrix} \alpha_\perp \cos^2\eta & \alpha_\perp \sin\eta\cos\eta & 0 \\ \alpha_\perp \sin\eta\cos\eta & \alpha_\perp \sin^2\eta & 0 \\ 0 & 0 & 0 \end{pmatrix} \tag{25.87}$$

and the second-order term

$$\begin{aligned}
\frac{1}{2}\mathbf{E}^t \alpha_n \mathbf{E} &= \frac{E_\perp^2 \alpha_\perp}{4}\left\{1 + \cos(2\eta - 2\xi + n4\pi/N) \right. \\
&\quad + \cos(-2\eta - 2\xi + n4\pi/N)\}
\end{aligned} \tag{25.88}$$

now has only Fourier components $\Delta k = 0, \pm 4\pi/N$.

25.2.3 Diagonal Disorder

Let us now consider a static distribution of site energies [110–112]. The Hamiltonian

$$H = \sum_{n=0}^{N-1} |n\rangle(E_0 + \delta E_n)\langle n| + \sum_{n=0}^{N-1}\sum_{n'=0, n'\neq n}^{N-1} |n\rangle V_{nn'}\langle n'| \tag{25.89}$$

contains energy shifts δE_n which are assumed to have a Gaussian distribution function:

$$P(\delta E_n) = \frac{1}{\Delta\sqrt{\pi}}\exp(-\delta E_n^2/\Delta^2). \tag{25.90}$$

Transforming to the delocalized states, the Hamiltonian becomes

$$\langle k'|H|k\rangle = \frac{1}{N}\sum_{n,n'} e^{-ik'n'+ikn}\langle n'|H|n\rangle$$

$$= \frac{1}{N}\sum_{n} e^{i(k-k')n}(E_0 + \delta E_n) + \frac{1}{N}\sum_{n}\sum_{n'\neq n} e^{-ik'n'+ikn}V_{nn'}$$

$$= \delta_{k,k'}E_k + \frac{1}{N}\sum_{n} e^{i(k-k')n}\delta E_n \tag{25.91}$$

and due to the disorder the k-states are mixed. The energy change of a non-degenerate state (for instance $k = 0$) is in lowest-order perturbation theory given by the diagonal matrix element

$$\delta E_k = \frac{1}{N}\sum_{n}\delta E_n \tag{25.92}$$

as the average of the local energy fluctuations. Obviously, the average is zero

$$\langle\delta E_k\rangle = \langle\delta E_n\rangle = 0 \tag{25.93}$$

and

$$\langle\delta E_k^2\rangle = \frac{1}{N^2}\sum_{n,n'}\langle\delta E_n\delta E_{n'}\rangle. \tag{25.94}$$

If the fluctuations of different sites are uncorrelated

$$\langle\delta E_n\delta E_{n'}\rangle = \delta_{n,n'}\langle\delta E^2\rangle, \tag{25.95}$$

the width of the k-state

$$\langle\delta E_k^2\rangle = \frac{\langle\delta E^2\rangle}{N}$$

is smaller by a factor $1/\sqrt{N}$.[5] If the fluctuations are fully correlated, on the other hand, the k-states have the same width as the site energies.

For the pairs of degenerate states $\pm k$, we have to consider the secular matrix ($\Delta k = k - k' = 2k$)

$$H_1 = \begin{pmatrix} \frac{1}{N}\sum_n \delta E_n & \frac{1}{N}\sum_n e^{in\Delta k}\delta E_n \\ \frac{1}{N}\sum_n e^{-in\Delta k}\delta E_n & \frac{1}{N}\sum_n \delta E_n \end{pmatrix}, \tag{25.96}$$

which has the eigenvalues

$$\delta E_{\pm k} = \frac{1}{N}\sum_{n}\delta E_n \pm \frac{1}{N}\sqrt{\sum_{n} e^{in\Delta k}\delta E_n \sum_{n'} e^{-in'\Delta k}\delta E_{n'}}. \tag{25.97}$$

---

[5] This is known as exchange or motional narrowing.

Obviously, the average is not affected

$$\left\langle \frac{\delta E_{+k} + \delta E_{-k}}{2} \right\rangle = 0 \tag{25.98}$$

and the width is given by

$$\left\langle \frac{\delta E_{+k}^2 + \delta E_{-k}^2}{2} \right\rangle = \frac{1}{N^2} \sum_{nn'} \langle \delta E_n \delta E_{n'} \rangle + \frac{1}{N^2} \sum_{nn'} e^{i(n-n')\Delta k} \langle \delta E_n \delta E_{n'} \rangle$$

$$= 2 \frac{\langle \delta E^2 \rangle}{N}. \tag{25.99}$$

25.2.4 Off-Diagonal Disorder

Consider now fluctuations also of the coupling matrix elements, for instance due to fluctuations of orientation and distances. The Hamiltonian contains an additional perturbation

$$H'_{kk'} = \frac{1}{N} \sum_{n'n} e^{-ik'n'+ikn} \delta V_{nn'}. \tag{25.100}$$

For uncorrelated fluctuations with the properties[6]

$$\langle \delta V_{nn'} \rangle = 0, \tag{25.101}$$

$$\langle \delta V_{nn'} \delta V_{mm'} \rangle = (\delta_{nm}\delta_{n'm'} + \delta_{nm'}\delta_{n'm} - \delta_{nm}\delta_{n'm'}\delta_{nn'})\langle \delta V_{|n-n'|}^2 \rangle, \tag{25.102}$$

it follows

$$\langle H'_{kk'} \rangle = 0 \tag{25.103}$$

$$\langle |H'_{kk'}|^2 \rangle = \frac{1}{N^2} \sum_{n'n} \sum_{m'm} e^{ik(n-m)+ik'(m'-n')} \langle \delta V_{nn'} \delta V_{mm'} \rangle$$

$$= \frac{1}{N^2} \sum_{n'n} \langle \delta V_{nn'}^2 \rangle + \frac{1}{N^2} \sum e^{i(k+k')(n-n')} \langle \delta V_{nn'}^2 \rangle$$

$$= \frac{\langle \delta E^2 \rangle}{N} + \frac{1}{N} \sum_{m \neq 0} \langle \delta V_{|m|}^2 \rangle \left[1 + \cos\left(m(k+k') \right) \right]. \tag{25.104}$$

If the dominant contributions come from the fluctuation of site energies and nearest neighbor couplings, this simplifies to

$$\langle |H'_{kk'}|^2 \rangle = \frac{1}{N} \left(\langle \delta E^2 \rangle + 2 \langle \delta V_{\pm 1}^2 \rangle (1 + \cos(k+k')) \right). \tag{25.105}$$

---

[6] We assume here that the fluctuation amplitudes obey the C_N-symmetry.

For the nondegenerate states, the width is

$$\langle \delta E_k^2 \rangle = \frac{\langle \delta E^2 \rangle + 4\langle \delta V_{\pm 1}^2 \rangle}{N} \tag{25.106}$$

and for the degenerate pairs the eigenvalues of the secular matrix become

$$\delta E_{\pm k} = H_{kk} \pm |H_{k,-k}| . \tag{25.107}$$

Again the average of the two is not affected

$$\left\langle \frac{\delta E_{+k} + \delta E_{-k}}{2} \right\rangle = 0, \tag{25.108}$$

whereas the width now is given by

$$\left\langle \frac{\delta E_{+k}^2 + \delta E_{-k}^2}{2} \right\rangle = \langle H_{kk}^2 \rangle + \langle H_{k,-k}^2 \rangle$$

$$= \frac{2}{N} \left(\langle \delta E^2 \rangle + \langle \delta V_{\pm 1}^2 \rangle \left(3 + \cos(2k) \right) \right) . \tag{25.109}$$

Off-diagonal disorder has a similar effect as diagonal disorder, but it influences the optically allowed $k = \pm 2\pi/N$ states stronger than the states in the center of the exciton band. More sophisticated investigations, including also partially correlated disorder and disorder of the transition dipoles, can be found in the literature [110–115].

Problems

25.1. Photosynthetic Reaction Center

The "special pair" in the photosynthetic reaction center of *Rps. viridis* is a dimer of two bacteriochlorophyll molecules whose centers of mass have a distance of 7 Å. The transition dipoles of the two molecules include an angle of 139°.

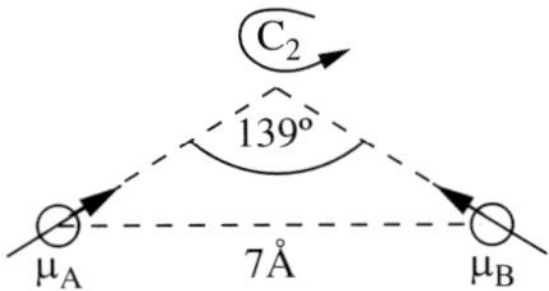

Calculate energies and intensities of the two dimer bands from a simple exciton model

$$\mathcal{H} = \begin{pmatrix} -\Delta/2 & V \\ V & \Delta/2 \end{pmatrix}$$

as a function of the energy difference Δ and the interaction V. The Hamiltonian is represented here in a basis spanned by the two localized excited states $|A^*B\rangle$ and $|B^*A\rangle$.

25.2. Light-Harvesting Complex

The circular light-harvesting complex of the bacterium *Rps. acidophila* consists of nine bacteriochlorophyll dimers in a C_9-symmetric arrangement. The two subunits of a dimer are denoted as α and β. The exciton Hamiltonian with nearest-neighbor and next-to-nearest-neighbor interactions only is (with the index n taken as modulo 9)

$$
\begin{aligned}
\mathcal{H} = \sum_{n=1}^{9} \{ & E_\alpha |n;\alpha\rangle\langle n;\alpha| + E_\beta |n;\beta\rangle\langle n;\beta| \\
& + V_{\text{dim}} (|n;\alpha\rangle\langle n;\beta| + |n;\beta\rangle\langle n;\alpha|) \\
& + V_{\beta\alpha,1} (|n;\alpha\rangle\langle n-1;\beta| + |n;\beta\rangle\langle n+1;\alpha|) \\
& + V_{\alpha\alpha,1} (|n;\alpha\rangle\langle n+1;\alpha| + |n;\alpha\rangle\langle n-1;\alpha|) \\
& + V_{\beta\beta,1} (|n;\beta\rangle\langle n+1;\beta| + |n;\beta\rangle\langle n-1;\beta|) \} .
\end{aligned}
$$

Transform the Hamiltonian to delocalized states

$$
|k;\alpha\rangle = \frac{1}{3} \sum_{n=1}^{9} e^{ikn} |n;\alpha\rangle, \quad |k;\beta\rangle = \frac{1}{3} \sum_{n=1}^{9} e^{ikn} |n;\beta\rangle,
$$

$$
k = l\,2\pi/9, \quad l = 0, \pm 1, \pm 2, \pm 3, \pm 4.
$$

(a) Show that states with different k-values do not interact

$$
\langle k', \alpha(\beta) | \mathcal{H} | k, \alpha(\beta)\rangle = 0 \quad \text{if} \quad k \neq k'.
$$

(b) Find the matrix elements

$$
H_{\alpha\alpha}(k) = \langle k;\alpha|\mathcal{H}|k;\alpha\rangle, \quad H_{\beta\beta}(k) = \langle k;\beta|\mathcal{H}|k;\beta\rangle, \quad H_{\alpha\beta}(k) = \langle k;\alpha|\mathcal{H}|k;\beta\rangle.
$$

(c) Solve the eigenvalue problem

$$
\begin{pmatrix} H_{\alpha\alpha}(k) & H_{\alpha\beta}(k) \\ H_{\alpha\beta}^{*}(k) & H_{\beta\beta}(k) \end{pmatrix} \begin{pmatrix} C_\alpha \\ C_\beta \end{pmatrix} = E_{1,2}(k) \begin{pmatrix} C_\alpha \\ C_\beta \end{pmatrix}.
$$

(d) The transition dipole moments are given by

$$
\boldsymbol{\mu}_{n,\alpha} = \mu \begin{pmatrix} \sin\theta \, \cos(\phi_\alpha - \nu + n\phi) \\ \sin\theta \, \sin(\phi_\alpha - \nu + n\phi) \\ \cos\theta \end{pmatrix},
$$

$$
\boldsymbol{\mu}_{n,\beta} = \mu \begin{pmatrix} \sin\theta \, \cos(\phi_\beta + \nu + n\phi) \\ \sin\theta \, \sin(\phi_\beta + \nu + n\phi) \\ \cos\theta \end{pmatrix},
$$

$\nu = 10.3°$, $\phi_\alpha = -112.5°$, $\phi_\beta = 63.2°$, and $\theta = 84.9°$.

Determine the optically allowed transitions from the ground state and calculate the relative intensities.

25.3. Exchange Narrowing

Consider excitons in a ring of chromophores with uncorrelated diagonal disorder. Show that in lowest order, the distribution function of E_k is Gaussian. *Hint.* Write the distribution function as

$$P(\delta E_k = X) = \int \mathrm{d}\delta E_1 \mathrm{d}\delta E_2 \cdots P(\delta E_1) P(\delta E_2) \cdots \delta \left(X - \frac{\sum \delta E_n}{N} \right)$$

and replace the δ-function with a Fourier integral.

Ultrafast Electron Transfer Processes in the Photosynthetic Reaction Center

In this chapter, we discuss a model for ultrafast electron transfer processes in biological systems where the density of vibrational states is very high. The vibrational modes which are coupled to the electronic transition[1] are treated within the model of displaced parallel (17.2) harmonic oscillators (Chap. 19). This gives the following approximation to the diabatic potential surfaces

$$E_i = E_i^0 + \sum_r \frac{\omega_r^2}{2} q_r^2,$$

$$E_f = E_f^0 + \sum_r \frac{\omega_r^2}{2} (q_r - \Delta q_r)^2, \tag{26.1}$$

and the energy gap

$$E_f - E_i = \Delta E + \sum_r \frac{(\omega_r \Delta q_r)^2}{2} - \sum_r \omega_r^2 \Delta q_r q_r. \tag{26.2}$$

The total reorganization energy is the sum of all partial reorganization energies

$$E_{\mathrm{R}} = \sum_r \frac{(\omega_r \Delta q_r)^2}{2} = \sum_r g_r^2 \hbar \omega_r \tag{26.3}$$

with the vibronical coupling strength

$$g_r = \sqrt{\frac{\omega_r}{2\hbar}} \Delta q_r. \tag{26.4}$$

The golden rule expression gives the rate for nonadiabatic electron transfer in analogy to (18.8) (Fig. 26.1)

---

[1] Which have different equilibrium positions in the two states $|i\rangle$ and $|f\rangle$.

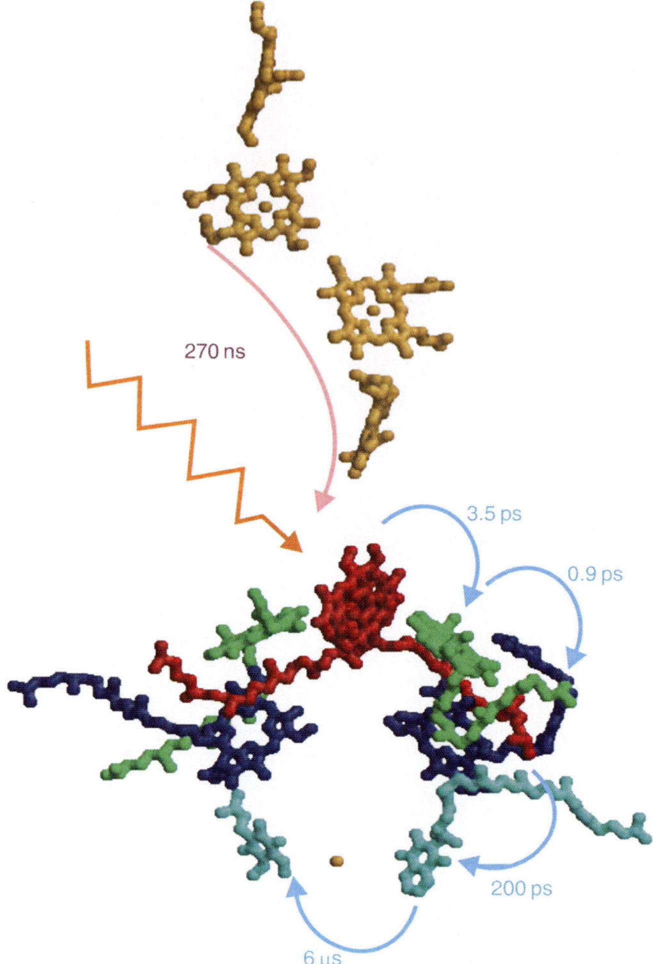

Fig. 26.1. Electron transfer processes in the reaction center of bacterial photosynthesis. After photoexcitation of the special pair dimer (*red*), an electron is transferred to a bacteriopheophytine (*blue*) in few picoseconds via another bacteriochlorophyll (*green*). The electron is then transferred on a longer time scale to the quinone Q_A and finally to Q_B via a nonheme ferreous ion. The electron hole in the special pair is filled by an electron from the hemes of the cytochrome c (*orange*). After a second excitation of the dimer, another electron is transferred via the same route to the semiquinone to form ubiquinone which then diffuses freely within the membrane. This figure was created with Rasmol [106] based on the structure of *Rps. viridis* [90–93] from the protein data bank [104, 105]

$$k_{\mathrm{et}} = \frac{2\pi V^2}{\hbar} \sum_{n_r n_{r'}} P(\{n_r\}) \prod_r (\mathrm{FC}_r(n_r, n_{r'}))^2 \delta\left(\Delta E + \sum_r \hbar\omega_r(n_{r'} - n_r)\right)$$

$$= \frac{2\pi V^2}{\hbar} \mathrm{FCD}(\Delta E), \qquad (26.5)$$

where the Franck–Condon-weighted density of states $\mathrm{FCD}(\Delta E)$ can be expressed with the help of the time correlation formalism (18.2) as

$$\mathrm{FCD}(\Delta E) = \frac{1}{2\pi\hbar} \int \mathrm{d}t\, \mathrm{e}^{-\mathrm{i}t\Delta E/\hbar} F(t) \qquad (26.6)$$

with

$$F(t) = \prod_r \exp\left(g_r^2 (\mathrm{e}^{\mathrm{i}\omega t} - 1)(\overline{n_r} + 1) + (\mathrm{e}^{-\mathrm{i}\omega t} - 1)\overline{n_r}\right). \qquad (26.7)$$

Quantum chemical calculations for the reaction center [116–119] put the electronic coupling in the range of $V = 5\text{–}50\,\mathrm{cm}^{-1}$ for the first step $\mathrm{P}^* \to \mathrm{P}^+\mathrm{B}^-$ and $V = 15\text{–}120\,\mathrm{cm}^{-1}$ for the second step $\mathrm{P}^+\mathrm{B}^- \to \mathrm{P}^+\mathrm{H}^-$. For a rough estimate, we take a reorganization energy of $2{,}000\,\mathrm{cm}^{-1}$ and a maximum of the Franck–Condon-weighted density of states $\mathrm{FCD} \approx 1/2{,}000\,\mathrm{cm}^{-1}$ yielding an electronic coupling of $V = 21\,\mathrm{cm}^{-1}$ for the first and $V = 42\,\mathrm{cm}^{-1}$ for the second step in quite good agreement with the calculations.

Proton Transfer in Biomolecules

Intraprotein proton transfer is perhaps the most fundamental flux in the biosphere [120]. It is essential for such important processes as photosynthesis, respiration, and ATP synthesis [121]. Within a protein, protons appear only in covalently bound states. Here, proton transfer is essentially the motion along a hydrogen bond, for instance peptide or DNA H-bonds (Fig. 27.1) or the more complicated pathways in proton-pumping proteins.

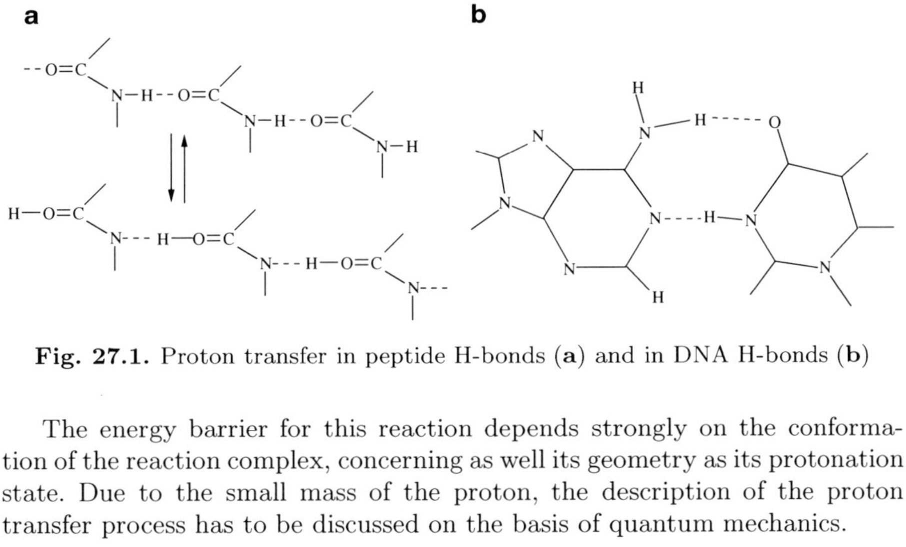

Fig. 27.1. Proton transfer in peptide H-bonds (**a**) and in DNA H-bonds (**b**)

The energy barrier for this reaction depends strongly on the conformation of the reaction complex, concerning as well its geometry as its protonation state. Due to the small mass of the proton, the description of the proton transfer process has to be discussed on the basis of quantum mechanics.

27.1 The Proton Pump Bacteriorhodopsin

Rhodopsin proteins collect the photon energy in the process of vision. Since the end of the 1950s, it has been recognized that the photoreceptor molecule retinal undergoes a structural change, the so-called *cis–trans* isomerization upon photoexcitation (G. Wald, R. Granit, H.K. Hartline, Nobel prize for

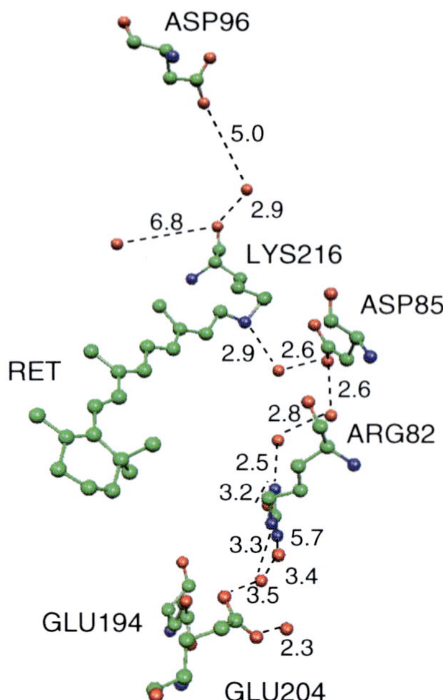

Fig. 27.2. X-ray structure of bR. The most important residues and structural waters are shown, together with possible pathways for proton transfer. *Numbers* show O–O and O–N distances in Å. Coordinates are from the structure 1c3w [122] in the protein data bank [104, 105]. Molekel graphics [89]

medicine 1961). Rhodopsin is not photostable and its spectroscopy remains difficult. Therefore, much more experimental work has been done on its bacterial analogue bacteriorhodopsin, which performs a closed photocycle (Figs. 27.2–27.4).

Bacteriorhodopsin is the simplest known natural photosynthetic system. Retinal is covalently bound to a lysine residue forming the so-called protonated Schiff base. This form of the protein absorbs sunlight very efficiently over a large spectral range (480–360 nm).

After photoinduced isomerization, a proton is transferred from the Schiff base to the negatively charged ASP85 ($L \rightarrow M$). This induces a large blue shift. In wild-type bR, under physiological conditions, a proton is released to the extracellular medium on the same time scale of 10 μs. The proton release group is believed to consist of GLU194, GLU204, and some structural waters. During the M-state, a large rearrangement on the cytoplasmatic side appears, which is not seen spectroscopically and which induces the possibility to reprotonate the Schiff base from ASP96 subsequently ($M \rightarrow N$). ASP96 is then reprotonated from the cytoplasmatic side and the retinal reisomerizes thermally to the all *trans*-configuration ($N \rightarrow O$). Finally, a proton is transferred from ASP85 via ARG82 to the release group and the ground state bR is recovered.

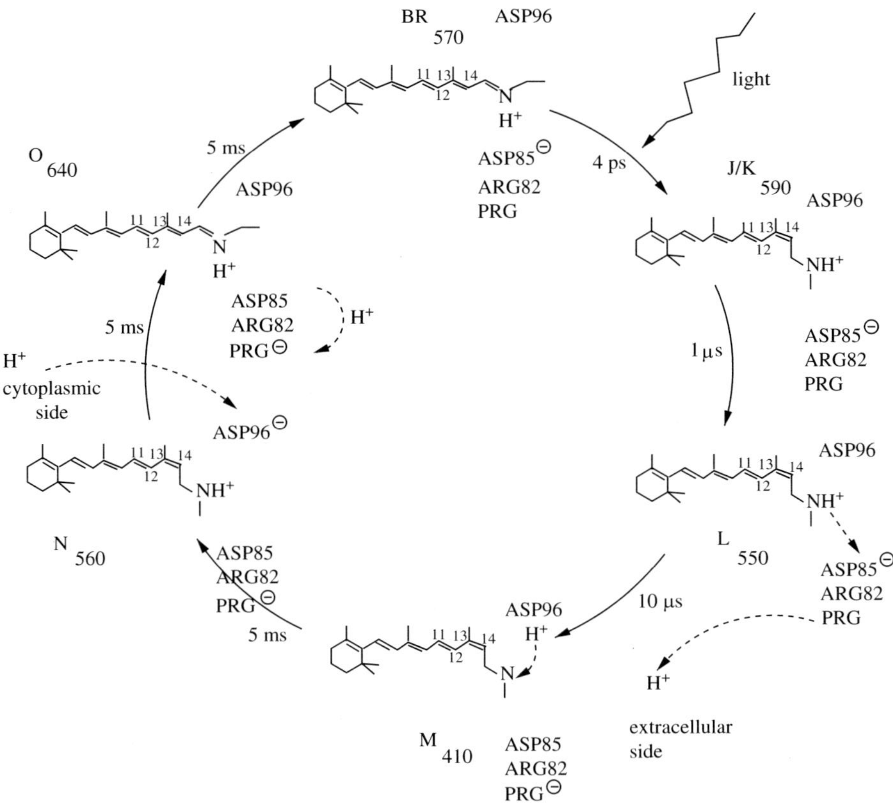

Fig. 27.3. The photocycle of bacteriorhodopsin [123]. Different states are labeled by the corresponding absorption maximum (nm)

Fig. 27.4. Photoisomerization of bR

27.2 Born–Oppenheimer Separation

Protons move on a faster time scale than the heavy nuclei but are slower than the electrons. Therefore, we use a double Born–Oppenheimer approximation for the wave function[1]

---

[1] The Born–Oppenheimer approximation and the nonadiabatic corrections to it are discussed more systematically in Sect. 17.1.

$$\psi(r, R_{\mathrm{P}}, Q)\chi(R_{\mathrm{p}}, Q)\Phi(Q). \tag{27.1}$$

The protonic wave function χ depends parametrically on the coordinates Q of the heavy atoms and the electronic wave function ψ depends parametrically on all nuclear coordinates. The Hamiltonian consists of the kinetic energy contributions and a potential energy term:

$$H = T_{\mathrm{el}} + T_{\mathrm{p}} + T_{\mathrm{N}} + V. \tag{27.2}$$

The Born–Oppenheimer approximation leads to the following hierarchy of equations. First, all nuclear coordinates (R_{p}, Q) are fixed and the electronic wave function of the state s is obtained from

$$(T_{\mathrm{el}} + V(r; R_{\mathrm{p}}, Q))\psi_s(r; R_{\mathrm{p}}, Q) = E_{\mathrm{el},s}(R_{\mathrm{p}}, Q)\psi_s(r; R_{\mathrm{p}}, Q). \tag{27.3}$$

In the second step, only the heavy atoms (Q) are fixed but the action of the kinetic energy of the proton on the electronic wave function is neglected. Then the wave function of proton state n obeys

$$(T_{\mathrm{p}} + E_{\mathrm{el},s}(R_{\mathrm{p}}; Q))\chi_{s,n}(R_{\mathrm{p}}; Q) = \varepsilon_{s,n}(Q)\chi_{s,n}(R_{\mathrm{p}}; Q). \tag{27.4}$$

The electronic energy $E_{\mathrm{el}}(R_{\mathrm{p}}, Q)$ plays the role of a potential energy surface for the nuclear motion. It is shown schematically in Fig. 27.5 for one proton

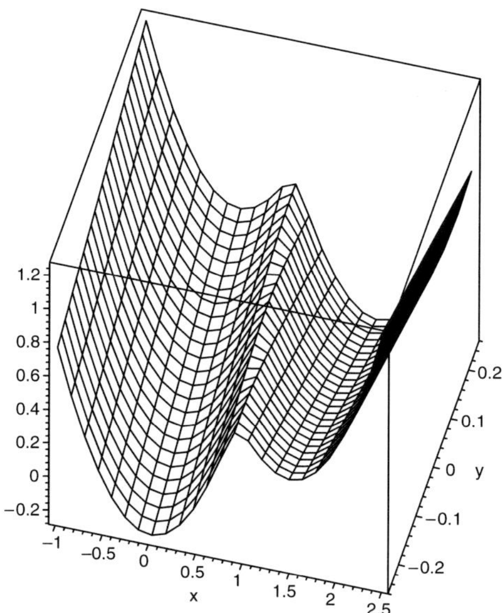

Fig. 27.5. Proton transfer potential. The figure shows schematically the potential which, as a function of the proton coordinate x, has two local minima corresponding to the structures $\cdots$ N–H $\cdots$ O $\cdots$ and $\cdots$ N $\cdots$ H–O $\cdots$ and which is modulated by a coordinate y representing the heavy atoms

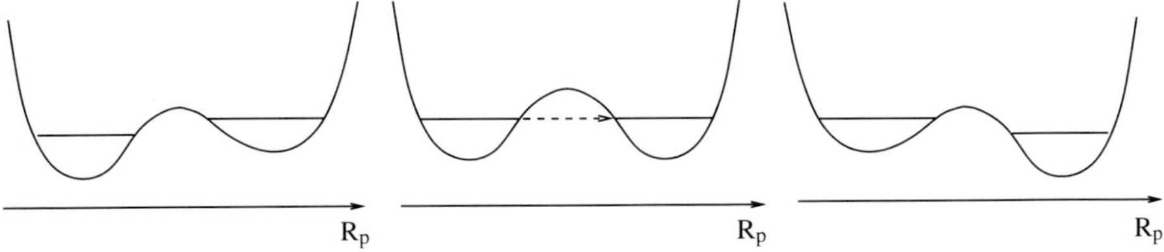

Fig. 27.6. Proton tunneling. The double well of the proton along the H-bond is shown schematically for different configurations of the reaction complex

coordinate (for instance, the O–H bond length) and one promoting mode of the heavy atoms which modulates the energy gap between the two minima.

For fixed positions Q of the heavy nuclei, $E_{el}(R_p, Q)$ has a double-well structure as shown in Fig. 27.6. The rate of proton tunneling depends on the configuration and is most efficient if the two localized states are in resonance. Finally, the wave function of the heavy nuclei in state α is the approximate solution of

$$(T_N + \varepsilon_{s,n}(Q))\Phi_{s,n,\alpha}(Q) = E_{s,n,\alpha}\Phi_{s,n,\alpha}(Q). \tag{27.5}$$

27.3 Nonadiabatic Proton Transfer (Small Coupling)

Initial and final states (s, n) and (s', m) are stabilized by the reorganization of the heavy nuclei. The transfer occurs whenever fluctuations bring the two states in resonance (Fig. 27.7).

We use the harmonic approximation

$$\varepsilon_{s,n}(Q) = \varepsilon_{s,n}^{(0)} + \frac{a}{2}(Q - Q_s^{(0)})^2 + \cdots = \varepsilon_{s,0}(Q) + (\varepsilon_{s,n} - \varepsilon_{s,0}) \tag{27.6}$$

similar to nonadiabatic electron transfer theory (Chap. 16). The activation free energy for the $(s, 0) \to (s', 0)$ transition is given by

$$\Delta G_{00}^{\ddagger} = \frac{(\Delta G_0 + E_R)^2}{4E_R} \tag{27.7}$$

and for the transition $(s, n) \to (s', m)$, we have approximately

$$\begin{aligned}
\Delta G_{nm}^{\ddagger} &= \Delta G_0^{\ddagger} + (\varepsilon_{sm} - \varepsilon_{s0}) - (\varepsilon_{s'n} - \varepsilon_{s'0}) \\
&= \frac{(\Delta G_0 + E_R + \varepsilon_{sm} - \varepsilon_{s'n} - \varepsilon_{s0} + \varepsilon_{s'0})^2}{4E_R}.
\end{aligned} \tag{27.8}$$

The partial rate k_{nm} is given in analogy to the ET rate (16.23) by

$$k_{nm} = \frac{\omega}{2\pi}\kappa_{nm}^P \exp\left\{-\frac{\Delta G_{nm}^{\ddagger}}{kT}\right\}. \tag{27.9}$$

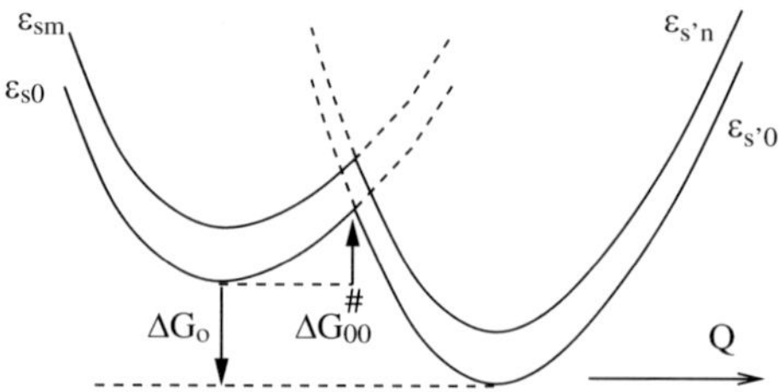

Fig. 27.7. Harmonic model for proton transfer

The interaction matrix element will be calculated using the BO approximation
and treating the heavy nuclei classically

$$
\begin{aligned}
V_{sns'm} &= \int dr\, dR_{\mathrm{p}}\, \psi_s(r, R_{\mathrm{P}}, Q)\chi_{sn}(R_{\mathrm{p}}, Q)V\chi_{s'm}(R_{\mathrm{p}}, Q)\psi_{s'}(r, R_{\mathrm{P}}, Q) \\
&= \int dR_{\mathrm{p}} \left(\int dr\, \psi_s(r, R_{\mathrm{P}}, Q)V\psi_{s'}(r, R_{\mathrm{P}}, Q) \right) \chi_{sn}(R_{\mathrm{p}}, Q)\chi_{s'm}(R_{\mathrm{p}}, Q) \\
&= \int dR_{\mathrm{p}}V_{ss'}^{\mathrm{el}}(R_{\mathrm{p}}, Q)\, \chi_{sn}(R_{\mathrm{p}}, Q)\chi_{s'm}(R_{\mathrm{p}}, Q).
\end{aligned}
\tag{27.10}
$$

The electronic matrix element

$$
V_{ss'}^{\mathrm{el}} = \int \psi_s(r, R_{\mathrm{p}}, Q)V\psi_{s'}(r, R_{\mathrm{p}}, Q)dr
\tag{27.11}
$$

is expanded around the equilibrium position of the proton R_{p}^*

$$
V_{ss'}^{\mathrm{el}}(R_{\mathrm{p}}, Q) \approx V_{ss'}^{\mathrm{el}}(R_{\mathrm{P}}^*, Q) + \cdots
\tag{27.12}
$$

and we have approximately

$$
V_{sns'm} \approx V_{ss'}^{\mathrm{el}}(R_{\mathrm{P}}^*, Q)S_{ss'}^{nm}(Q)
\tag{27.13}
$$

with the overlap integral of the protonic wave functions (Franck–Condon
factor)

$$
S_{ss'}^{nm}(Q) = \int \chi_{sn}\chi_{s'm}dR_{\mathrm{p}}.
\tag{27.14}
$$

27.4 Strongly Bound Protons

The frequencies of O–H or N–H bond vibrations are typically in the range of
$3{,}000\,\mathrm{cm}^{-1}$ which is much larger than thermal energy. If, on the other hand,
the barrier for proton transfer is on the same order, only the transition between
the vibrational ground states is involved. For small electronic matrix element
V^{el}, the situation is very similar to electron transfer described by Marcus

theory (Chap. 16) and the reaction rate is given by Borgis and Hynes [124] for symmetrical proton transfer ($E_\mathrm{R} \gg \Delta G_0, \Delta G_{00}^{\ddagger} = E_\mathrm{R}/4$):

$$k = \frac{2\pi V_0^2}{\hbar} \sqrt{\frac{1}{4\pi kT E_\mathrm{R}}} \, \exp\left(-\frac{E_\mathrm{R}}{4kT}\right). \tag{27.15}$$

In the derivation of (27.15), the Condon approximation has been used which corresponds to application of (27.12) and approximation of the overlap integral (27.14) by a constant value. However, for certain modes (which influence the distance of the two potential minima), the modulation of the coupling matrix element can be much stronger than in the case of electron transfer processes due to the larger mass of the proton. The dependence on the transfer distance $\delta Q = Q_{s'}^{(0)} - Q_{s}^{(0)}$ is approximately exponential:

$$V_{s0s'0} \sim \mathrm{e}^{-\alpha \delta Q}, \tag{27.16}$$

where typically $\alpha \approx 25\text{–}35 A^{-1}$, whereas for electron transfer processes typical values are around $\alpha \approx 0.5\text{–}1 A^{-1}$. Borgis and Hynes [124] found the following result for the rate when the low-frequency substrate mode with frequency Ω and reduced mass M is thermally excited

$$k = \frac{2\pi \langle V^2 \rangle}{\hbar} \sqrt{\frac{1}{4\pi kT 4\Delta G_\mathrm{tot}}} \, \exp\left(-\frac{\Delta G_\mathrm{tot}}{kT}\right) \chi_\mathrm{SC}, \tag{27.17}$$

where the average coupling square can be expanded as

$$\langle V^2 \rangle = V_0^2 \exp\left(\frac{\hbar \alpha^2}{2M\Omega} \left\{ \frac{4kT}{\hbar\Omega} + \frac{\hbar\Omega}{3kT} + O\left(\left(\frac{\hbar\Omega}{kT}\right)^2\right) \right\}\right). \tag{27.18}$$

The activation energy is

$$\Delta G_\mathrm{tot} = \Delta G^{\ddagger} + \frac{\hbar^2 \alpha^2}{8M} \tag{27.19}$$

and the additional factor χ_SC is given by

$$\chi_\mathrm{SC} = \exp\left(\gamma_\mathrm{SC}\left(1 - \frac{\gamma_\mathrm{SC} kT}{4\Delta G_\mathrm{tot}}\right)\right) \tag{27.20}$$

with

$$\gamma_\mathrm{SC} = \frac{\alpha \langle D \rangle}{M\Omega^2}, \tag{27.21}$$

where

$$D = \frac{\partial \Delta H}{\partial Q}(Q_0) \tag{27.22}$$

measures the modulation of the energy difference by the promoting mode Ω.

27.5 Adiabatic Proton Transfer

If V^{el} is large and the tunnel factor approaches unity (for $s = s'$), then the application of the low-friction result (8.29) gives

$$k_{nm} = \frac{\omega}{2\pi} \exp\left\{-\frac{\Delta G_{nm}^{\ddagger\,\mathrm{ad}}}{kT}\right\},\tag{27.23}$$

where the activation energy has to be corrected by the tunnel splitting

$$\Delta G_{nm}^{\ddagger\,\mathrm{ad}} = \Delta G_{nm}^{\ddagger} - \frac{1}{2}\Delta\varepsilon_{nm}^{\mathrm{P}}.\tag{27.24}$$

Part VIII

Molecular Motor Models

Continuous Ratchet Models

A particular class of proteins (molecular motors) can move linearly along its designated track, against an external force, by using chemical energy. The movement of a single protein within a cell is subject to thermal agitation from its environment and is, therefore, a Brownian motion with drift.

The following discussion is based on the pioneering work by Jülicher and Prost [125–128].

28.1 Transport Equations

Treating the center of mass of the motor, as subject to Brownian motion within a periodic potential, we apply a Smoluchowski equation with periodic boundary conditions:

$$\frac{\partial}{\partial t} W(x,t) = \frac{k_{\mathrm{B}}T}{m\gamma} \frac{\partial^2}{\partial x^2} W(x,t) + \frac{1}{m\gamma} \frac{\partial}{\partial x} \left(\frac{\partial U}{\partial x} W(x,t) \right)$$

$$= \frac{k_{\mathrm{B}}T}{m\gamma} \frac{\partial^2}{\partial x^2} W(x,t) - \frac{1}{m\gamma} \frac{\partial}{\partial x} \left(F(x) W(x,t) \right), \qquad (28.1)$$

$$U(x+L) = U(x), \qquad (28.2)$$

which can be written in terms of the probability flux

$$\frac{\partial}{\partial t} W(x,t) = -\frac{\partial}{\partial x} S(x,t), \qquad (28.3)$$

$$S(x,t) = -\frac{k_{\mathrm{B}}T}{m\gamma} \frac{\partial}{\partial x} W(x,t) + \frac{1}{m\gamma} F(x) W(x,t). \qquad (28.4)$$

By comparison with the continuity equation for the density

$$\frac{\partial}{\partial t} \rho = -\mathrm{div}(\rho v), \qquad (28.5)$$

we find

$$\frac{\partial W}{\partial t} = -\frac{\partial}{\partial x}S = -\frac{\partial}{\partial x}(Wv) \tag{28.6}$$

hence the drift velocity at x is

$$v_{\mathrm{d}}(x) = \frac{S(x)}{W(x)} \tag{28.7}$$

and the average velocity of systems in the interval $0, L$ is

$$\langle v \rangle = \frac{\int_0^L W(x)v_{\mathrm{d}}(x)\mathrm{d}x}{\int_0^L W(x)\mathrm{d}x} = \frac{\int_0^L S(x)\mathrm{d}x}{\int_0^L W(x)\mathrm{d}x}. \tag{28.8}$$

Equation (28.6) shows that for a stationary solution, the flux has to be constant. However, the stationary distribution is easily found by integration and has the form

$$W_{\mathrm{st}} = \mathrm{e}^{-U(x)/k_{\mathrm{B}}T}\left(C - \frac{m\gamma S}{k_{\mathrm{B}}T}\int_0^x \mathrm{e}^{U(x')/k_{\mathrm{B}}T}\mathrm{d}x'\right). \tag{28.9}$$

From the periodic boundary condition, we find

$$W_{\mathrm{st}}(L) = \mathrm{e}^{-U(0)/k_{\mathrm{B}}T}\left(C - \frac{m\gamma}{k_{\mathrm{B}}T}S\int_0^L \mathrm{e}^{U(x')/k_{\mathrm{B}}T}\mathrm{d}x'\right)$$

$$= W_{\mathrm{st}}(0) = \mathrm{e}^{-U(0)/k_{\mathrm{B}}T}C, \tag{28.10}$$

which shows that in a stationary state, there is zero flux (Fig. 28.1)

$$S = 0 \tag{28.11}$$

and no average velocity

$$\langle v \rangle = 0. \tag{28.12}$$

If an external force is applied to the system

$$U(x) \rightarrow U(x) - xF_{\mathrm{ext}}, \tag{28.13}$$

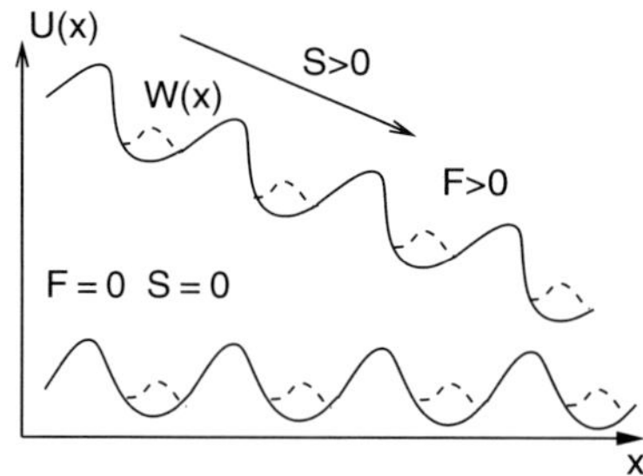

Fig. 28.1. Periodic potential with external force

then the stationary solution is

$$W(x) = e^{(xF - U(x))/k_BT} \left(C - \frac{m\gamma}{k_BT} S \int e^{(U(x) - xF)/k_BT} dx \right). \qquad (28.14)$$

Assuming periodicity for $W(x)$

$$W(0) = W(L),$$

$$Ce^{-U(0)/k_BT} = e^{(LF - U(0))/k_BT} \left(C - \frac{m\gamma}{k_BT} S \int_0^L e^{(U(x) - xF)/k_BT} dx \right), \qquad (28.15)$$

there will be a net flux which is a function of the external force:

$$S = \frac{k_BT}{m\gamma} W(0) \frac{e^{LF/k_BT} - 1}{\int_0^L e^{(U(x) - U(0) - (x-L)F)/k_BT} dx} = S(F). \qquad (28.16)$$

From

$$S_{st} = -\frac{k_BT}{m\gamma} \frac{\partial}{\partial x} W_{st}(x) - \frac{1}{m\gamma} W_{st}(x) \frac{\partial}{\partial x} \left(U(x) - xF_{ext} \right), \qquad (28.17)$$

we find

$$m\gamma S \frac{1}{W(x)} = -\frac{\partial}{\partial x} (k_BT \ln W(x) + U + U_{ext}), \qquad (28.18)$$

which can be interpreted in terms of a chemical potential [129]

$$\mu(x) = k_BT \ln W(x) + U(x) + U_{ext}(x) \qquad (28.19)$$

as

$$\frac{\partial}{\partial x} \mu(x) = -\frac{m\gamma S}{W(x)} = -m\gamma v_d(x), \qquad (28.20)$$

where the right-hand side gives the energy which is dissipated by a particle moving in a viscous medium.

28.2 Chemical Transitions

We assume that the chemistry of the motor can be described by a number m of discrete states or conformations $i = 1, \ldots, m$. Many biochemical models focus on a model with four states (Fig. 28.2).

Transitions between these states are fast compared to the motion of the motor and will be described by chemical kinetics

$$i \underset{k_{ji}}{\overset{k_{ij}}{\rightleftharpoons}} j. \qquad (28.21)$$

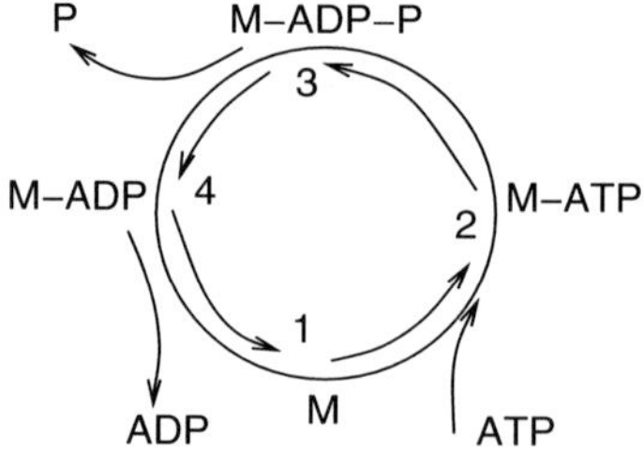

Fig. 28.2. Chemical cycle of the four-state model

The motion in each conformation will be described by a Smoluchowski equation in a corresponding potential $U_i(x)$ which is assumed to be periodic and asymmetric

$$U_i(x + L) = U_i(x), \quad U_i(x) \neq U_i(-x) \tag{28.22}$$

and where the potential of an external force may be added

$$U_i \rightarrow U_i(x) - xF_{\text{ext}}. \tag{28.23}$$

Together, we have the equations

$$\frac{\partial}{\partial t}W_i(x,t) + \frac{\partial}{\partial x}S_i(x,t) = \sum_{i \neq j}\left(k_{ji}(x)W_i(x,t) - k_{ij}(x)W_j(x,t)\right), \tag{28.24}$$

$$S_i(x,t) = -\frac{k_\mathrm{B}T}{m\gamma}\frac{\partial}{\partial x}W_i(x,t) - \frac{1}{\gamma m}W_i(x,t)\frac{\partial}{\partial x}U_i(x). \tag{28.25}$$

28.3 The Two-State Model

We discuss in the following a simple two-state model which has only a small number of parameters, but which is very useful to understand the general behavior. The model is a system of two coupled differential equations:

$$\frac{\partial}{\partial t}W_1 + \frac{\partial}{\partial x}S_1 = -k_{12}W_1 + k_{21}W_2,$$

$$\frac{\partial}{\partial t}W_2 + \frac{\partial}{\partial x}S_2 = -k_{21}W_2 + k_{12}W_2. \tag{28.26}$$

28.3.1 The Chemical Cycle

The energy source is the hydrolysis reaction

$$\mathrm{ATP} \rightarrow \mathrm{ADP+P}. \tag{28.27}$$

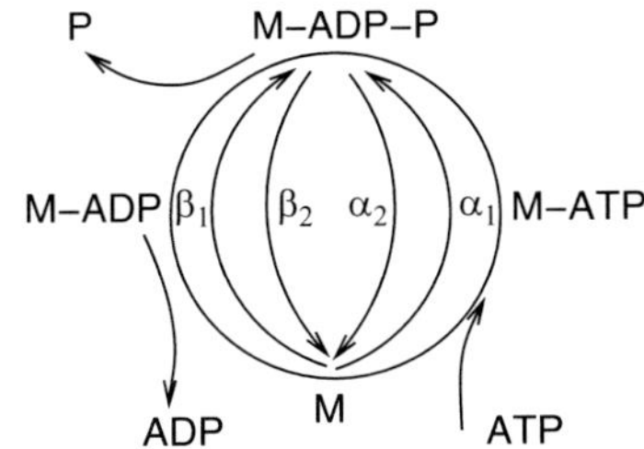

Fig. 28.3. Simplified chemical cycle

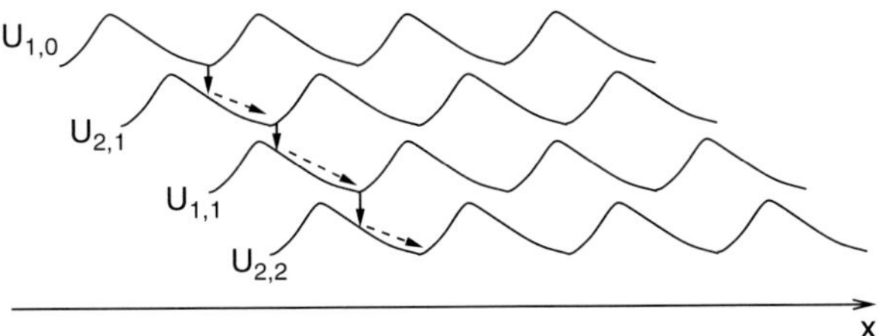

Fig. 28.4. Hierarchy of potentials

The motor protein undergoes a chemical cycle which can be divided into four states. In a simplified two-state model, this cycle has to be further divided into two substeps. One option is to combine ATP binding and hydrolysis (Fig. 28.3)

$$\text{M+ATP} \underset{\alpha_2}{\overset{\alpha_1}{\rightleftharpoons}} \text{M–ADP–P} \tag{28.28}$$

as well as the release of ADP and P

$$\text{M+ADP+P} \underset{\beta_2}{\overset{\beta_1}{\rightleftharpoons}} \text{M–ADP–P}. \tag{28.29}$$

Since each cycle consumes the energy from hydrolysis of one ATP, we consider a sequence of states characterized by the motor state and the number of total cycles (Fig. 28.4)

$$\cdots \text{M}_{1,0} \underset{\alpha_2}{\overset{\alpha_1}{\rightleftharpoons}} \text{M}_{2,1} \underset{\beta_1}{\overset{\beta_2}{\rightleftharpoons}} \text{M}_{1,1} \underset{\alpha_2}{\overset{\alpha_1}{\rightleftharpoons}} \text{M}_{2,2} \cdots \tag{28.30}$$

The potential energies $U_{s,n}$ and $U_{s,n+m}$ only differ by a vertical shift corresponding to the consumption of m energy units

$$\Delta\mu = \mu_{\text{ATP}} - \mu_{\text{ADP}} - \mu_{\text{P}} \tag{28.31}$$

and therefore

$$U_{s,n} = U_s - n\Delta\mu. \tag{28.32}$$

Detailed balance of the chemical reactions implies

$$\frac{\alpha_1}{\alpha_2} = e^{(U_{1,n}-U_{2,n+1})/k_{\mathrm{B}}T} = e^{(U_1-U_2+\Delta\mu)/k_{\mathrm{B}}T}, \tag{28.33}$$

$$\frac{\beta_1}{\beta_2} = e^{(U_{1,n}-U_{2,n})/k_{\mathrm{B}}T} = e^{(U_1-U_2)/k_{\mathrm{B}}T}. \tag{28.34}$$

The time evolution of the motor is described by a hierarchy of equations

$$\frac{\partial}{\partial t}W_{1,n} + \frac{\partial}{\partial x}S_{1,n} = -(\alpha_1 + \beta_1)W_{1,n} + \alpha_2 W_{2,n+1} + \beta_2 W_{2,n},$$
$$\frac{\partial}{\partial t}W_{2,n} + \frac{\partial}{\partial x}S_{2,n} = -(\alpha_2 + \beta_2)W_{2,n} + \beta_1 W_{1,n} + \alpha_1 W_{1,n-1}. \tag{28.35}$$

Obviously, we can sum over the hierarchy

$$W_s = \sum_n W_{s,n}, \quad S_s = \sum_n S_{s,n} \tag{28.36}$$

to obtain the two-state model:

$$\frac{\partial}{\partial t}W_1 + \frac{\partial}{\partial x}S_1 = -(\alpha_1 + \beta_1)W_1 + (\alpha_2 + \beta_2)W_2, \tag{28.37}$$

$$\frac{\partial}{\partial t}W_2 + \frac{\partial}{\partial x}S_2 = -(\alpha_2 + \beta_2)W_2 + (\alpha_1 + \beta_1)W_1, \tag{28.38}$$

where the transition rates are superpositions:

$$k_{21} = \alpha_2 + \beta_2, \tag{28.39}$$
$$k_{12} = \alpha_1 + \beta_1 = \alpha_2 e^{(U_1-U_2+\Delta\mu)/k_{\mathrm{B}}T} + \beta_2 e^{(U_1-U_2)/k_{\mathrm{B}}T}. \tag{28.40}$$

If $\Delta\mu = 0$, then these rates obey the detailed balance

$$\frac{k_{12}(x)}{k_{21}(x)} = e^{-(U_2-U_1)/k_{\mathrm{B}}T}, \tag{28.41}$$

which indicates that the system is not chemically driven and is only subject to thermal fluctuations. The steady state is then again given by

$$W_i = N e^{-U_i(x)/k_{\mathrm{B}}T} \tag{28.42}$$

for which the flux is zero. For $\Delta\mu > 0$, the system is chemically driven and spontaneous motion with $v \neq 0$ can occur. However, for a symmetric system with $U_{1,2}(-x) = U_{1,2}(x)$, the steady-state distributions are also symmetric

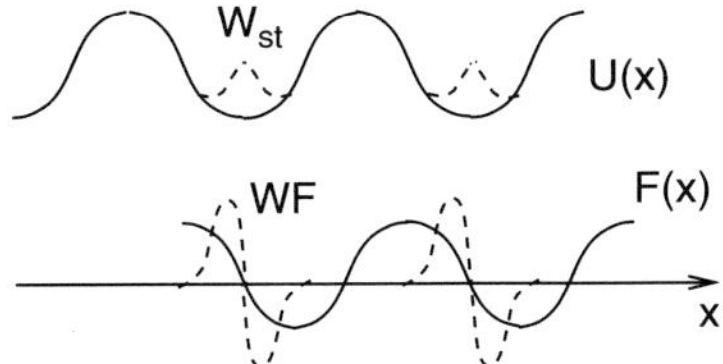

Fig. 28.5. Symmetric system

and the velocity

$$\langle v \rangle = -\frac{1}{m\gamma} \int_0^L dx \left(W_1 \frac{\partial U_1}{\partial x} + W_2 \frac{\partial U_2}{\partial x} \right) \qquad (28.43)$$

vanishes, since the forces are antisymmetric functions (Fig. 28.5).

The ATP hydrolysis rate

$$r = \int_0^L dx (\alpha_1 W_1 - \alpha_2 W_2), \qquad (28.44)$$

on the other hand, is generally not zero in this case since $\alpha_{1,2}$ are symmetric functions. Therefore, two conditions have to be fulfilled for spontaneous motion to occur: detailed balance has to be broken ($\Delta\mu \neq 0$) and the system must have polar symmetry.

28.3.2 The Fast Reaction Limit

If the chemical reactions are faster than the diffusive motion, then there is always a local equilibrium

$$\frac{W_1(x,t)}{W_2(x,t)} = \frac{k_{21}(x)}{k_{12}(x)} \qquad (28.45)$$

and the total probability $W = W_1 + W_2$ obeys the equation

$$\begin{aligned}
\frac{\partial}{\partial t} W &= \frac{k_{\mathrm{B}}T}{m\gamma} \frac{\partial^2}{\partial x^2} W + \frac{1}{m\gamma} \frac{\partial}{\partial x} \left(\frac{\partial}{\partial x} (U_1 W_1 + U_2 W_2) \right) \\
&= \frac{k_{\mathrm{B}}T}{m\gamma} \frac{\partial^2}{\partial x^2} W + \frac{1}{m\gamma} \frac{\partial}{\partial x} \left(\frac{\partial}{\partial x} \left(\frac{k_{21}}{k_{12} + k_{21}} U_1 + \frac{k_{12}}{k_{12} + k_{21}} U_2 \right) W \right),
\end{aligned}$$
$$(28.46)$$

which describes motion in the average potential.

28.3.3 The Fast Diffusion Limit

If the diffusive motion is faster than the chemical reactions, the motor can work as a Brownian ratchet. In the time between chemical reactions, equilibrium is established in both states

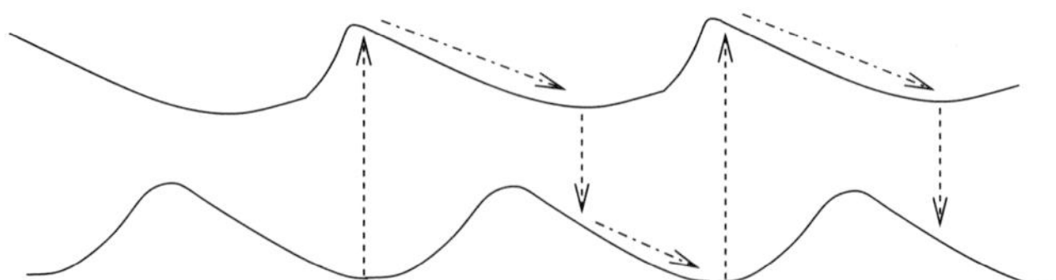

Fig. 28.6. Brownian ratchet

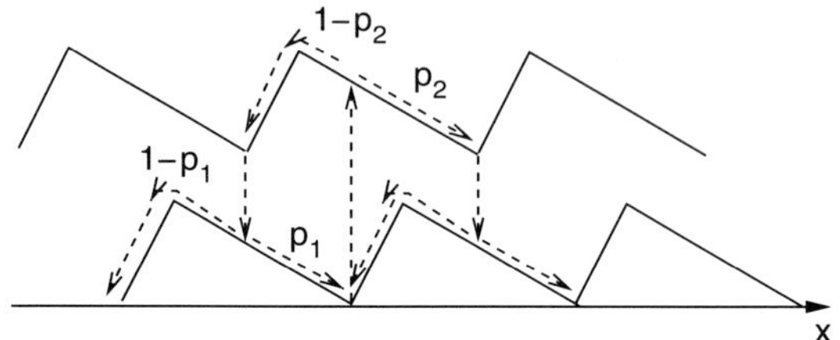

Fig. 28.7. Fast diffusion limit far from equilibrium

$$W_{\mathrm{st}}^0 = c_{1,2}\mathrm{e}^{-U_{1,2}(x)/k_{\mathrm{B}}T}. \tag{28.47}$$

If the barrier height is large and the equilibrium distribution is confined around the minima, the diffusion process can be approximated by a simpler rate process. Every ATP consumption corresponds to a transition to the second state. Then the system moves to one of the two neighboring minima with probabilities p_2 and $1 - p_2$, respectively. From here it makes a transition back to the first state. Finally, it moves right or left with probability p_1 and $1 - p_1$, respectively (Fig. 28.6).

During this cycle, the system proceeds in a forward direction with probability (Fig. 28.7)

$$p_+ = p_1 p_2 \tag{28.48}$$

backwards with probability

$$p_- = (1 - p_1)(1 - p_2) \tag{28.49}$$

or it returns to its initial position with probability

$$p_0 = p_2(1 - p_1) + p_1(1 - p_2). \tag{28.50}$$

The average displacement per consumed ATP is

$$\langle x \rangle \approx L(p_+ - p_-) = L(p_1 + p_2 - 1) \tag{28.51}$$

and the average velocity is

$$v = r\langle x \rangle. \tag{28.52}$$

In the presence of an external force, the efficiency can thus be estimated as

$$\eta = \frac{-F_{\mathrm{ext}}}{\Delta\mu}\langle x \rangle = \frac{-F_{\mathrm{ext}}L}{\Delta\mu}(p_1 + p_2 - 1). \tag{28.53}$$

28.4 Operation Close to Thermal Equilibrium

For constant T and μ_i, the local heat dissipation is given by

$$T\sigma(x) = \sum_k \left(F_{\text{ext}} - \frac{\partial U_k}{\partial x}\right) S_k + r(x)\Delta\mu. \tag{28.54}$$

In a stationary state, the total heat dissipation is

$$\Pi = \int T\sigma(x)\mathrm{d}x = F_{\text{ext}}v + r\Delta\mu, \tag{28.55}$$

since

$$\int \frac{\partial U}{\partial x}S\,\mathrm{d}x = -\int U(x)\frac{\partial S}{\partial x}\mathrm{d}x = 0. \tag{28.56}$$

Under the action of an external force F_{ext}, the velocity and the hydrolysis rate are functions of the external force and $\Delta\mu$:

$$v = v(F_{\text{ext}}, \Delta\mu),$$
$$r = r(F_{\text{ext}}, \Delta\mu). \tag{28.57}$$

Close to equilibrium, linearization gives

$$v = \lambda_{11}F_{\text{ext}} + \lambda_{12}\Delta\mu,$$
$$r = \lambda_{21}F_{\text{ext}} + \lambda_{22}\Delta\mu. \tag{28.58}$$

The rate of energy dissipation is given by mechanical plus chemical work

$$\Pi = F_{\text{ext}}v + r\Delta\mu = \lambda_{11}F_{\text{ext}}^2 + \lambda_{22}\Delta\mu^2 + (\lambda_{12} + \lambda_{21})F_{\text{ext}}\Delta\mu \tag{28.59}$$

which must be always positive due to the second law of thermodynamics. This is the case if

$$\lambda_{11} > 0, \quad \lambda_{22} > 0, \quad \text{and} \quad \lambda_{11}\lambda_{22} - \lambda_{12}\lambda_{21} > 0. \tag{28.60}$$

The efficiency of the motor can be defined as

$$\eta = \frac{-F_{\text{ext}}v}{r\Delta\mu} = \frac{-\lambda_{11}F_{\text{ext}}^2 - \lambda_{12}F_{\text{ext}}\Delta\mu}{\lambda_{22}\Delta\mu^2 + \lambda_{21}F_{\text{ext}}\Delta\mu}$$

or shorter

$$\eta = -\frac{\lambda_{11}a^2 + \lambda_{12}a}{\lambda_{22} + \lambda_{21}a}, \quad a = \frac{F_{\text{ext}}}{\Delta\mu}. \tag{28.61}$$

It vanishes for zero force but also for $v = 0$ and has a maximum at (Figs. 28.8 and 28.9)

$$a = \frac{\sqrt{1 - (\lambda_{12}\lambda_{21}/\lambda_{11}\lambda_{22})} - 1}{\lambda_{21}/\lambda_{22}}, \quad \eta_{\max} = \frac{(1 - \sqrt{1 - \Lambda})^2}{\Lambda}, \quad \Lambda = \frac{\lambda_{12}\lambda_{21}}{\lambda_{11}\lambda_{22}}. \tag{28.62}$$

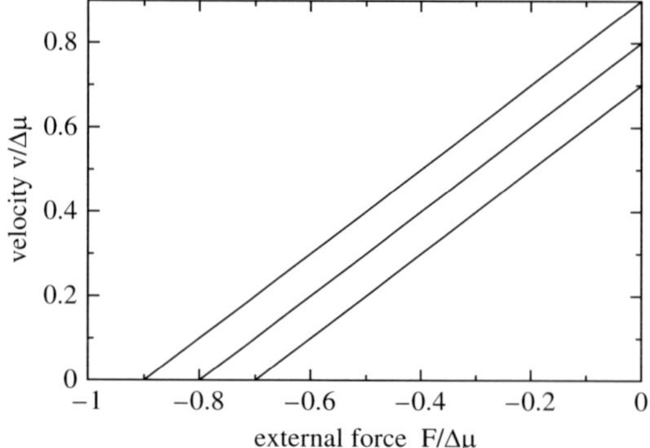

Fig. 28.8. Velocity–force relation

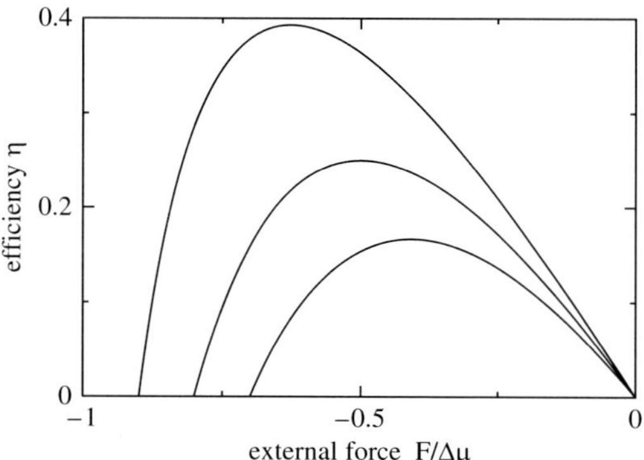

Fig. 28.9. Efficiency of the motor

Problems

28.1. Deviation from Equilibrium
Make use of the detailed balance conditions

$$\frac{\alpha_1}{\alpha_2} = e^{(\Delta\mu - \Delta U)/k_B T},$$

$$\frac{\beta_1}{\beta_2} = e^{-\Delta U/k_B T},$$

and calculate the quantity

$$\Omega = \frac{k_{12}}{k_{21}} - e^{-\Delta U/k_B T} = \frac{\alpha_1 + \beta_1}{\alpha_2 + \beta_2} - e^{-\Delta U/k_B T}.$$

Use the approximation

$$\Delta\mu = \Delta\mu^0 + k_B T \ln \frac{C(\mathrm{ATP})}{C(\mathrm{ADP})C(\mathrm{P})}$$

and express Ω in terms of the the equilibrium constant

$$K_{\mathrm{eq}} = \mathrm{e}^{\Delta\mu^0/k_{\mathrm{B}}T}.$$

Consider the limiting cases

$$\Delta\mu \neq 0 \quad \text{but} \quad \Delta\mu \ll k_{\mathrm{B}}T$$

and

$$\Delta\mu \gg k_{\mathrm{B}}T.$$

Discrete Ratchet Models

Simplified motor models describe the motion of the protein as a hopping process between positions x_n combined with a cyclical transition between internal motor states:

$$\mathrm{M}_{1,x_n} \rightleftharpoons \mathrm{M}_{2,x_n} \cdots \rightleftharpoons \mathrm{M}_{N,n_1} \rightleftharpoons \mathrm{M}_{1,x_{n+1}} \cdots \tag{29.1}$$

For example, the four-state model of the ATP hydrolysis can be easily translated into such a scheme (Fig. 29.1).

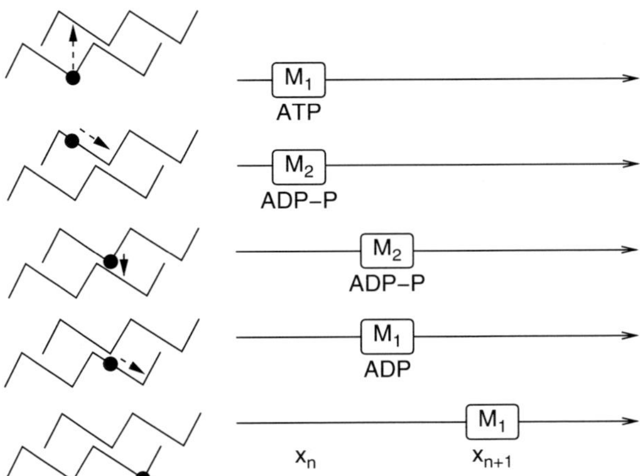

Fig. 29.1. Discrete motor model

29.1 Linear Model with Two Internal States

Fisher [130] considers a linear periodic process with two internal states:

$$\cdots \mathrm{M}_{1,x_n} \underset{\alpha_2}{\overset{\alpha_1}{\rightleftharpoons}} \mathrm{M}_{2,x_n} \underset{\beta_2}{\overset{\beta_1}{\rightleftharpoons}} \mathrm{M}_{1,x_{n+1}} \cdots \tag{29.2}$$

The master equation is

$$\frac{\mathrm{d}}{\mathrm{d}t} P_{1n} = -(\alpha_1 + \beta_2) P_{1n} + \alpha_2 P_{2n} + \beta_1 P_{2,n-1},$$

$$\frac{\mathrm{d}}{\mathrm{d}t} P_{2n} = -(\alpha_2 + \beta_1) P_{2n} + \alpha_1 P_{1n} + \beta_2 P_{1,n+1}, \tag{29.3}$$

and the stationary solution corresponds to the zero eigenvalue of the matrix

$$\begin{pmatrix} -\alpha_1 - \beta_2 & \alpha_2 + \beta_1 \\ \alpha_1 + \beta_2 & -\alpha_2 - \beta_1 \end{pmatrix}. \tag{29.4}$$

With the help of the left and right eigenvectors

$$L_0 = \begin{pmatrix} 1 & 1 \end{pmatrix}, \quad R_0 = \begin{pmatrix} \alpha_2 + \beta_1 \\ \alpha_1 + \beta_2 \end{pmatrix}, \tag{29.5}$$

we find

$$P_{1,\mathrm{st}} = \frac{\alpha_2 + \beta_1}{\alpha_2 + \beta_1 + \alpha_1 + \beta_2}, \quad P_{2,\mathrm{st}} = \frac{\alpha_1 + \beta_2}{\alpha_1 + \beta_2 + \alpha_2 + \beta_1} \tag{29.6}$$

and the stationary current is

$$S = \alpha_1 P_{1,\mathrm{st}} - \alpha_2 P_{2,\mathrm{st}} = \frac{\alpha_1 \beta_1 - \alpha_2 \beta_2}{\alpha_1 + \beta_2 + \alpha_2 + \beta_1}. \tag{29.7}$$

With the definition of

$$\omega = \frac{\alpha_2 \beta_2}{\alpha_1 + \alpha_2 + \beta_1 + \beta_2} \tag{29.8}$$

and

$$\Gamma = \frac{\alpha_1 \beta_1}{\alpha_2 \beta_2} = \mathrm{e}^{\Delta \mu / k_\mathrm{B} T}, \tag{29.9}$$

the current can be written as

$$S = (\Gamma - 1)\omega. \tag{29.10}$$

Consider now application of an external force corresponding to an additional potential energy

$$U_n = -n\,d\,F_{\mathrm{ext}}. \tag{29.11}$$

Assuming that the motion corresponds to the transition $\mathrm{M}_{2,n} \rightleftharpoons \mathrm{M}_{1,n+1}$, the corresponding rates will become dependent on the force (Fig. 29.2)

$$\beta_1 = \beta_1^0 \mathrm{e}^{\Theta \mathrm{F} d / k_\mathrm{B} T},$$

$$\beta_2 = \beta_2^0 \mathrm{e}^{-(1-\Theta) \mathrm{F} d / k_\mathrm{B} T}. \tag{29.12}$$

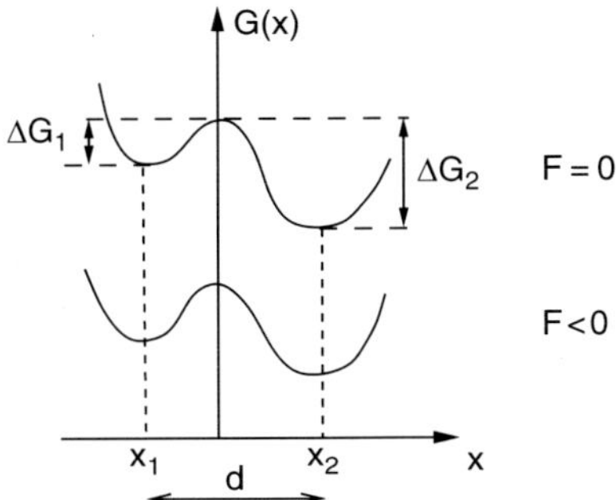

Fig. 29.2. Influence of the external force. The external force changes the activation barriers which become approximately $\Delta G_1(F) = \Delta G_1(0) - x_1 F$ and $\Delta G_2(F) = \Delta G_2(0) + x_2 F$

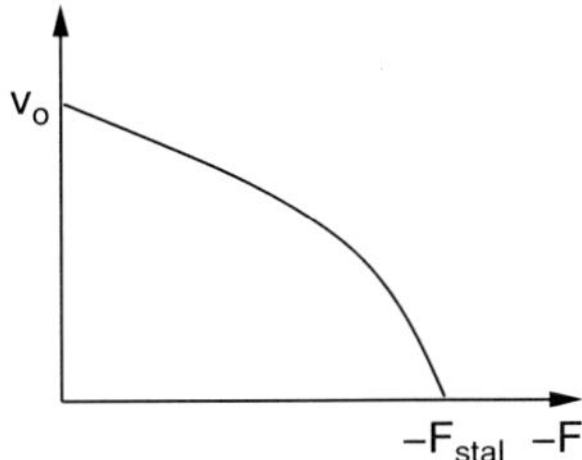

Fig. 29.3. Velocity–force relation

Θ is the so-called splitting parameter. The force-dependent current is

$$S = (e^{\Delta\mu/k_{\rm B}T + {\rm F}d/k_{\rm B}T} - 1)\frac{\alpha_2\beta_2^0 e^{-(1-\Theta){\rm F}d/k_{\rm B}T}}{\alpha_1 + \alpha_2 + \beta_1^0 e^{\Theta {\rm F}d/k_{\rm B}T} + \beta_2^0 e^{-(1-\Theta){\rm F}d/k_{\rm B}T}},$$

(29.13)

which vanishes for the stalling force:

$$F_{\rm stal} = -\frac{\Delta\mu}{d}.$$

(29.14)

Close to equilibrium, we expand (Fig. 29.3)

$$\Gamma - 1 = e^{\Delta\mu/k_{\rm B}T + {\rm F}d/k_{\rm B}T} - 1$$

$$= e^{(1-F/F_{\rm stal})\Delta\mu/k_{\rm B}T} - 1 \approx (1 - F/F_{\rm stal})\frac{\Delta\mu}{k_{\rm B}T},$$

(29.15)

$$\omega \approx \omega^0\left(1 - \frac{{\rm F}d}{k_{\rm B}T}\frac{(\beta_1^0 + (\alpha_1 + \alpha_2)(1 - \Theta))}{(\alpha_1 + \alpha_2 + \beta_1^0 + \beta_2^0)}\right).$$

(29.16)

Part IX

Appendix

A

The Grand Canonical Ensemble

Consider an ensemble of M systems which can exchange energy as well as particles with a reservoir. For large M, the total number of particles $N_{\text{tot}} = M\overline{N}$ and the total energy $E_{\text{tot}} = M\overline{E}$ have well-defined values since the relative widths decrease as

$$\frac{\sqrt{\overline{N_{\text{tot}}^2} - \overline{N_{\text{tot}}}^2}}{\overline{N_{\text{tot}}}} \sim \frac{1}{\sqrt{M}}, \quad \frac{\sqrt{\overline{E_{\text{tot}}^2} - \overline{E_{\text{tot}}}^2}}{\overline{E_{\text{tot}}}} \sim \frac{1}{\sqrt{M}}. \tag{A.1}$$

Hence also the average number and energy can be assumed to have well-defined values (Fig. A.1).

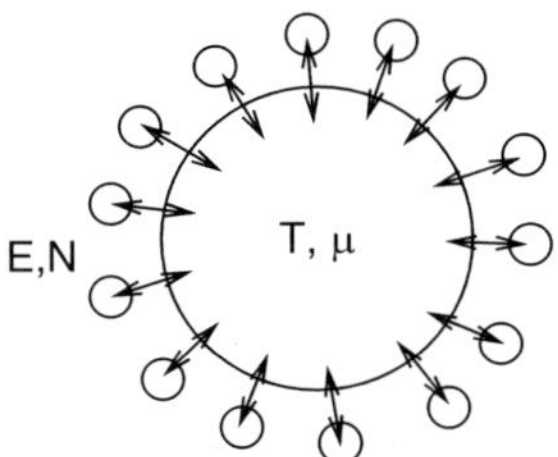

Fig. A.1. Ensemble of systems exchanging energy and particles with a reservoir

A.1 Grand Canonical Distribution

We distinguish different microstates j which are characterized by the number of particles N_j and the energy E_j. The number of systems in a certain microstate j will be denoted by n_j. The total number of particles and the total energy of the ensemble in a macrostate with n_j systems in the state j are

$$N_{\text{tot}} = \sum_j n_j N_j, \quad E_{\text{tot}} = \sum_j n_j E_j \tag{A.2}$$

and the number of systems is

$$M = \sum_j n_j. \tag{A.3}$$

The number of possible representations is given by a multinomial coefficient:

$$W(\{n_j\}) = \frac{M!}{\prod_j n_j!}. \tag{A.4}$$

From Stirling's formula, we have

$$\ln W \approx \ln(M!) - \sum_j (n_j \ln n_j - n_j). \tag{A.5}$$

We search for the maximum of (A.5) under the restraints imposed by (A.2) and (A.3). To this end, we use the method of undetermined factors (Lagrange method) and consider the variation

$$\delta\left(\ln W - \alpha\left(\sum_j n_j - M\right) - \beta\left(\sum_j n_j E_j - E_{\text{tot}}\right) - \gamma\left(\sum_j n_j N_j - N_{\text{tot}}\right)\right)$$

$$= \sum_j \delta n_j \left(-\ln n_j - \alpha - \beta E_j - \gamma N_j\right) = 0. \tag{A.6}$$

Since the n_j now can be varied independently, we find

$$n_j = \exp(-\alpha - \beta E_j - \gamma N_j). \tag{A.7}$$

The unknown factors α, β, and γ have to be determined from the restraints.
First we have

$$\sum_j n_j = e^{-\alpha} \sum_j e^{-\beta E_j - \gamma N_j} = M. \tag{A.8}$$

With the grand canonical partition function

$$\Xi = \sum_j e^{-\beta E_j - \gamma N_j}, \tag{A.9}$$

the probability of a certain microstate is given by

$$P(E_j, N_j) = \frac{n_j}{\sum_j n_j} = \frac{e^{-\beta E_j - \gamma N_j}}{\Xi} \tag{A.10}$$

and further

$$e^{-\alpha} = \frac{M}{\Xi}. \tag{A.11}$$

From

$$\sum_j n_j E_j = \frac{M}{\Xi} \sum_j E_j \mathrm{e}^{-\beta E_j - \gamma N_j} = E_{\text{tot}}, \qquad (A.12)$$

we find the average energy per system

$$\overline{E} = \frac{\sum n_j E_j}{M} = -\frac{\partial}{\partial \beta} \ln \Xi \qquad (A.13)$$

and similarly from

$$\sum_j n_j N_j = \frac{M}{\Xi} \sum_j N_j \mathrm{e}^{-\beta E_j - \gamma N_j} = N_{\text{tot}} \qquad (A.14)$$

the average particle number of a system

$$\overline{N} = \frac{\sum n_j N_j}{M} = -\frac{\partial}{\partial \gamma} \ln \Xi. \qquad (A.15)$$

Equations (A.13) and (A.15) determine the parameters β, γ implicitly and then α follows from (A.11).

A.2 Connection to Thermodynamics

Entropy is given by

$$\begin{aligned} S &= -k \sum P(E_j, N_j) \ln P(E_j, N_j) \\ &= -k \sum P(E_j, N_j)(-\beta E_j - \gamma N_j - \ln \Xi) \\ &= k\beta \overline{E} + k\gamma \overline{N} + k \ln \Xi. \end{aligned} \qquad (A.16)$$

From thermodynamics, the Duhem–Gibbs relation is known which states for the free enthalpy

$$G = U - TS + pV = \mu N$$

where $U = \overline{E}$ and S, V, N are the thermodynamic averages. Solving for the entropy, we have

$$S = \frac{U + pV - \mu N}{T} \qquad (A.17)$$

and comparison with (A.16) shows that

$$\beta = \frac{1}{k_B T}, \quad \gamma = -\frac{\mu}{k_B T}, \qquad (A.18)$$

$$k_B T \ln \Xi = pV. \qquad (A.19)$$

The summation over microstates j with energy E_j and N_j particles can be replaced by a double sum over energies $E_i(N)$ and particle number N to give

$$\Xi = \sum_{E,N} e^{-\beta(E(N)-\mu N)},$$

which can be written as a sum over canonical partition functions with different particle numbers

$$\Xi = \sum_N e^{\beta\mu N} \sum_E e^{-\beta E(N)} = \sum_N e^{\beta\mu N} Q(N). \tag{A.20}$$

B

Time Correlation Function of the Displaced Harmonic Oscillator Model

In the following, we evaluate the time correlation function of the displaced harmonic oscillator model (19.7)

$$f_r(t) = \left\langle \mathrm{e}^{-it\omega_r b_r^\dagger b_r} \mathrm{e}^{it\omega_r(b_r^\dagger + g_r)(b_r + g_r)} \right\rangle. \tag{B.1}$$

B.1 Evaluation of the Time Correlation Function

To proceed, we need some theorems which will be derived afterwards.

Theorem 1. *A displacement of the oscillator potential energy minimum can be formulated as a canonical transformation*

$$\hbar\omega_r(b_r^\dagger + g_r)(b_r + g_r) = \mathrm{e}^{-g_r(b_r^\dagger - b_r)} \hbar\omega_r b_r^\dagger b_r \mathrm{e}^{g_r(b_r^\dagger - b_r)}. \tag{B.2}$$

With the help of this relation, the single mode correlation function becomes

$$F_r(t) = \left\langle \mathrm{e}^{-it\omega_r b_r^\dagger b_r} \mathrm{e}^{-g_r(b_r^\dagger - b_r)} \mathrm{e}^{it\omega b_r^\dagger b_r} \mathrm{e}^{g_r(b_r^\dagger - b_r)} \right\rangle. \tag{B.3}$$

The first three factors can be interpreted as another canonical transformation. To this end, we apply the following theorem.

Theorem 2. *The time-dependent boson operators are given by*

$$\mathrm{e}^{-i\omega t b^\dagger b} b^\dagger \mathrm{e}^{i\omega t b^\dagger b} = b^\dagger \mathrm{e}^{-i\omega t},$$
$$\mathrm{e}^{-i\omega t b^\dagger b} b \, \mathrm{e}^{i\omega t b^\dagger b} = b \, \mathrm{e}^{i\omega t}. \tag{B.4}$$

we find

$$F_r(t) = \left\langle \exp\left(-g_r(b_r^\dagger \mathrm{e}^{-i\omega_r t} - b_r \mathrm{e}^{i\omega t})\right) \exp\left(g_r(b_r^\dagger - b_r)\right) \right\rangle. \tag{B.5}$$

The two exponentials can be combined due to the following theorem.

Theorem 3. *If the commutator of two operators A and B is a c-number then*

$$e^{A+B} = e^A e^B e^{-(1/2)[A,B]}. \tag{B.6}$$

The commutator is

$$-g_r^2[b_r^\dagger e^{-i\omega_r t} - b_r e^{i\omega t}, b_r^\dagger - b_r] = -g_r^2(e^{-i\omega t} - e^{i\omega t}) \tag{B.7}$$

and we have

$$F_r(t) = \exp\left(-\frac{1}{2}g_r^2(e^{-i\omega t} - e^{i\omega t})\right)$$
$$\times \left\langle \exp\left(-g_r b_r^\dagger(e^{-i\omega t} - 1) + g_r b_r(e^{i\omega t} - 1)\right)\right\rangle. \tag{B.8}$$

The remaining average is easily evaluated due to the following theorem.

Theorem 4. *For a linear combination of b_r and $b_r^\dagger$, the second-order cumulant expansion is exact*

$$\left\langle e^{\mu b^\dagger + \tau b}\right\rangle = e^{1/2\langle(\mu b^\dagger + \tau b)^2\rangle} = e^{\mu\tau\langle b^\dagger b + 1/2\rangle}. \tag{B.9}$$

The average square is

$$\left\langle\left(b_r^\dagger(e^{-i\omega t} - 1) - b_r(e^{i\omega t} - 1)\right)^2\right\rangle g_r^2$$
$$= -g_r^2(2 - e^{i\omega t} - e^{-i\omega t})\left\langle b_r^\dagger b_r + b_r b_r^\dagger\right\rangle$$
$$= -g_r^2(2 - e^{i\omega t} - e^{-i\omega t})(2\overline{n}_r + 1) \tag{B.10}$$

with the average phonon number[1]

$$\overline{n}_r = \frac{1}{e^{\beta\hbar\omega_r} - 1}. \tag{B.11}$$

Finally we have

$$F_r(t) = \exp\left(-\frac{1}{2}g_r^2(2 - e^{i\omega t} - e^{-i\omega t})(2\overline{n}_r + 1) - \frac{1}{2}g_r^2(e^{-i\omega t} - e^{i\omega t})\right)$$
$$= \exp\left(g_r^2(e^{i\omega t} - 1)(\overline{n}_r + 1) + (e^{-i\omega t} - 1)\overline{n}_r\right). \tag{B.12}$$

B.2 Boson Algebra

B.2.1 Derivation of Theorem 1

Consider the following unitary transformation

$$A = e^{-g(b^\dagger - b)}b^\dagger b e^{g(b^\dagger - b)} \tag{B.13}$$

---

[1] In the following, we use the abbreviation $\beta = 1/k_B T$.

and make a series expansion

$$A = A(0) + g\frac{\mathrm{d}A}{\mathrm{d}g} + \frac{1}{2}g^2\frac{\mathrm{d}^2 A}{\mathrm{d}g^2} + \cdots \tag{B.14}$$

The derivatives are

$$\left.\frac{\mathrm{d}A}{\mathrm{d}g}\right|_{g=0} = [b^\dagger b, b^\dagger - b] = b^\dagger[b, b^\dagger] + [b^\dagger, -b]b = (b + b^+),$$

$$\left.\frac{\mathrm{d}^2 A}{\mathrm{d}g^2}\right|_{g=0} = [[b^\dagger b, b^\dagger - b], b^\dagger - b] = [b + b^\dagger, b^\dagger - b] = 2,$$

$$\left.\frac{\mathrm{d}^n A}{\mathrm{d}g^n}\right|_{g=0} = 0 \quad \text{for} \quad n \geq 3, \tag{B.15}$$

and the series is finite

$$A = b^\dagger b + g(b^\dagger + b) + g^2 = (b^\dagger + g)(b + g). \tag{B.16}$$

Hence for any of the normal modes

$$\hbar\omega_r(b_r^\dagger + g_r)(b_r + g_r) = \mathrm{e}^{-g_r(b_r^\dagger - b_r)}\hbar\omega_r b_r^\dagger b_r \mathrm{e}^{g_r(b_r^\dagger - b_r)}. \tag{B.17}$$

B.2.2 Derivation of Theorem 2

Consider

$$A = \mathrm{e}^{\tau b^\dagger b}b\,\mathrm{e}^{-\tau b^\dagger b} \quad \text{with} \quad \tau = -\mathrm{i}\omega t. \tag{B.18}$$

Make again a series expansion

$$\frac{\mathrm{d}A}{\mathrm{d}\tau} = [b^+ b, b] = -b, \tag{B.19}$$

$$\frac{\mathrm{d}^2 A}{\mathrm{d}\tau^2} = [b^+ b, -b] = b, \quad \text{etc.,} \tag{B.20}$$

$$A = b\left(1 - \tau + \frac{\tau^2}{2} - \cdots\right) = b\left(1 + \mathrm{i}t\omega + \frac{(\mathrm{i}t\omega)^2}{2} + \cdots\right) = b\,\mathrm{e}^{\mathrm{i}\omega t}. \tag{B.21}$$

Hermitian conjugation gives

$$\mathrm{e}^{-\mathrm{i}\omega t b^\dagger b}b^+ \mathrm{e}^{\mathrm{i}\omega t b^\dagger b} = b^\dagger \mathrm{e}^{-\mathrm{i}\omega t}. \tag{B.22}$$

B.2.3 Derivation of Theorem 3

Consider the operator

$$f(\tau) = \mathrm{e}^{-B\tau}\mathrm{e}^{-A\tau}\mathrm{e}^{(A+B)\tau} \tag{B.23}$$

as a function of the c-number τ. Differentiation gives

$$\frac{\mathrm{d}f(\tau)}{\mathrm{d}\tau} = \mathrm{e}^{-B\tau}(-A-B)\mathrm{e}^{-A\tau}\mathrm{e}^{(A+B)\tau} + \mathrm{e}^{-B\tau}\mathrm{e}^{-A\tau}(A+B)\mathrm{e}^{(A+B)\tau}$$

$$= \mathrm{e}^{-B\tau}[\mathrm{e}^{-A\tau}, B]\mathrm{e}^{(A+B)\tau}. \tag{B.24}$$

Now if the commutator $[A, B]$ is a c-number, then

$$[A^n, B] = A[A^{n-1}, B] + [A, B]A^{n-1} = \cdots n[A, B]A^{n-1} \tag{B.25}$$

and therefore

$$[\mathrm{e}^{-A\tau}, B] = \sum_{n=0}^{\infty} \frac{(-\tau)^n}{n!}[A^n, B] = \sum_{n} \frac{(-\tau)^n}{(n-1)!}A^{n-1}[A, B]$$

$$= -\tau[A, B]\mathrm{e}^{-A\tau} \tag{B.26}$$

and (B.24) gives

$$\frac{\mathrm{d}f(\tau)}{\mathrm{d}\tau} = -\tau[A, B]\mathrm{e}^{-B\tau}\mathrm{e}^{-A\tau}\mathrm{e}^{(A+B)\tau} = -\tau[A, B]f(\tau) \tag{B.27}$$

which is for the initial condition $f(0) = 1$ solved by

$$f(\tau) = \exp\left(-\frac{\tau^2}{2}[A, B]\right). \tag{B.28}$$

Substituting $\tau = 1$ finally gives

$$\mathrm{e}^{-B}\mathrm{e}^{-A}\mathrm{e}^{(A+B)} = \mathrm{e}^{-(1/2)[A,B]}. \tag{B.29}$$

B.2.4 Derivation of Theorem 4

This derivation is based on [131]. For one single oscillator, consider the linear combination

$$\mu b^\dagger + \tau b = A + B, \tag{B.30}$$

$$[A, B] = -\mu\tau. \tag{B.31}$$

Application of (B.29) gives

$$\left\langle \mathrm{e}^{\mu b^\dagger + \tau b} \right\rangle = \left\langle \mathrm{e}^{\mu b^\dagger}\mathrm{e}^{\tau b} \right\rangle \mathrm{e}^{\mu\tau/2} \tag{B.32}$$

and after exchange of A and B

$$\left\langle \mathrm{e}^{\mu b^\dagger + \tau b} \right\rangle = \left\langle \mathrm{e}^{\tau b}\mathrm{e}^{\mu b^\dagger} \right\rangle \mathrm{e}^{-\mu\tau/2}. \tag{B.33}$$

Combination of the last two equations gives

$$\left\langle e^{\tau b} e^{\mu b^\dagger} \right\rangle = \left\langle e^{\mu b^\dagger} e^{\tau b} \right\rangle e^{\mu\tau}. \tag{B.34}$$

Using the explicit form of the averages, we find

$$Q^{-1}\, \mathrm{tr}\left(e^{-\beta\hbar\omega b^\dagger b} e^{\tau b} e^{\mu b^\dagger} \right) = Q^{-1}\, \mathrm{tr}\left(e^{-\beta\hbar\omega b^\dagger b} e^{\mu b^\dagger} e^{\tau b} \right) e^{\mu\tau} \tag{B.35}$$

and due to the cyclic invariance of the trace operation, the right-hand side becomes

$$Q^{-1}\mathrm{tr}(e^{\tau b} e^{-\beta\hbar\omega b^\dagger b} e^{\mu b^\dagger}) e^{\mu\tau} \tag{B.36}$$

which can be written as

$$Q^{-1}\mathrm{tr}(e^{-\beta\hbar\omega b^\dagger b + \beta\hbar\omega b^\dagger b} e^{\tau b} e^{-\beta\hbar\omega b^\dagger b} e^{\mu b^\dagger}) e^{\mu\tau}$$
$$= \left\langle e^{+\beta\hbar\omega b^\dagger b} e^{\tau b} e^{-\beta\hbar\omega b^\dagger b} e^{\mu b^\dagger} \right\rangle e^{\mu\tau}. \tag{B.37}$$

Application of (B.4) finally gives the relation

$$\left\langle \exp(\tau b) \exp(\mu b^\dagger) \right\rangle = \left\langle \exp\left(\tau e^{-\beta\hbar\omega} b \right) \exp\left(\mu b^\dagger \right) \right\rangle e^{\mu\tau} \tag{B.38}$$

which can be iterated to give

$$\left\langle \exp(\tau b) \exp(\mu b^\dagger) \right\rangle = \left\langle \exp\left(\tau e^{-2\beta\hbar\omega} b \right) \exp\left(\mu b^\dagger \right) \right\rangle e^{\mu\tau(1+e^{-\beta\hbar\omega})} = \cdots$$
$$= \left\langle \exp\left(\tau e^{-(n+1)\beta\hbar\omega} b \right) \exp\left(\mu b^\dagger \right) \right\rangle e^{\mu\tau(1+e^{-\beta\hbar\omega}+\cdots+e^{-n\beta\hbar\omega})} \tag{B.39}$$

and in the limit $n \to \infty$

$$\left\langle \exp(\tau b) \exp(\mu b^\dagger) \right\rangle = \left\langle \exp(\mu b^\dagger) \right\rangle \exp\left(\frac{\mu\tau}{1 - e^{-\beta\hbar\omega}} \right)$$
$$= \exp\left(\frac{\mu\tau}{1 - e^{-\beta\hbar\omega}} \right) \tag{B.40}$$

since only the zero-order term of the expansion of the exponential gives a nonzero contribution to the average. With the average number of vibrations

$$\tilde{n} = \frac{1}{e^{\beta\hbar\omega} - 1}, \tag{B.41}$$

we have

$$\left\langle \exp(\tau b) \exp(\mu b^\dagger) \right\rangle = \exp\left(\mu\tau(\tilde{n} + 1) \right) \tag{B.42}$$

and finally

$$\left\langle e^{\mu b^\dagger + \tau b} \right\rangle = e^{\mu\tau(\tilde{n}+1/2)}. \tag{B.43}$$

The average of the square is

$$\left\langle (\mu b^\dagger + \tau b)^2 \right\rangle = \mu\tau \left\langle b^\dagger b + b\, b^\dagger \right\rangle = \mu\tau(2\tilde{n} + 1), \tag{B.44}$$

which shows the validity of the theorem.

C

The Saddle Point Method

The saddle point method is an asymptotic method to calculate integrals of the type

$$\int_{-\infty}^{\infty} e^{n\phi(x)} \mathrm{d}x \tag{C.1}$$

for large n. If the function $\phi(x)$ has a maximum at x_0, then the integrand also has a maximum there which becomes very sharp for large n. Then the integral can be approximated by expanding the exponent around x_0

$$n\phi(x) = n\phi(x_0) + \frac{1}{2} \left. \frac{\mathrm{d}^2(n\phi(x))}{\mathrm{d}x^2} \right|_{x_0} (x - x_0)^2 + \cdots \tag{C.2}$$

as a Gaussian integral

$$\int_{-\infty}^{\infty} e^{n\phi(x)} \mathrm{d}x \approx e^{n\phi(x_0)} \sqrt{\frac{2\pi}{n|\phi''(x_0)|}}. \tag{C.3}$$

The method can be extended to integrals in the complex plane

$$\int_C e^{n\phi(z)} \mathrm{d}z = \int_C e^{n\Re(\phi(z))} e^{in\Im(\phi(z))} \mathrm{d}z. \tag{C.4}$$

If the integration contour is deformed such that the imaginary part is constant, then the Laplace method is applicable and gives

$$\int_C e^{n\phi(z)} \mathrm{d}z = e^{in\Im(\phi(z_0))} \int_{C'} e^{n\Re(\phi(z))} \mathrm{d}z. \tag{C.5}$$

The contour C' and the expansion point are determined from

$$\phi'(z_0) = 0, \tag{C.6}$$

$$\phi(z) = u(z) + iv(z) = u(z) + iv(z_0). \tag{C.7}$$

Now consider the imaginary part as a function in $\mathbb{R}^2$. The gradient is

$$\nabla v(x,y) = \left(\frac{\partial v}{\partial x}, \frac{\partial v}{\partial y} \right). \tag{C.8}$$

For an analytic function $\phi(x + iy)$, we have

$$u(x + dx + iy + idy) + iv(x + dx + iy + idy) - (u(x + iy) + iv(x + iy))$$

$$= \frac{\partial(u + iv)}{\partial x} dx + \frac{\partial(u + iv)}{\partial y} dy, \tag{C.9}$$

which can be written as

$$\frac{d\phi}{dz} dz = \frac{d\phi}{dz}(dx + idy) \tag{C.10}$$

only if

$$i \frac{\partial(u + iv)}{\partial x} = \frac{\partial(u + iv)}{\partial y} \tag{C.11}$$

or

$$\frac{\partial u}{\partial x} = \frac{\partial v}{\partial y}, \quad \frac{\partial v}{\partial x} = -\frac{\partial u}{\partial y}. \tag{C.12}$$

Hence the gradient of the imaginary part

$$\nabla v(x,y) = \left(-\frac{\partial u}{\partial y}, \frac{\partial u}{\partial x} \right) \tag{C.13}$$

is perpendicular to the gradient of the real part

$$\nabla u(x,y) = \left(\frac{\partial u}{\partial x}, \frac{\partial u}{\partial y} \right) \tag{C.14}$$

which gives the direction of steepest descent. The method is known as saddle point method since a maximum of the real part always is a saddle point which can be seen from the expansion

$$\phi(z) = \phi(z_0) + \frac{1}{2}\phi''(z_0)(z - z_0)^2 + \cdots \tag{C.15}$$

We substitute

$$dz^2 = dx^2 - dy^2 + 2idx\,dy \tag{C.16}$$

and find for the real part

$$\Re(\phi(z)) = \Re(\phi(z_0)) - \frac{1}{2}\left(\Re(\phi''(z_0))(dx^2 - dy^2) - 2\Im(\phi''(z_0))dx\,dy \right)$$

$$= \Re(\phi(z_0)) - \frac{1}{2}(dx, dy) \begin{pmatrix} \Re(\phi'') & -\Im(\phi'') \\ -\Im(\phi'') & -\Re(\phi'') \end{pmatrix} \begin{pmatrix} dx \\ dy \end{pmatrix}. \tag{C.17}$$

The eigenvalues of the matrix are

$$\pm \sqrt{\Re(\phi'')^2 + \Im(\phi'')^2} = \pm|\phi''|. \tag{C.18}$$

Solutions

Problems of Chap. 1

1.1. Gaussian Polymer Model

(a)
$$\Delta \mathbf{r}_i = \mathbf{r}_i - \mathbf{r}_{i-1}, \quad i = 1, \ldots, N,$$

$$P(\Delta \mathbf{r}_i) = \frac{1}{b^3} \sqrt{\frac{27}{8\pi^3}} \exp\left\{ -\frac{3(\Delta \mathbf{r}_i)^2}{2b^2} \right\}$$

is the normalized probability function with

$$\int_{-\infty}^{\infty} P(\Delta \mathbf{r}_i) \mathrm{d}^3 \Delta \mathbf{r}_i = 1, \qquad \int_{-\infty}^{\infty} \Delta \mathbf{r}_i^2 P(\Delta \mathbf{r}_i) \mathrm{d}^3 \Delta \mathbf{r}_i = b^2.$$

(b)
$$\mathbf{r}_N - \mathbf{r}_0 = \sum_{i=1}^{N} \Delta \mathbf{r}_i,$$

$$P\left(\sum_{i=1}^{N} \Delta \mathbf{r}_i = \mathbf{R} \right)$$

$$= \int_{-\infty}^{\infty} \mathrm{d}^3 \Delta \mathbf{r}_1 \cdots \int_{-\infty}^{\infty} \mathrm{d}^3 \Delta \mathbf{r}_N \prod_{i=1}^{N} P(\Delta \mathbf{r}_i) \delta\left(\mathbf{R} - \sum_{i=1}^{N} \Delta \mathbf{r}_i \right)$$

$$= \int \frac{1}{(2\pi)^3} \mathrm{d}^3 \mathbf{k} \, \mathrm{e}^{\mathrm{i}\mathbf{k}\mathbf{R}} \int_{-\infty}^{\infty} \mathrm{d}^3 \Delta \mathbf{r}_1 \cdots \int_{-\infty}^{\infty} \mathrm{d}^3 \Delta \mathbf{r}_N \left(\frac{1}{b^3} \sqrt{\frac{27}{8\pi^3}} \right)^N$$

$$\times \prod \exp\left\{ -\frac{3\Delta \mathbf{r}_i^2}{2b^2} - \mathrm{i}\mathbf{k}\Delta \mathbf{r}_i \right\}$$

$$= \int \frac{1}{(2\pi)^3} \mathrm{d}^3 \mathbf{k} \, \mathrm{e}^{\mathrm{i}\mathbf{k}\mathbf{R}}$$

$$\times \left[\left(\frac{1}{b^3} \sqrt{\frac{27}{8\pi^3}} \right) \int_{-\infty}^{\infty} d^3\Delta\mathbf{r} \exp\left\{ -\frac{3\Delta r^2}{2b^2} - i\mathbf{k}\Delta\mathbf{r} \right\} \right]^N$$

$$= \int \frac{1}{(2\pi)^3} d^3\mathbf{k}\, e^{i\mathbf{k}\mathbf{R}} \exp\left(-\frac{Nb^2}{6}\mathbf{k}^2 \right)$$

$$= \frac{1}{b^3} \sqrt{\frac{27}{8\pi^3 N^3}} \exp\left\{ -\frac{3\mathbf{R}^2}{2Nb^2} \right\}.$$

(c)
$$\exp\left\{ -\frac{1}{k_{\mathrm{B}}T} \left(\frac{f}{2} \sum \Delta\mathbf{r}_i^2 - \kappa \sum \Delta\mathbf{r}_i \right) \right\}$$

$$= \prod_i \exp\left\{ -\frac{1}{k_{\mathrm{B}}T} \left(\frac{f}{2} \Delta\mathbf{r}_i^2 - \kappa\Delta\mathbf{r}_i \right) \right\}$$

$$\frac{f}{2k_{\mathrm{B}}T} = \frac{3}{2b^2} \rightarrow f = \frac{3k_{\mathrm{B}}T}{b^2}.$$

(d)
$$x_N - x_0 = \frac{N\kappa}{f}, \quad y_N - y_0 = z_N - z_0 = 0,$$

$$L = x_N - x_0 = \frac{N\kappa b^2}{3k_{\mathrm{B}}T}.$$

1.2. Three-Dimensional Polymer Model

(a) Nb^2.

(b) $b^2 \left(N\dfrac{1+x}{1-x} + \dfrac{2x(x^2-1)}{(1-x)^2} \right) \quad \text{with} \quad x = \cos\theta$

$\approx Nb^2 \dfrac{1+\cos x}{1-\cos x}.$

(c) $Nb^2 \dfrac{(1+\cos\theta_1)(1+\cos\theta_2)}{1-\cos\theta_1\cos\theta_2}.$

(d) $N \gg a/b$ with the coherence length
$$a = b\frac{1+\cos x}{(1-\cos x)\cos(x/2)}.$$

(e) $Nb^2 \left(\dfrac{4}{\theta^2} - 1 \right) - b^2 \left(\dfrac{4}{\theta^2} - \dfrac{8}{\theta^4} \right).$

1.3. Two-Component Model

(a)
$$\kappa = -k_{\mathrm{B}}T\frac{1}{l_\alpha - l_\beta} \ln\left(\frac{Ml_\alpha - L}{L - Ml_\beta} \right)$$

$$-k_{\mathrm{B}}T \left(\frac{1}{2Ml_\alpha - 2L} + \frac{1}{2Ml_\beta - L} + \frac{l_\alpha - l_\beta}{12(L - Ml_\beta)^2} - \frac{l_\alpha - l_\beta}{12(Ml_\alpha - L)^2} \right).$$

The exact solution can be written with the digamma function Ψ which is well known by algebra programs as

$$\kappa = -k_{\mathrm{B}}T \frac{1}{l_\alpha - l_\beta} \left(-\Psi \left(\frac{L - Ml_\beta}{l_\alpha - l_\beta} + 1 \right) + \Psi \left(\frac{Ml_\alpha - L}{l_\alpha - l_\beta} \right) \right).$$

The error of the asymptotic expansion is largest for $L \approx Ml_\alpha$ or $L \approx Ml_\beta$. The following table compares the relative errors of the Stirling's approximation and the higher-order asymptotic expansion for $M = 1,000$ and $l_\beta/l_\alpha = 2$:

L/l_α	Stirling	Asympt. expansion
1,000.2	0.18	0.13
1,000.5	0.11	0.009
1,001	0.065	0.00094
1,005	0.019	2.5×10^{-6}

(b)
$$Z(\kappa, M, T) = \left(e^{\kappa l_\alpha/k_{\mathrm{B}}T} + e^{\kappa l_\beta/k_{\mathrm{B}}T} \right)^M,$$

$$\overline{L} = M \frac{l_\alpha e^{\kappa l_\alpha/k_{\mathrm{B}}T} + l_\beta e^{\kappa l_\beta/k_{\mathrm{B}}T}}{e^{\kappa l_\alpha/k_{\mathrm{B}}T} + e^{\kappa l_\beta/k_{\mathrm{B}}T}},$$

$$\overline{L^2} = \overline{L}^2 + M e^{\kappa(l_\alpha + l_\beta)/k_{\mathrm{B}}T} \left(\frac{l_\alpha - l_\beta}{e^{\kappa l_\alpha/k_{\mathrm{B}}T} + e^{\kappa l_\beta/k_{\mathrm{B}}T}} \right)^2,$$

$$\sigma^2 = M e^{\kappa(l_\alpha + l_\beta)/k_{\mathrm{B}}T} \left(\frac{l_\alpha - l_\beta}{e^{\kappa l_\alpha/k_{\mathrm{B}}T} + e^{\kappa l_\beta/k_{\mathrm{B}}T}} \right)^2,$$

$$\frac{\sigma}{\overline{L}} \sim \frac{1}{\sqrt{N}},$$

$$\frac{\partial \sigma}{\partial \kappa} = 0 \quad \text{for} \quad (l_\alpha + l_\beta) = 2 \frac{l_\alpha e^{\kappa l_\alpha/k_{\mathrm{B}}T} + l_\beta e^{\kappa l_\beta/k_{\mathrm{B}}T}}{e^{\kappa l_\alpha/k_{\mathrm{B}}T} + e^{\kappa l_\beta/k_{\mathrm{B}}T}},$$

hence for

$$\kappa = 0$$

$$\frac{\partial^2 \sigma^2}{\partial \kappa^2}(\kappa = 0) = -\frac{M}{k(k_{\mathrm{B}}T)^2}(l_\alpha - l_\beta)^2 < 0 \rightarrow \text{maximum}$$

also a maximum of σ since the square root is monotonous.

Problems of Chap. 2

2.1. Osmotic Pressure of a Polymer Solution

$$\mu_\alpha(P,T) - \mu_\alpha^0(P,T) = k_\mathrm{B}T\left(\ln(1-\phi_\beta) + \left(1 - \frac{1}{M}\right)\phi_\beta + \chi\phi_\beta^2\right),$$

$$\mu_\alpha^0(P',T) - \mu_\alpha^0(P,T) = \mu_\alpha(P,T) - \mu_\alpha^0(P,T) = -\Pi\frac{\partial\mu_\alpha^0(P,T)}{\partial P},$$

$$\Pi = -\left(\frac{\partial\mu_\alpha^0(P,T)}{\partial P}\right)^{-1} k_\mathrm{B}T\left(\ln(1-\phi_\beta) + \left(1 - \frac{1}{M}\right)\phi_\beta + \chi\phi_\beta^2\right).$$

For the pure solvent,

$$\mu_\alpha^0 = \frac{G}{N_\alpha},$$

$$dG = -S\,dT + V\,dP + \mu_\alpha^0(P,T)dN,$$

$$\left.\frac{\partial\mu_\alpha^0}{\partial P}\right|_{T,N_\alpha} = \frac{V}{N_\alpha},$$

$$\Pi = -\frac{N_\alpha}{V}k_\mathrm{B}T\left(-\phi_\beta - \frac{1}{2}\phi_\beta^2 - \frac{1}{3}\phi_\beta^3 + \cdots + \left(1 - \frac{1}{M}\right)\phi_\beta + \chi\phi_\beta^2\right)$$

$$= \frac{N_\alpha k_\mathrm{B}T}{V}\left(\frac{1}{M}\phi_\beta + \left(\frac{1}{2} - \chi\right)\phi_\beta^2 + \cdots\right),$$

$$\chi = \frac{\chi_0 T_0}{T},$$

high T:

$$\frac{1}{2} - \chi > 0, \quad \Pi > 0 \text{ good solvent}$$

low T:

$$\Pi < 0 \text{ bad solvent, possibly phase separation.}$$

2.2. Polymer Mixture

$$\Delta F = N k_\mathrm{B}T\left(\frac{\phi_1}{M_1}\ln\phi_1 + \frac{\phi_2}{M_2}\ln\phi_2 + \chi\phi_1\phi_2\right),$$

$$\phi_{2,\mathrm{c}} = \frac{1}{1 + \sqrt{M_2/M_1}},$$

$$\chi_\mathrm{c} = \frac{1}{2}\left(\sqrt{M_1} + \sqrt{M_2}\right)\left(\frac{1}{M_2\sqrt{M_1}} + \frac{1}{M_1\sqrt{M_2}}\right),$$

symmetric case

$$\phi_{\mathrm{c}} = \frac{1}{2}$$

$$\chi_{\mathrm{c}} = \frac{2}{M} \quad \text{can be small, demixing possible.}$$

Problems of Chap. 4

4.1. Membrane Potential

$$\Phi_{\mathrm{I}} = B e^{\kappa x}, \quad \Phi_{\mathrm{II}} = B\left(1 + \frac{\epsilon_W}{\epsilon_M}\kappa x\right), \quad \Phi_{\mathrm{III}} = V - B e^{-\kappa(x-L)},$$

$$B = \frac{V}{2 + (\epsilon_w/\epsilon_M)\kappa L},$$

$$Q/A = \epsilon_W \kappa B \quad \text{per area} \ \ A,$$

$$C/A = \frac{Q/A}{V} = \frac{\epsilon_W \kappa}{2 + (\epsilon_W/\epsilon_M)\kappa L} = \frac{1}{(2/\epsilon_w \kappa) + (L/\epsilon_M)}.$$

4.2. Ion Activity

$$k_{\mathrm{B}}T \ln \gamma^{\mathrm{c}} = k_{\mathrm{B}}T \ln \frac{a}{c} = -\frac{Z^2 \mathrm{e}^2}{8\pi\epsilon}\frac{\kappa}{1 + \kappa R},$$

$$\ln \gamma^{\mathrm{c}}_+ = \ln \gamma^{\mathrm{c}}_- = -\frac{1}{k_{\mathrm{B}}T}\frac{Z^2 \mathrm{e}^2}{8\pi\epsilon}\frac{\kappa}{1 + \kappa R},$$

$$\ln \gamma^{\mathrm{c}}_\pm = -\frac{1}{k_{\mathrm{B}}T}\frac{Z^2 \mathrm{e}^2}{16\pi\epsilon}\left(\frac{\kappa}{1 + \kappa R_+} + \frac{\kappa}{1 + \kappa R_-}\right),$$

$\kappa \to 0$ for dilute solution,

$$\ln \gamma^{\mathrm{c}}_\pm \to -\frac{1}{k_{\mathrm{B}}T}\frac{Z^2 \mathrm{e}^2 \kappa}{8\pi\epsilon}.$$

Problems of Chap. 5

5.1. Abnormal Titration Curve

$$\begin{aligned}
\Delta G(\mathrm{B}, \mathrm{B}) &= 0, \\
\Delta G(\mathrm{BH}^+, \mathrm{B}) &= \Delta G_{1,\mathrm{int}}, \\
\Delta G(\mathrm{B}, \mathrm{BH}^+) &= \Delta G_{2,\mathrm{int}}, \\
\Delta G(\mathrm{BH}^+, \mathrm{BH}^+) &= \Delta G_{1,\mathrm{int}} + \Delta G_{2,\mathrm{int}} + W_{1,2},
\end{aligned}$$

$$\Xi = 1 + e^{-\beta(\Delta G_{1,\mathrm{int}}-\mu)} + e^{-\beta(\Delta G_{2,\mathrm{int}}-\mu)} + e^{-\beta(\Delta G_{1,\mathrm{int}}+\Delta G_{2,\mathrm{int}}-W+2\mu)},$$

$$\overline{s_1} = \frac{e^{-\beta(\Delta G_{1,\mathrm{int}}-\mu)} + e^{-\beta(\Delta G_{1,\mathrm{int}}+\Delta G_{2,\mathrm{int}}-W+2\mu)}}{\Xi},$$

$$\overline{s_2} = \frac{e^{-\beta(\Delta G_{2,\mathrm{int}}-\mu)} + e^{-\beta(\Delta G_{1,\mathrm{int}}+\Delta G_{2,\mathrm{int}}-W+2\mu)}}{\Xi}.$$

Problems of Chap. 6

6.1. pH Dependence of Enzyme Activity

$$\frac{r}{r_{\max}} = \frac{1}{1 + (1 + (c_{\mathrm{H+}}/K))(K_M/c_{\mathrm{S}})},$$

$$K = \frac{c_{\mathrm{H+}} c_{\mathrm{S-}}}{c_{\mathrm{HS}}}, \quad c_s = c_{\mathrm{S-}} + c_{\mathrm{HS}}.$$

6.2. Polymerization at the End of a Polymer

$$c_{iM} = K c_M c_{(i-1)M} = \cdots = \frac{(K c_M)^i}{K},$$

$$\langle i \rangle = \frac{\sum_{i=1}^{\infty} i(K c_M)^i}{\sum_{i=1}^{\infty} (K c_M)^i} = \frac{1}{1 - K c_M} \quad \text{for } K c_M < 1.$$

6.3. Primary Salt Effect

$$r = k_1 c_{\mathrm{X}},$$

$$K = \frac{c_{\mathrm{X}}}{c_{\mathrm{A}} c_{\mathrm{B}}} \exp\left\{ -\frac{Z_{\mathrm{A}} Z_{\mathrm{B}} e^2 \kappa}{4\pi \epsilon k_{\mathrm{B}} T} \right\},$$

$$c_{\mathrm{X}} = K c_{\mathrm{A}} c_{\mathrm{B}} \exp\left\{ \frac{Z_{\mathrm{A}} Z_{\mathrm{B}} e^2 \kappa}{4\pi \epsilon k_{\mathrm{B}} T} \right\}.$$

Problems of Chap. 7

7.1. Smoluchowski Equation

$$P(t + \Delta t, x) = e^{\Delta t(\partial/\partial t)} P(t, x),$$

$$w^{\pm}(x \pm \Delta x) P(t, x \pm \Delta x) = e^{\pm \Delta x(\partial/\partial x)} w^{\pm}(x) P(t, x),$$

$$e^{\Delta t(\partial/\partial t)} P(t, x) = e^{\Delta x(\partial/\partial x)} w^{+}(x) P(t, x) + e^{-\Delta x(\partial/\partial x)} w^{-}(x) P(t, x),$$

$$P(t, x) + \Delta t \frac{\partial}{\partial t} P(t, x) + \cdots = (w^{+}(x) + w^{-}(x)) P(x, t)$$

$$+\Delta x \frac{\partial}{\partial x}(w^{+}(x) - w^{-}(x)) P(x, t) + \frac{\Delta x^2}{2} \frac{\partial^2}{\partial x^2}(w^{+}(x) + w^{-}(x)) P(x, t) + \cdots,$$

$$\frac{\partial}{\partial t}P(t,x) = \frac{\Delta x}{\Delta t}\frac{\partial}{\partial x}(w^+(x) - w^-(x))P(x,t) + \frac{\Delta x^2}{\Delta t}\frac{\partial^2}{\partial x^2}P(x,t) + \cdots,$$

$$D = \frac{\Delta x^2}{\Delta t}, \quad K(x) = -\frac{k_{\mathrm{B}}T}{\Delta x}(w^+(x) - w^-(x)).$$

7.2. Eigenvalue Solution to the Smoluchowski Equation

$$-\frac{k_{\mathrm{B}}T}{m\gamma}\mathrm{e}^{-U/k_{\mathrm{B}}T}\left(\frac{\partial}{\partial x}\mathrm{e}^{U/k_{\mathrm{B}}T}W\right)$$

$$= -\frac{k_{\mathrm{B}}T}{m\gamma}\mathrm{e}^{-U/k_{\mathrm{B}}T}\left(\mathrm{e}^{U/k_{\mathrm{B}}T}\frac{\partial W}{\partial x} + \mathrm{e}^{U/k_{\mathrm{B}}T}W\frac{1}{k_{\mathrm{B}}T}\frac{\partial U}{\partial x}\right)$$

$$= -\frac{k_{\mathrm{B}}T}{m\gamma}\frac{\partial W}{\partial x} - \frac{1}{m\gamma}\frac{\partial U}{\partial x}W = S,$$

$$\mathfrak{L}_{\mathrm{FP}}W = -\frac{\partial}{\partial x}S = \frac{\partial}{\partial x}\frac{k_{\mathrm{B}}T}{m\gamma}\mathrm{e}^{-U/k_{\mathrm{B}}T}\left(\frac{\partial}{\partial x}\mathrm{e}^{U/k_{\mathrm{B}}T}W\right),$$

$$\mathfrak{L}_{\mathrm{FP}} = \frac{k_{\mathrm{B}}T}{m\gamma}\frac{\partial}{\partial x}\mathrm{e}^{-U/k_{\mathrm{B}}T}\frac{\partial}{\partial x}\mathrm{e}^{U/k_{\mathrm{B}}T},$$

$$\mathfrak{L} = \mathrm{e}^{U/2k_{\mathrm{B}}T}\mathfrak{L}_{\mathrm{FP}}\mathrm{e}^{-U/2k_{\mathrm{B}}T} = \mathrm{e}^{U/2k_{\mathrm{B}}T}\frac{k_{\mathrm{B}}T}{m\gamma}\frac{\partial}{\partial x}\mathrm{e}^{-U/k_{\mathrm{B}}T}\frac{\partial}{\partial x}\mathrm{e}^{U/2k_{\mathrm{B}}T},$$

$$\mathfrak{L}^{\mathrm{H}} = \mathrm{e}^{U/2k_{\mathrm{B}}T}\left(-\frac{\partial}{\partial x}\right)\mathrm{e}^{-U/k_{\mathrm{B}}T}\left(-\frac{\partial}{\partial x}\right)\frac{k_{\mathrm{B}}T}{m\gamma}\mathrm{e}^{U/2k_{\mathrm{B}}T} = \mathfrak{L},$$

since

$$\frac{k_{\mathrm{B}}T}{m\gamma} = \text{constant.}$$

For an eigenfunction ψ of $\mathfrak{L}$, we have

$$\lambda\psi = \mathfrak{L}\psi = \mathrm{e}^{U/2k_{\mathrm{B}}T}\mathfrak{L}_{\mathrm{FP}}\mathrm{e}^{-U/2k_{\mathrm{B}}T}\psi.$$

Hence

$$\mathfrak{L}_{\mathrm{FP}}\left(\mathrm{e}^{-U/2k_{\mathrm{B}}T}\psi\right) = \lambda\left(\mathrm{e}^{-U/2k_{\mathrm{B}}T}\psi\right)$$

gives an eigenfunction of the Fokker–Planck operator to the same eigenvalue λ. A solution of the Smoluchowski equation is then given by

$$W(x,t) = \mathrm{e}^{\lambda t}\mathrm{e}^{-U/2k_{\mathrm{B}}T}\psi(x).$$

The Hermitian operator is very similar to the harmonic oscillator in second quantization

$$\mathfrak{L} = \frac{k_{\mathrm{B}}T}{m\gamma}\left(\mathrm{e}^{U/2k_{\mathrm{B}}T}\frac{\partial}{\partial x}\mathrm{e}^{-U/2k_{\mathrm{B}}T}\right)\left(\mathrm{e}^{-U/2k_{\mathrm{B}}T}\frac{\partial}{\partial x}\mathrm{e}^{U/2k_{\mathrm{B}}T}\right)$$

$$= \frac{k_{\mathrm{B}}T}{m\gamma}\left(\frac{\partial}{\partial x} - \frac{1}{2k_{\mathrm{B}}T}\frac{\partial U}{\partial x}\right)\left(\frac{\partial}{\partial x} + \frac{1}{2k_{\mathrm{B}}T}\frac{\partial U}{\partial x}\right)$$

$$= \frac{k_{\mathrm{B}}T}{m\gamma}\left(\frac{\partial}{\partial x} - \frac{m\omega^2}{2k_{\mathrm{B}}T}x\right)\left(\frac{\partial}{\partial x} + \frac{m\omega^2}{2k_{\mathrm{B}}T}x\right)$$

$$= -\frac{\omega^2}{\gamma}\left(\sqrt{\frac{k_{\mathrm{B}}T}{m\omega^2}}\frac{\partial}{\partial x} - \frac{1}{2}\sqrt{\frac{m\omega^2}{k_{\mathrm{B}}T}}x\right)\left(\sqrt{\frac{k_{\mathrm{B}}T}{m\omega^2}}\frac{\partial}{\partial x} + \frac{1}{2}\sqrt{\frac{m\omega^2}{k_{\mathrm{B}}T}}x\right)$$

$$= -\frac{\omega^2}{\gamma}\left(\frac{\partial}{\partial \xi} - \frac{1}{2}\xi\right)\left(\frac{\partial}{\partial \xi} + \frac{1}{2}\xi\right) = -\frac{\omega^2}{\gamma}b^+ b$$

with Boson operators

$$b^+ b - bb^+ = 1.$$

From comparison with the harmonic oscillator, we know the eigenvalues

$$\lambda_n = -\frac{\omega^2}{\gamma}n, \quad n = 0, 1, 2, \ldots$$

The ground state obeys

$$a\psi_0 = \left(\frac{\partial}{\partial x} + \frac{1}{2k_{\mathrm{B}}T}\frac{\partial U}{\partial x}\right)\psi_0 = 0$$

with the solution

$$\psi_0 = \mathrm{e}^{-U(x)/2k_{\mathrm{B}}T}.$$

This corresponds to the stationary solution of the Smoluchowski equation:

$$W = \sqrt{\frac{m\omega^2}{2\pi k_{\mathrm{B}}T}}\mathrm{e}^{-U(x)/k_{\mathrm{B}}T}.$$

7.3. Diffusion Through a Membrane

$$k_{\mathrm{AB}} = k_{\mathrm{A}} + k_{\mathrm{B}},$$

$$0 = \frac{\mathrm{d}\overline{N}}{\mathrm{d}t} = \sum_{N=1}^{M}\frac{\mathrm{d}P_N}{\mathrm{d}t}N = -\sum_{N=1}^{M}k_{\mathrm{AB}}MNP_N + \sum_{N=1}^{N}(k_{\mathrm{AB}} - 2k_m)N^2 P_N$$

$$+ \sum_{N=2}^{M}k_{\mathrm{AB}}MNP_{N-1} - \sum_{N=2}^{M}k_{\mathrm{AB}}(N-1)NP_{N-1}$$

$$+ \sum_{N=1}^{M-1}2k_m N(N+1)P_{N+1}$$

$$\approx -k_{\mathrm{AB}}M\overline{N} + (k_{\mathrm{AB}} - 2k_m)\overline{N^2} + k_{\mathrm{AB}}M(1 + \overline{N}) - k_{\mathrm{AB}}(\overline{N^2} + \overline{N})$$

$$+ 2k_m(\overline{N^2} - \overline{N})$$

$$= k_{\mathrm{AB}}M - k_{\mathrm{AB}}\overline{N} - 2k_m\overline{N},$$

$$\overline{N} = M\frac{k_{\mathrm{A}} + k_{\mathrm{B}}}{k_{\mathrm{A}} + k_{\mathrm{B}} + 2k_m},$$

$$0 = \frac{\mathrm{d}\overline{N^2}}{\mathrm{d}t} = \sum N^2 \frac{\mathrm{d}P_N}{\mathrm{d}t} = -\sum_{N=1}^{M} k_{\mathrm{AB}}M N^2 P_N + \sum_{N=1}^{N} (k_{\mathrm{AB}} - 2k_m)N^3 P_N$$

$$+ \sum_{N=2}^{M} k_{\mathrm{AB}}M N^2 P_{N-1} - \sum_{N=2}^{M} k_{\mathrm{AB}}(N-1)N^2 P_{N-1}$$

$$+ \sum_{N=1}^{M-1} 2k_m N^2 (N+1) P_{N+1}$$

$$\approx -k_{\mathrm{AB}}M\overline{N^2} + (k_{\mathrm{AB}} - 2k_m)\overline{N^3} + k_{\mathrm{AB}}M(\overline{N^2} + 2\overline{N} + 1)$$

$$- k_{\mathrm{AB}}(\overline{N^3} + 2\overline{N^2} + \overline{N}) + 2k_m(\overline{N^3} - 2\overline{N^2} + \overline{N})$$

$$= k_{\mathrm{AB}}M(2\overline{N} + 1) - k_{\mathrm{AB}}(2\overline{N^2} + \overline{N}) + 2k_m(-2\overline{N^2} + \overline{N})$$

$$= k_{\mathrm{AB}}M + \overline{N}(2k_{\mathrm{AB}}M - k_{\mathrm{AB}} + 2k_m) - \overline{N^2}(2k_{\mathrm{AB}} + 4k_m)$$

$$\overline{N^2} = \frac{k_{\mathrm{AB}}M + (2k_{\mathrm{AB}}M - k_{\mathrm{AB}} + 2k_m)M\frac{k_{\mathrm{AB}}}{k_{\mathrm{AB}}+2k_m}}{2k_{\mathrm{AB}} + 4k_m}$$

$$= \frac{2k_{\mathrm{AB}}k_m}{(k_{\mathrm{AB}} + 2k_m)^2}M + \frac{k_{\mathrm{AB}}^2}{(k_{\mathrm{AB}} + 2k_m)^2}M^2,$$

and the variance is

$$\overline{N^2} - \overline{N}^2 = \frac{2k_m k_{\mathrm{AB}}}{(k_{\mathrm{AB}} + 2k_m)^2}M = \frac{2k_m}{k_{\mathrm{AB}}M}\overline{N}^2.$$

The diffusion current from A $\rightarrow$ B is

$$J = \frac{\mathrm{d}N_{\mathrm{A}}}{\mathrm{d}t} - \frac{\mathrm{d}N_{\mathrm{B}}}{\mathrm{d}t} = \sum_N \left(-k_{\mathrm{A}}(M - N)P_N + k_m N P_N\right)$$

$$- \sum_N \left(-k_{\mathrm{B}}(M - N)P_N + k_m N P_N\right)$$

$$= \sum_N (k_{\mathrm{B}} - k_{\mathrm{A}})(M - N)P_N = (k_{\mathrm{B}} - k_{\mathrm{A}})(M - \overline{N}).$$

Problems of Chap. 9

9.1. Dichotomous Model

$$\lambda_1 = 0,$$

$$\mathbf{L}_1 = (\,1\;1\;1\;1\,), \quad \mathbf{R}_1 = \begin{pmatrix} 0 \\ 0 \\ \beta \\ \alpha \end{pmatrix}, \quad \frac{(\mathbf{L}_1\mathbf{P}_0)}{(\mathbf{L}_1\mathbf{R}_1)} = \frac{1}{\alpha + \beta},$$

$$\lambda_2 = -(\alpha + \beta),$$

$$\mathbf{L}_2 = (\,\alpha\;-\beta\;\alpha\;-\beta\,), \mathbf{R}_2 = \begin{pmatrix} 0 \\ 0 \\ -1 \\ 1 \end{pmatrix}, \quad \frac{(\mathbf{L}_2\mathbf{P}_0)}{(\mathbf{L}_2\mathbf{R}_2)} = 0,$$

$$\lambda_{3,4} = -\frac{k_- + k_+ + \alpha + \beta}{2}$$

$$\pm \frac{1}{2}\sqrt{(\alpha + \beta)^2 + (k_+ - k_-)^2 + 2(\beta - \alpha)(k_- - k_+)},$$

Fast fluctuations:

$$\lambda_3 = -\frac{\alpha}{\alpha + \beta}k_- - \frac{\beta}{\alpha + \beta}k_+ + O(k^2),$$

$$\mathbf{L}_3 \approx (1, 1, 0, 0), \quad \mathbf{R}_3 \approx \begin{pmatrix} \beta \\ \alpha \\ -\beta \\ -\alpha \end{pmatrix},$$

$$\frac{(\mathbf{L}_3\mathbf{P}_0)}{(\mathbf{L}_3\mathbf{R}_3)} \approx \frac{1}{\alpha + \beta},$$

$$\lambda_4 = -(\alpha + \beta) - \frac{\alpha}{\alpha + \beta}k_+ - \frac{\beta}{\alpha + \beta}k_- + O(k^2),$$

$$\mathbf{L}_4 \approx (-\alpha, \beta, 0, 0), \quad \mathbf{R}_4 \approx \begin{pmatrix} 1 \\ -1 \\ 1 \\ -1 \end{pmatrix}, \quad \frac{(\mathbf{L}_4\mathbf{P}_0)}{(\mathbf{L}_4\mathbf{R}_1)} \approx 0,$$

$$\mathbf{P}(t) \approx \begin{pmatrix} 0 \\ 0 \\ \frac{\beta}{\alpha+\beta} \\ \frac{\alpha}{\alpha+\beta} \end{pmatrix} + \begin{pmatrix} \frac{\beta}{\alpha+\beta} \\ \frac{\alpha}{\alpha+\beta} \\ -\frac{\beta}{\alpha+\beta} \\ -\frac{\alpha}{\alpha+\beta} \end{pmatrix} e^{\lambda_3 t} \to P(D^*) = e^{\lambda_3 t}.$$

Slow fluctuations:

$$\lambda_3 \approx -k_+ - \alpha,$$

$$\mathbf{L}_3 \approx (k_+ - k_-, -\beta, 0, 0), \quad \mathbf{R}_3 \approx \begin{pmatrix} k_+ - k_- \\ -\alpha \\ -(k_+ - k_-) \\ \alpha \end{pmatrix},$$

$$\frac{\mathbf{L}_3 \mathbf{P}_0}{\mathbf{L}_3 \mathbf{R}_3} \approx \frac{\beta}{\alpha + \beta} \frac{1}{k_+ - k_-},$$

$$\lambda_4 \approx -k_- - \beta,$$

$$\mathbf{L}_4 \approx (\alpha, k_+ - k_-, 0, 0), \ \mathbf{R}_4 \approx \begin{pmatrix} \beta \\ k_+ - k_- \\ -\beta \\ -(k_+ - k_-) \end{pmatrix}, \ \frac{\mathbf{L}_4 \mathbf{P}_0}{\mathbf{L}_4 \mathbf{R}_4} \approx \frac{\alpha}{\alpha + \beta} \frac{1}{k_+ - k_-},$$

$$\mathbf{P}(t) \approx \begin{pmatrix} 0 \\ 0 \\ \frac{\beta}{\alpha+\beta} \\ \frac{\alpha}{\alpha+\beta} \end{pmatrix} + \frac{\beta}{\alpha + \beta} \begin{pmatrix} 1 \\ 0 \\ -1 \\ 0 \end{pmatrix} \mathrm{e}^{-(k_+ + \alpha)t} + \frac{\alpha}{\alpha + \beta} \begin{pmatrix} 0 \\ 1 \\ 0 \\ -1 \end{pmatrix} \mathrm{e}^{-(k_- + \beta)t},$$

$$P(\mathrm{D}^*) \approx \frac{\beta}{\alpha + \beta} \mathrm{e}^{-(k_+ + \alpha)t} + \frac{\alpha}{\alpha + \beta} \mathrm{e}^{-(k_- + \beta)t}.$$

Problems of Chap. 10

10.1. Entropy Production

$$0 = \mathrm{d}H = T\,\mathrm{d}S + V\,\mathrm{d}p + \sum_k \mu_k \mathrm{d}N_k,$$

$$T\,\mathrm{d}S = -\sum_k \mu_k \mathrm{d}N_k = -\sum_j \sum_k \mu_k \nu_{kj} d\xi_j = \sum_j A_j d\xi_j,$$

$$\frac{\mathrm{d}S}{\mathrm{d}t} = \sum_j \frac{A_j}{T} r_j.$$

Problems of Chap. 11

11.1. ATP Synthesis

At chemical equilibrium,

$$0 = A = -\sum \nu_k \mu_k$$

$$= \mu^0(\text{ADP}) + k_\text{B}T \ln c(\text{ADP}) + \mu^0(\text{POH}) + k_\text{B}T \ln c(\text{POH})$$
$$+ 2k_\text{B}T \ln c(H^+_\text{out}) + 2e\Phi_\text{out}$$
$$- \mu^0(\text{ATP}) - k_\text{B}T \ln c(\text{ATP}) - \mu^0(\text{H}_2\text{O}) - k_\text{B}T \ln c(\text{H}_2\text{O})$$
$$- 2k_\text{B}T \ln c(H^+_\text{in}) - 2e\Phi_\text{in},$$
$$k_\text{B}T \ln K = -\Delta G^0 = \mu^0(\text{ADP}) - \mu^0(\text{ATP}) + \mu^0(\text{POH}) - \mu^0(\text{H}_2\text{O})$$
$$= k_\text{B}T \ln \frac{c(\text{ATP})c(\text{H}_2\text{O})c^2(H^+_\text{in})}{c(\text{ADP})c(\text{POH})c^2(H^+_\text{out})} + 2e(\Phi_\text{in} - \Phi_\text{out})$$
$$= k_\text{B}T \ln \frac{c(\text{ATP})c(\text{H}_2\text{O})}{c(\text{ADP})c(\text{POH})} + 2k_\text{B}T \ln \frac{c(H^+_\text{in})}{c(H^+_\text{out})} + 2e(\Phi_\text{in} - \Phi_\text{out}).$$

Problems of Chap. 15

15.1. Transition State Theory

$$\text{A} + \text{B} \overset{K^\ddagger}{\rightleftharpoons} [\text{AB}]^\ddagger \overset{k^\ddagger}{\to} \text{P},$$

$$k = k^\ddagger c_{[\text{AB}]^\ddagger} = k^\ddagger K^\ddagger c_\text{A} c_\text{B},$$

$$K^\ddagger = \frac{c_{[\text{AB}]^\ddagger}}{c_\text{A} c_\text{B}} = \frac{q_{[\text{AB}]^\ddagger}}{q_\text{A} q_\text{B}} e^{-\Delta H^\ddagger/k_\text{B}T} = q_x \frac{q^\ddagger_{[\text{AB}]^\ddagger}}{q_\text{A} q_\text{B}} e^{-\Delta H^\ddagger/k_\text{B}T},$$

$$q_x = \frac{\sqrt{2\pi m k_\text{B}T}}{h} \delta x,$$

$$k^\ddagger = \frac{v^\ddagger}{\delta x} = \frac{1}{\delta x}\sqrt{\frac{k_\text{B}T}{2\pi m}},$$

$$k = \frac{1}{\delta x}\sqrt{\frac{k_\text{B}T}{2\pi m}} \frac{\sqrt{2\pi m k_\text{B}T}}{h} \delta x \frac{q^\ddagger_{[\text{AB}]^\ddagger}}{q_\text{A} q_\text{B}} e^{-\Delta H^\ddagger/k_\text{B}T} c_\text{A} c_\text{B}$$

$$= \frac{k_\text{B}T}{h} \frac{q^\ddagger_{[\text{AB}]^\ddagger}}{q_\text{A} q_\text{B}} e^{-\Delta H^\ddagger/k_\text{B}T} c_\text{A} c_\text{B}.$$

15.2. Harmonic Transition State Theory

$$k = v\langle \delta(x - x^\ddagger)\rangle = \sqrt{\frac{k_\text{B}T}{2\pi m}} \frac{\int_{-\infty}^{\infty} e^{-m\omega^2 x^2/2k_\text{B}T}\delta(x - x^\ddagger)}{\int_{-\infty}^{\infty} e^{-m\omega^2 x^2/2k_\text{B}T}}$$

$$= \sqrt{\frac{k_\text{B}T}{2\pi m}} \frac{e^{-m\omega^2 x^{\ddagger 2}/2k_\text{B}T}}{\sqrt{2\pi k_\text{B}T/m\omega^2}}$$

$$= \frac{\omega}{2\pi} e^{-\Delta E/k_\text{B}T}.$$

Problems of Chap. 16

16.1. Marcus Cross Relation

$$A+A^- \to A^-+A, \qquad\qquad \lambda_A = 2\Delta E(A_{eq} \to A_{eq}^-),$$

$$D+D^+ \to D^++D, \qquad\qquad \lambda_D = 2\Delta E(D_{eq} \to D_{eq}^+),$$

$$A+D \to A^-+D^+, \quad \lambda_{AD} = \Delta E(A_{eq} \to A_{eq}^-) + \Delta E(D_{eq} \to D_{eq}^+) = \frac{\lambda_A+\lambda_D}{2},$$

$$k_A = \frac{\omega_A}{2\pi} e^{-\lambda_A/4k_B T},$$

$$k_B = \frac{\omega_B}{2\pi} e^{-\lambda_B/4k_B T},$$

$$K_{AD} = e^{-\Delta G/k_B T},$$

$$k_{AD} = \frac{\omega_{AD}}{2\pi} \exp\left\{ -\frac{(\lambda_{AD} + \Delta G)^2}{4\lambda_{AD}k_B T} \right\}$$

$$= \frac{\omega_{AD}}{2\pi} \exp\left\{ -\frac{\lambda_A + \lambda_D}{8k_B T} - \frac{\Delta G}{2k_B T} - \frac{\Delta G^2}{4\lambda_{AD}k_B T} \right\}$$

$$= \sqrt{k_A k_B K_{AD}} \sqrt{\frac{\omega_{AD}^2}{\omega_A \omega_D}} \exp\left\{ -\frac{\Delta G^2}{4\lambda_{AD}k_B T} \right\}.$$

Problems of Chap. 18

18.1. Absorption Spectrum

$$\alpha = \frac{1}{2\pi\hbar} \int dt \sum_{i,f} e^{i\omega t} \left\langle i \middle| \sum \frac{e^{-\beta H}}{Q} e^{-i\omega_i t} \mu f \right\rangle e^{i\omega_f t} \langle f \mu i \rangle$$

$$= \frac{1}{2\pi\hbar} \int dt\, e^{i\omega t} \langle e^{-iHt/\hbar} \mu e^{iHt/\hbar} \mu \rangle$$

$$= \frac{1}{2\pi\hbar} \int dt\, e^{i\omega t} \langle \mu(0)\mu(t) \rangle$$

$$\approx \frac{|\mu_{eg}|^2}{2\pi\hbar} \int dt\, e^{i\omega t} \langle e^{-iH_g t/\hbar} e^{iH_e t/\hbar} \rangle_g.$$

Problems of Chap. 20

20.1. Motional Narrowing

$$(s + i\omega_1)(s + i\omega_2) + (\alpha + \beta)(s + i\overline{\omega})$$

$$= -\left(\Omega + \frac{\Delta\omega}{2}\right)\left(\Omega - \frac{\Delta\omega}{2}\right) - i\omega_c\Omega,$$

$$\Omega = \omega - \overline{\omega},$$

$$\Omega^2 - \frac{\Delta\omega^2}{4} + i\omega_c\Omega = 0,$$

$$\left(\Omega + \frac{i\omega_c}{2}\right)^2 = \frac{\Delta\omega^2}{4} - \frac{\omega_c^2}{4}.$$

For $\omega_c \ll \Delta\omega$, the poles are approximately at

$$\Omega_{\mathrm{p}} = -\frac{i\omega_c}{2} \pm \frac{\Delta\omega}{2}$$

and two lines are observed centered at the unperturbed frequencies $\overline{\omega} \pm \Delta\omega/2$ and with their width determined by ω_c. For $\omega_c = \Delta\omega$, the two poles coincide at

$$\Omega_{\mathrm{p}} = -\frac{i\omega_c}{2}$$

and a single line at the average frequency $\overline{\omega}$ appears. For $\omega_c \gg \Delta\omega$, one pole approaches zero according to

$$\Omega_{\mathrm{p}} = -i\frac{\Delta\omega^2}{4\omega_c},$$

which corresponds to a sharp line at the average frequency $\overline{\omega}$. The other pole approaches infinity as

$$\Omega_{\mathrm{p}} = -i\omega_c.$$

It contributes a broad line at $\overline{\omega}$ which vanishes in the limit of large ω_c (Fig. 32.1).

Problems of Chap. 21

21.1. Crude Adiabatic Model

$$\frac{\partial C}{\partial Q} = \begin{pmatrix} -s & c \\ -c & -s \end{pmatrix}\frac{\partial \zeta}{\partial Q}, \quad C^\dagger\frac{\partial C}{\partial Q} = \begin{pmatrix} 0 & 1 \\ -1 & 0 \end{pmatrix}\frac{\partial \zeta}{\partial Q},$$

$$\frac{\partial^2 C}{\partial Q^2} = \begin{pmatrix} -s & c \\ -c & -s \end{pmatrix}\frac{\partial^2 \zeta}{\partial Q^2} - C\left(\frac{\partial \zeta}{\partial Q}\right)^2,$$

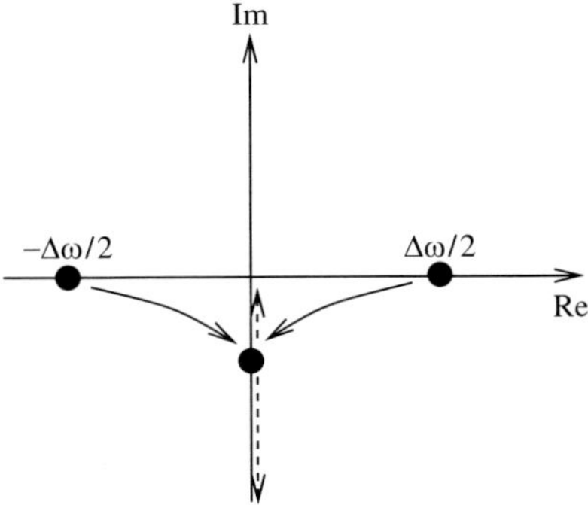

Fig. 32.1. Poles of the lineshape function

$$C^\dagger \frac{\partial^2 C}{\partial Q^2} = \begin{pmatrix} 0 & 1 \\ -1 & 0 \end{pmatrix} \frac{\partial^2 \zeta}{\partial Q^2} - \left(\frac{\partial \zeta}{\partial Q}\right)^2,$$

$$\int dr\, C^\dagger \Phi^\dagger \frac{\partial^2}{\partial Q^2} \Phi C = C^\dagger \frac{\partial^2 C}{\partial Q^2} + 2 C^\dagger \frac{\partial C}{\partial Q} \frac{\partial}{\partial Q} + \frac{\partial^2}{\partial Q^2}$$

$$= \frac{\partial^2}{\partial Q^2} + \begin{pmatrix} 0 & 1 \\ -1 & 0 \end{pmatrix} \frac{\partial^2 \zeta}{\partial Q^2} - \left(\frac{\partial \zeta}{\partial Q}\right)^2 + 2 \begin{pmatrix} 0 & 1 \\ -1 & 0 \end{pmatrix} \frac{\partial \zeta}{\partial Q} \frac{\partial}{\partial Q},$$

$$\int dr\, C^\dagger \Phi^\dagger (T_{\mathrm{el}} + V_0 + \Delta V) \Phi C$$

$$= C^\dagger E C + C^\dagger \int dr\, \Phi^\dagger \Delta V \Phi, \quad C = C^\dagger \begin{pmatrix} \overline{E}(Q) - \frac{\Delta E(Q)}{2} & V(Q) \\ V(Q) & \overline{E}(Q) + \frac{\Delta E(Q)}{2} \end{pmatrix} C,$$

$$\int dr\, C^\dagger \Phi^\dagger H \Phi C$$

$$= -\frac{\hbar^2}{2m} \frac{\partial^2}{\partial Q^2} + C^\dagger \begin{pmatrix} \overline{E}(Q) - \frac{\Delta E(Q)}{2} & V(Q) \\ V(Q) & \overline{E}(Q) + \frac{\Delta E(Q)}{2} \end{pmatrix} C + \frac{\hbar^2}{2m} \left(\frac{\partial \zeta}{\partial Q}\right)^2$$

$$- \frac{\hbar^2}{2m} \begin{pmatrix} 0 & 1 \\ -1 & 0 \end{pmatrix} \left(\frac{\partial^2 \zeta}{\partial Q^2} + 2 \frac{\partial \zeta}{\partial Q} \frac{\partial}{\partial Q}\right),$$

$$\cos\zeta \sin\zeta = \frac{V(Q)}{\sqrt{4V(Q)^2 + \Delta E(Q)^2}},$$

$$\cos\zeta^2 - \sin\zeta^2 = \frac{\Delta E(Q)}{\sqrt{4V(Q)^2 + \Delta E(Q)^2}},$$

$$\frac{\partial}{\partial Q}(cs)^2 = 2cs(c^2 - s^2)\frac{\partial \zeta}{\partial Q} = \frac{2V\Delta E}{4V^2 + \Delta E^2} \frac{\partial \zeta}{\partial Q},$$

$$\frac{\partial}{\partial Q}(cs)^2 = \frac{\partial}{\partial Q}\frac{V^2}{4V^2 + \Delta E^2}$$

$$= \frac{2V}{4V^2 + \Delta E^2}\frac{\partial V}{\partial Q} - \frac{V^2}{(4V^2 + \Delta E^2)^2}\left(2\Delta E\frac{\partial \Delta E}{\partial Q} + 8V\frac{\partial V}{\partial Q}\right),$$

$$\frac{\partial \zeta}{\partial Q} = \frac{\Delta E}{4V^2 + \Delta E^2}\frac{\partial V}{\partial Q} - \frac{V}{4V^2 + \Delta E^2}\frac{\partial \Delta E}{\partial Q} \approx \frac{1}{\Delta E}\frac{\partial V}{\partial Q},$$

$$\frac{\partial^2 \zeta}{\partial Q^2} = \frac{\Delta E\frac{\partial^2 V}{\partial Q^2} - V\frac{\partial^2 \Delta E}{\partial Q^2}}{4V^2 + \Delta E^2} - \left(\Delta E\frac{\partial V}{\partial Q} - V\frac{\partial \Delta E}{\partial Q}\right)\frac{2\Delta E\frac{\partial \Delta E}{\partial Q} + 8V\frac{\partial V}{\partial Q}}{(4V^2 + \Delta e^2)^2}$$

$$\approx \frac{1}{\Delta E}\frac{\partial^2 V}{\partial Q^2} - \frac{2}{\Delta E^2}\frac{\partial \Delta E}{\partial Q}\frac{\partial V}{\partial Q},$$

$$\widetilde{H} \approx -\frac{\hbar^2}{2m}\frac{\partial^2}{\partial Q^2} + \left(\begin{matrix}\overline{E} - \sqrt{4V^2 + \Delta E^2} & \\ & \overline{E} + \sqrt{4V^2 + \Delta E^2}\end{matrix}\right)$$

$$+ \frac{\hbar^2}{2m}\frac{1}{\Delta E^2}\left(\frac{\partial V}{\partial Q}\right)^2$$

$$- \frac{\hbar^2}{2m}\begin{pmatrix}0 & 1 \\ -1 & 0\end{pmatrix}\left(\frac{1}{\Delta E}\frac{\partial^2 V}{\partial Q^2} - \frac{2}{\Delta E^2}\frac{\partial \Delta E}{\partial Q}\frac{\partial V}{\partial Q} + \frac{2}{\Delta E}\frac{\partial V}{\partial Q}\frac{\partial}{\partial Q}\right).$$

Problems of Chap. 22

22.1. Ladder Model

$$\mathrm{i}\hbar \dot{C}_0 = V\sum_{j=1}^{n} C_j,$$

$$\mathrm{i}\hbar \dot{C}_j = E_j C_j + VC_0,$$

$$C_j = u_j \mathrm{e}^{(E_j/\mathrm{i}\hbar)t},$$

$$\mathrm{i}\hbar \dot{u}_j \mathrm{e}^{(E_j/\mathrm{i}\hbar)t} = VC_0,$$

$$u_j = \frac{V}{\mathrm{i}\hbar}\int_0^t \mathrm{e}^{-(E_j/\mathrm{i}\hbar)t'}C_0(t')\mathrm{d}t',$$

$$C_j = \frac{V}{\mathrm{i}\hbar}\int_0^t \mathrm{e}^{\mathrm{i}(E_j/\hbar)(t-t')}C_0(t')\mathrm{d}t',$$

$$E_j = \alpha + j^* \hbar \Delta\omega,$$

$$\dot{C}_0 = \frac{V}{\mathrm{i}\hbar}\sum_{j=1}^{n} C_j = -\frac{V^2}{\hbar^2}\sum\int_0^t \mathrm{e}^{\mathrm{i}(j\Delta\omega + \alpha/\hbar)(t-t')}C_0(t')\mathrm{d}t',$$

$$\omega = j\Delta\omega + \frac{\alpha}{\hbar},$$

$$\sum_{j=-\infty}^{\infty} e^{i(j\Delta\omega+\alpha/\hbar)(t-t')}\Delta j \;\rightarrow\; \int_{-\infty}^{\infty} e^{i\omega(t-t')}\frac{d\omega}{\Delta\omega} = \frac{2\pi}{\Delta\omega}\delta(t-t'),$$

$$\dot{C}_0 = -\frac{2\pi V^2}{\Delta\omega}C_0 = -\frac{2\pi V^2}{\hbar}\rho(E)C_0,$$

$$\rho(E) = \frac{1}{\hbar\Delta\omega} = \frac{1}{\Delta E}.$$

Problems of Chap. 23

23.1. Hückel Model with Alternating Bonds

(a)
$$\alpha e^{ikn} + \beta e^{i(kn+\chi)} + \beta' e^{i(kn+k+\chi)} = e^{ikn}\left(\alpha + \beta e^{i\chi} + \beta' e^{i(k+\chi)}\right),$$

$$\alpha e^{i(kn+\chi)} + \beta' e^{i(kn-k)} + \beta e^{ikn} = e^{i(kn+\chi)}\left(\alpha + \beta' e^{-i(k+\chi)} + \beta e^{-i\chi}\right).$$

(b)
$$\beta e^{i\chi} + \beta' e^{i(k+\chi)} = \beta' e^{-i(k+\chi)} + \beta e^{-i\chi},$$

$$e^{2i\chi} = \frac{\beta' e^{-ik} + \beta}{\beta' e^{ik} + \beta} = e^{-ik}\frac{\beta' e^{-ik/2} + \beta e^{ik/2}}{\beta' e^{ik/2} + \beta e^{-ik/2}} = e^{-ik}\frac{\left(\beta' e^{-ik/2} + \beta e^{ik/2}\right)^2}{\beta'^2 + \beta^2 + 2\beta\beta'\cos k},$$

$$e^{i\chi} = \pm e^{-ik/2}\frac{\beta' e^{-ik/2} + \beta e^{ik/2}}{\sqrt{\beta'^2 + \beta^2 + 2\beta\beta'\cos k}},$$

$$\lambda = \alpha + \beta e^{i\chi} + \beta' e^{i(k+\chi)} = \alpha \pm \frac{\beta\beta' e^{-ik} + \beta^2 + \beta'^2 + \beta\beta' e^{ik}}{\sqrt{\beta'^2 + \beta^2 + 2\beta\beta'\cos k}}$$

$$= \alpha \pm \sqrt{\beta'^2 + \beta^2 + 2\beta\beta'\cos k}.$$

(c)
$$0 = \Im\left(e^{i\chi+i(N+1)k}\right) = \Im\left(\pm e^{-ik/2}\frac{\beta' e^{-ik/2} + \beta e^{ik/2}}{\sqrt{\beta'^2 + \beta^2 + 2\beta\beta'\cos k}}e^{i(N+1)k}\right)$$

$$= \pm\Im\left(\frac{\beta' e^{iNk} + \beta e^{i(N+1)k}}{\sqrt{\beta'^2 + \beta^2 + 2\beta\beta'\cos k}}\right),$$

$$0 = \beta'\sin(Nk) + \beta\sin(N+1)k.$$

(d) For a linear polyene with $2N-1$ carbon atoms, use again eigenfunctions
$$c_{2n} = \sin(kn) = \Im(e^{ikn}),$$
$$c_{2n-1} = \sin(kn+\chi) = \Im(e^{i(kn+\chi)}),$$

and chose the k-values such that
$$\Im(e^{iNk}) = \sin(Nk) = 0.$$

Problems of Chap. 25

25.1. Special Pair Dimer

$$\begin{pmatrix} c & -s \\ s & c \end{pmatrix} \begin{pmatrix} -\Delta/2 & V \\ V & \Delta/2 \end{pmatrix} \begin{pmatrix} c & s \\ -s & c \end{pmatrix}$$

is diagonalized if

$$(c^2 - s^2)V = cs\Delta$$

or with $c = \cos\chi$, $s = \sin\chi$

$$\tan(2\chi) = \frac{2V}{\Delta},$$

$$c^2 - s^2 = \cos 2\chi = \frac{1}{\sqrt{1 + (4V^2/\Delta^2)}} \geq 0,$$

$$2cs = \sin(2\chi) = \frac{1}{\sqrt{1 + (\Delta^2/4V^2)}} \geq 0.$$

The eigenvalues are (Fig. 32.2)

$$E_\pm = \pm\left(\frac{\Delta}{2}(c^2 - s^2) + 2csV\right)$$

$$= \pm\left(\frac{\Delta}{2}\frac{1}{\sqrt{1 + \frac{4V^2}{\Delta^2}}} + V\frac{1}{\sqrt{1 + \frac{\Delta^2}{4V^2}}}\right) = \pm\frac{1}{2}\sqrt{\Delta^2 + 4V^2}.$$

The transition dipoles are

$$\mu_+ = s\mu_{\mathrm{a}} + c\mu_{\mathrm{b}},$$

$$\mu_- = c\mu_{\mathrm{a}} - s\mu_{\mathrm{b}},$$

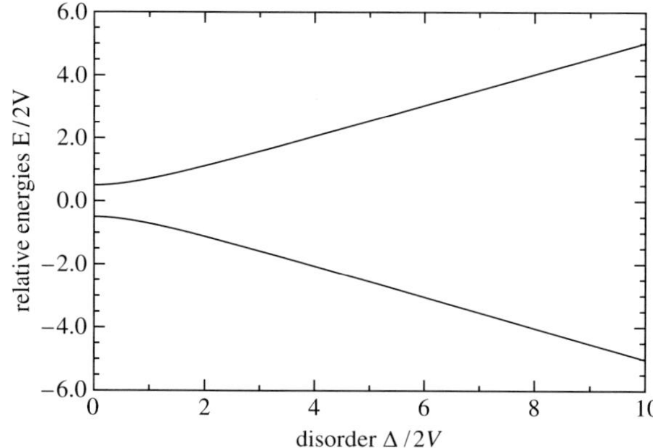

Fig. 32.2. Energy splitting of the two dimer bands

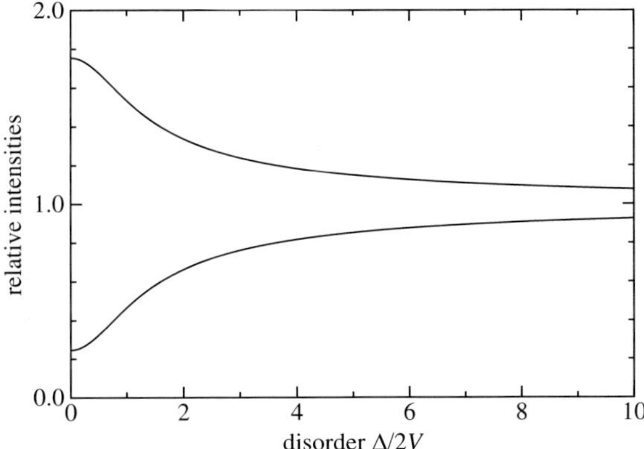

Fig. 32.3. Intensities of the two dimer bands

and the intensities (for $|\mu_a| = |\mu_b| = \mu$)

$$|\mu_\pm|^2 = \mu^2(1 \pm 2cs\cos\alpha) = \mu^2\left(1 \pm \frac{\cos\alpha}{\sqrt{1 + (\Delta^2/4V^2)}}\right)$$

with (Fig. 32.3)

$$\cos\alpha = -0.755.$$

25.2. LH2

$$|n;\alpha\rangle = \frac{1}{3}\sum_k e^{-ikn}|k;\alpha\rangle,$$

$$\sum_{n=1}^{9} E_\alpha|n;\alpha\rangle\langle n;\alpha| = \sum_{n=1}^{9} E_\alpha \frac{1}{9}\sum_{k,k'} e^{-i(k-k')n}|k;\alpha\rangle\langle k';\alpha|$$

$$= \delta_{k,k'} E_\alpha|k;\alpha\rangle\langle k;\alpha|,$$

$$\sum_{n=1}^{9} E_\alpha|n;\beta\rangle\langle n;\beta| \sum_{n=1}^{9} E_\alpha \frac{1}{9}\sum_{k,k'} e^{-i(k-k')n}|k;\beta\rangle\langle k';\beta|$$

$$= \delta_{k,k'} E_\alpha|k;\beta\rangle\langle k;\beta|,$$

$$\sum_{n=1}^{9} V_{\mathrm{dim}}|n;\alpha\rangle\langle n;\beta| = \sum_{n=1}^{9} V_{\mathrm{dim}} \frac{1}{9}\sum_{k,k'} e^{-i(k-k')n}|k;\alpha\rangle\langle k';\beta|$$

$$= \delta_{k,k'} V_{\mathrm{dim}}|k;\alpha\rangle\langle k;\beta|,$$

$$\sum_{n=1}^{9} V_{\beta\alpha,1}|n;\alpha\rangle\langle n-1;\beta| = \sum_{n=1}^{9} V_{\beta\alpha,1}\mathrm{e}^{-\mathrm{i}k}\frac{1}{9}\sum_{k,k'}\mathrm{e}^{-\mathrm{i}(k-k')n}|k;\alpha\rangle\langle k';\beta|$$

$$= \delta_{k,k'}V_{\beta\alpha,1}\mathrm{e}^{-\mathrm{i}k}|k;\alpha\rangle\langle k;\beta|,$$

$$\sum_{n=1}^{9} V_{\beta\alpha,1}|n;\beta\rangle\langle n+1;\alpha| = \sum_{n=1}^{9} V_{\beta\alpha,1}\mathrm{e}^{\mathrm{i}k}\frac{1}{9}\sum_{k,k'}\mathrm{e}^{-\mathrm{i}(k-k')n}|k;\beta\rangle\langle k';\alpha|$$

$$= \delta_{k,k'}V_{\beta\alpha,1}\mathrm{e}^{\mathrm{i}k}|k;\beta\rangle\langle k;\alpha|,$$

$$\sum_{n=1}^{9} V_{\alpha\alpha,1}(|n;\alpha\rangle\langle n+1;\alpha| + \text{h.c.})$$

$$= \sum_{n=1}^{9} V_{\alpha\alpha,1}\mathrm{e}^{\mathrm{i}k}\frac{1}{9}\sum_{k,k'}\mathrm{e}^{-\mathrm{i}(k-k')n}|k;\alpha\rangle\langle k';\alpha| + \text{h.c.}$$

$$= \delta_{k,k'}2V_{\alpha\alpha,1}\cos k|k;\alpha\rangle\langle k;\alpha|,$$

$$H_{\alpha\alpha}(k) = E_\alpha + 2V_{\alpha\alpha1}\cos k,$$

$$H_{\beta\beta}(k) = E_\beta + 2V_{\beta\beta1}\cos k,$$

$$H_{\alpha\beta}(k) = V_{\mathrm{dim}} + \mathrm{e}^{-\mathrm{i}k}V_{\beta\alpha,1},$$

$$H_{\beta\alpha}(k) = V_{\mathrm{dim}} + \mathrm{e}^{\mathrm{i}k}V_{\beta\alpha,1},$$

$$H(k) = \begin{pmatrix} E_\alpha + 2V_{\alpha\alpha1}\cos k & V + \mathrm{e}^{-\mathrm{i}k}W \\ V + \mathrm{e}^{\mathrm{i}k}W & E_\beta + 2V_{\beta\beta1}\cos k \end{pmatrix}$$

$$= \overline{E}_k + \begin{pmatrix} -\Delta_k/2 & V + \mathrm{e}^{-\mathrm{i}k}W \\ V + \mathrm{e}^{\mathrm{i}k}W & \Delta_k/2 \end{pmatrix}.$$

Perform a canonical transformation with

$$S = \begin{pmatrix} c & -s\mathrm{e}^{-\mathrm{i}\chi} \\ s\mathrm{e}^{\mathrm{i}\chi} & c \end{pmatrix}.$$

$S^\dagger H S$ becomes diagonal if

$$c^2(V + \mathrm{e}^{\mathrm{i}k}W) - s^2\mathrm{e}^{2\mathrm{i}\chi}(V + \mathrm{e}^{-\mathrm{i}k}W) + cs\mathrm{e}^{\mathrm{i}\chi}\Delta_k = 0.$$

Chose χ such that

$$V + \mathrm{e}^{\mathrm{i}k}W = \mathrm{sign}\,V|V + \mathrm{e}^{\mathrm{i}k}W|\mathrm{e}^{\mathrm{i}\chi} = U(k)\mathrm{e}^{\mathrm{i}\chi}$$

and solve

$$(c^2 - s^2)U + cs\Delta_k = 0$$

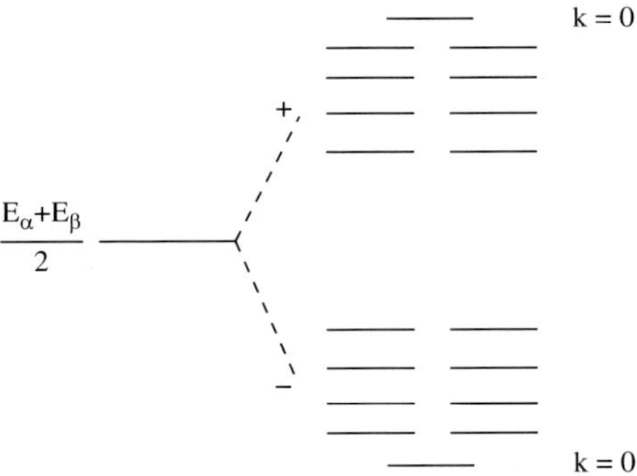

Fig. 32.4. Energy levels of LH2

by[1]

$$c^2 - s^2 = -\text{sign}\left(\frac{U}{\Delta}\right)\frac{1}{\sqrt{1 + (4U^2/\Delta^2)}}, \quad cs = \frac{|U/\Delta|}{\sqrt{1 + (4U^2/\Delta^2)}}.$$

The eigenvalues are (Fig. 32.4)

$$E_\pm(k) = \overline{E}_k \pm \text{sign}\, V\frac{1}{2}\sqrt{\Delta^2 + 4U^2}$$

$$= \frac{E_\alpha + E_\beta}{2} + (V_{\alpha\alpha 1} + V_{\beta\beta 1})\cos k$$

$$\pm \text{sign}\, V\sqrt{\left(\frac{E_\alpha - E_\beta + 2(V_{\alpha\alpha 1} - V_{\beta\beta 1})\cos k}{2}\right)^2 + V^2 + W^2 + 2VW\cos k}$$

$$\mu_{k,+} = c\frac{1}{3}\sum_n e^{ikn}\mu_{n\alpha} + se^{i\chi}\frac{1}{3}\sum_n e^{ikn}\mu_{n\beta}$$

$$= c\frac{1}{3}\sum_n e^{ikn}S_9^n\mu_{0\alpha} + se^{i\chi}\frac{1}{3}\sum_n e^{ikn}S_9^n R_z(2\nu + \phi_\beta - \phi_\alpha)\mu_{0,\alpha}$$

$$= \frac{1}{3}\sum_n e^{ikn}\begin{pmatrix}\cos\frac{2\pi}{9}n & -\sin\frac{2\pi}{9}n & \\ \sin\frac{2\pi}{9}n & \cos\frac{2\pi}{9}n & \\ & & 1\end{pmatrix}$$

$$\times \left(c + se^{i\chi}\begin{pmatrix}\cos\epsilon & -\sin\epsilon & \\ \sin\epsilon & \cos\epsilon & \\ & & 1\end{pmatrix}\right)\mu_0\begin{pmatrix}\sin\theta \\ 0 \\ \cos\theta\end{pmatrix}.$$

---

[1] The sign is chosen such that for $|\Delta| \ll |U|$, the solution becomes $c = s = 1/\sqrt{2}$.

The sum over n again gives the selection rules

$$k = 0 \quad z\text{-polarization,}$$

$$k = \pm\frac{2\pi}{9} \quad \text{circular } xy\text{-polarization.}$$

The second factor gives for z-polarization

$$\mu = 3\mu_0(c + se^{i\chi})\cos\theta,$$

$$|\mu|^2 = 9\mu_0^2(1 + 2cs\,\cos\chi)\cos^2\theta,$$

with

$$\cos^2\theta \approx 0.008$$

and for polarization in the xy-plane

$$\mu = 3\mu_0\sin\theta\left(\begin{array}{c} c + se^{i\chi}\cos\epsilon \\ se^{i\chi}\sin\epsilon \end{array}\right),$$

$$|\mu|^2 = 9\mu_0^2\sin^2\theta(1 + 2\cos\epsilon\,cs\cos\chi),$$

with

$$\sin^2\theta \approx 0.99, \quad \cos\epsilon \approx -0.952.$$

The intensities of the $(k, -)$ states are

$$|\mu_z|^2 = 9\mu_0^2(1 - 2cs\,\cos\chi)\cos^2\theta,$$

$$|\mu_\perp|^2 = 9\mu_0^2\sin^2\theta(1 - 2\cos\epsilon\,cs\cos\chi).$$

25.3. Exchange Narrowing

$$P(\delta E_k = X) = \int d\delta E_1 d\delta E_2 \cdots P(\delta E_1)P(\delta E_2)\cdots\delta\left(X - \frac{\sum \delta E_n}{N}\right),$$

$$P(\delta E_k = X) = \frac{1}{2\pi}\int dt \int \delta E_1 d\delta E_2 \cdots \frac{1}{\Delta\sqrt{\pi}}e^{-\delta E_1^2/\Delta^2}\cdots e^{it(X - \sum \delta E_n/N)}$$

$$= \frac{1}{2\pi}\int dt\, e^{itX}\left(\frac{1}{\Delta\sqrt{\pi}}\int dt\,\delta E_1 e^{-\delta E_1^2/\Delta^2 - it\delta E_1/N}\right)^N$$

$$= \frac{1}{2\pi}\int dt\, e^{itX}e^{-\Delta^2 t^2/4N}$$

$$= \frac{\sqrt{N}}{\sqrt{\pi}\Delta}e^{-X^2 N/\Delta^2}. \tag{32.19}$$

Problems of Chap. 28

28.1. Deviation from Equilibrium

$$\Omega = \mathrm{e}^{-\Delta U/k_{\mathrm{B}}T}\left(1 - \mathrm{e}^{\Delta\mu/k_{\mathrm{B}}T}\right)\frac{\alpha_2}{\alpha_2 + \beta_2},$$

$$\Omega(x) = \mathrm{e}^{(\Delta\mu^0 - \Delta U(x))/k_{\mathrm{B}}T}\frac{\alpha_2(x)}{\alpha_2(x) + \beta_2(x)}\left(K_{\mathrm{eq}} - \frac{C(\mathrm{ATP})}{C(\mathrm{ADP})C(\mathrm{P})}\right).$$

For $\Delta\mu \neq 0$ but $\Delta\mu \ll k_{\mathrm{B}}T$, Ω becomes a linear function of $\Delta\mu$

$$\Omega(x) \to -\mathrm{e}^{-\Delta U(x)/k_{\mathrm{B}}T}\frac{\alpha_2(x)}{\alpha_2(x) + \beta_2(x)}\frac{\Delta\mu}{k_{\mathrm{B}}T},$$

whereas in the opposite limit $\Delta\mu \gg k_{\mathrm{B}}T$ it becomes proportional to the concentration ratio:

$$\Omega(x) \to -\mathrm{e}^{-\Delta U(x)/k_{\mathrm{B}}T}\frac{\alpha_2(x)}{\alpha_2(x) + \beta_2(x)}\mathrm{e}^{\Delta\mu^0/k_{\mathrm{B}}T}\frac{C(\mathrm{ATP})}{C(\mathrm{ADP})C(\mathrm{P})}.$$

References

1. T.L. Hill, *An Introduction to Statistical Thermodynamics* (Dover, New York, 1986)
2. S. Cocco, J.F. Marko, R. Monasson, arXiv:cond-mat/0206238v1
3. C. Storm, P.C. Nelson, Phys. Rev. E **67**, 51906 (2003)
4. K.K. Mueller-Niedebock, H.L. Frisch, Polymer **44**, 3151 (2003)
5. C. Leubner, Eur. J. Phys. **6**, 299 (1985)
6. P.J. Flory, J. Chem. Phys. **10**, 51 (1942)
7. M.L. Huggins, J. Phys. Chem. **46**, 151 (1942)
8. M. Feig, C.L. Brooks III, Curr. Opin. Struct. Biol. **14**, 217 (2004)
9. B. Roux, T. Simonson, Biophys. Chem. **78**, 1 (1999)
10. A. Onufriev, Annu. Rep. Comput. Chem. **4**, 125 (2008)
11. M. Born, Z. Phys. **1**, 45 (1920)
12. D. Bashford, D. Case, Annu. Rev. Phys. Chem. **51**, 29 (2000)
13. W.C. Still, A. Tempczyk, R.C. Hawley, T. Hendrickson, JACS **112**, 6127 (1990)
14. G.D. Hawkins, C.J. Cramer, D.G. Truhlar, Chem. Phys. Lett. **246**, 122 (1995)
15. M. Schaefer, M. Karplus, J. Phys. Chem. **100**, 1578 (1996)
16. I.M. Wonpil, M.S. Lee, C.L. Brooks III, J. Comput. Chem. **24**, 1691 (2003)
17. F. Fogolari, A. Brigo, H. Molinari, J. Mol. Recognit. **15**, 377 (2002)
18. A.I. Shestakov, J.L. Milovich, A. Noy, J. Colloid Interface Sci. **247**, 62 (2002)
19. B. Lu, D. Zhang, J.A. McCammon, J. Chem. Phys. **122**, 214102 (2005)
20. P. Koehl, Curr. Opin. Struct. Biol. **16**, 142 (2006)
21. P. Debye, E. Hueckel, Phys. Z. **24**, 305 (1923)
22. G. Goüy, Comt. Rend. **149**, 654 (1909)
23. G. Goüy, J. Phys. **9**, 457 (1910)
24. D.L. Chapman, Philos. Mag. **25**, 475 (1913)
25. O. Stern, Z. Elektrochem. **30**, 508 (1924)
26. A.-S. Yang, M. Gunner, R. Sampogna, K. Sharp, B. Honig, Proteins **15**, 252 (1993)
27. P.W. Atkins, *Physical Chemistry* (Freeman, New York, 2006)
28. W.J. Moore, *Basic Physical Chemistry* (Prentice-Hall, New York, 1983)
29. M. Schaefer, M. Sommer, M. Karplus, J. Phys. Chem. B **101**, 1663 (1997)
30. R.A. Raupp-Kossmann, C. Scharnagl, Chem. Phys. Lett. **336**, 177 (2001)
31. H. Risken, *The Fokker–Planck Equation* (Springer, Berlin, 1989)
32. H.A. Kramers, Physica **7**, 284 (1941)

33. P.O.J. Scherer, Chem. Phys. Lett. **214**, 149 (1993)
34. E.W. Montroll, H. Scher, J. Stat. Phys. **9**, 101 (1973)
35. E.W. Montroll, G.H. Weiss, J. Math. Phys. **6**, 167 (1965)
36. A.A. Zharikov, P.O.J. Scherer, S.F. Fischer, J. Phys. Chem. **98**, 3424 (1994)
37. A.A. Zharikov, S.F. Fischer, Chem. Phys. Lett. **249**, 459 (1995)
38. S.R. de Groot, P. Mazur, *Irreversible Thermodynamics* (Dover, New York, 1984)
39. D.G. Miller, Faraday Discuss. Chem. Soc. **64**, 295 (1977)
40. R. Paterson, Faraday Discuss. Chem. Soc. **64**, 304 (1977)
41. D.E. Goldman, J. Gen. Physiol. **27**, 37 (1943)
42. A.L. Hodgkin, A.F. Huxley, J. Physiol. **117**, 500 (1952)
43. J. Kenyon, How to solve and program the Hodgkin–Huxley equations (http://134.197.54.225/department/Faculty/kenyon/Hodgkin&Huxley/pdfs/HH.Program.pdf)
44. J. Vreeken, A friendly introduction to reaction–diffusion systems, Internship Paper, AILab, Zurich, July 2002
45. E. Pollak, P. Talkner, Chaos **15**, 026116 (2005)
46. F.T. Gucker, R.L. Seifert, *Physical Chemistry* (W.W. Norton, New York, 1966)
47. S. Glasstone, K.J. Laidler, H. Eyring, *The Theory of Rate Processes* (McGraw-Hill, New York, 1941)
48. K.J. Laidler, *Chemical Kinetics*, 3rd edn. (Harper & Row, New York, 1987)
49. G.A. Natanson, J. Chem. Phys. **94**, 7875 (1991)
50. R.A. Marcus, Annu. Rev. Phys. Chem. **15**, 155 (1964)
51. R.A. Marcus, N. Sutin, Biochim. Biophys. Acta **811**, 265 (1985)
52. R.A. Marcus, Angew. Chem. Int. Ed. Engl. **32**, 1111 (1993)
53. A.M. Kuznetsov, J. Ulstrup, *Electron Transfer in Chemistry and Biology* (Wiley, Chichester, 1998), p. 49
54. P.W. Anderson, J. Phys. Soc. Jpn. **9**, 316 (1954)
55. R. Kubo, J. Phys. Soc. Jpn. **6**, 935 (1954)
56. R. Kubo, in *Fluctuations, Relaxations and Resonance in Magnetic Systems*, ed. by D. ter Haar (Plenum, New York, 1962)
57. T.G. Heil, A. Dalgarno, J. Phys. B **12**, 557 (1979)
58. L.D. Landau, Phys. Z. Sowjetun. **1**, 88 (1932)
59. C. Zener, Proc. R. Soc. Lond. A **137**, 696 (1932)
60. E.S. Kryachko, A.J.C. Varandas, Int. J. Quant. Chem. **89**, 255 (2001)
61. E.N. Economou, *Green's Functions in Quantum Physics* (Springer, Berlin, 1978)
62. H. Scheer, W.A. Svec, B.T. Cope, M.H. Studler, R.G. Scott, J.J. Katz, JACS **29**, 3714 (1974)
63. A. Streitwieser, *Molecular Orbital Theory for Organic Chemists* (Wiley, New York, 1961)
64. J.E. Lennard-Jones, Proc. R. Soc. Lond. A **158**, 280 (1937)
65. B.S. Hudson, B.E. Kohler, K. Schulten, in *Excited States*, ed. by E.C. Lin (Academic, New York, 1982), pp. 1–95
66. B.E. Kohler, C. Spangler, C. Westerfield, J. Chem. Phys. **89**, 5422 (1988)
67. T. Polivka, J.L. Herek, D. Zigmantas, H.-E. Akerlund, V. Sundstrom, Proc. Natl. Acad. Sci. USA **96**, 4914 (1999)
68. B. Hudson, B. Kohler, Synth. Met. **9**, 241 (1984)
69. B.E. Kohler, J. Chem. Phys. **93**, 5838 (1990)

70. W.T. Simpson, J. Chem. Phys. **17**, 1218 (1949)
71. H. Kuhn, J. Chem. Phys. **17**, 1198 (1949)
72. M. Gouterman, J. Mol. Spectrosc. **6**, 138 (1961)
73. M. Gouterman, J. Chem. Phys. **30**, 1139 (1959)
74. M. Gouterman, G.H. Wagniere, L.C. Snyder, J. Mol. Spectrosc. **11**, 108 (1963)
75. C. Weiss, *The Porphyrins*, vol. III (Academic, New York, 1978), p. 211
76. D. Spangler, G.M. Maggiora, L.L. Shipman, R.E. Christofferson, J. Am. Chem. Soc. **99**, 7470 (1977)
77. B.R. Green, D.G. Durnford, Annu. Rev. Plant Physiol. Plant Mol. Biol. **47**, 685 (1996)
78. H.A. Frank et al., Pure Appl. Chem. **69**, 2117 (1997)
79. R.J. Cogdell et al., Pure Appl. Chem. **66**, 1041 (1994)
80. H. van Amerongen, R. van Grondelle, J. Phys. Chem. B **105**, 604 (2001)
81. T. Förster, Ann. Phys. **2**, 55 (1948)
82. T. Förster, Disc. Faraday Trans. **27**, 7 (1965)
83. D.L. Dexter, J. Chem. Phys. **21**, 836 (1953)
84. F.J. Kleima, M. Wendling, E. Hofmann, E.J.G. Peterman, R. van Grondelle, H. van Amerongen, Biochemistry **39**, 5184 (2000)
85. T. Pullerits, M. Chachisvilis, V. Sundström, J. Phys. Chem. **100**, 10787 (1996)
86. R.C. Hilborn, Am. J. Phys. **50**, 982 (1982), revised 2002
87. S.H. Lin, Proc. R. Soc. Lond. A **335**, 51 (1973)
88. S.H. Lin, W.Z. Xiao, W. Dietz, Phys. Rev. E **47**, 3698 (1993)
89. MOLEKEL 4.0, P. Fluekiger, H.P. Luethi, S. Portmann, J. Weber, Swiss National Supercomputing Centre CSCS, Manno Switzerland, 2000
90. J. Deisenhofer, H. Michel, Science **245**, 1463 (1989)
91. J. Deisenhofer, O. Epp, K. Miki, R. Huber, H. Michel, Nature **318**, 618 (1985)
92. J. Deisenhofer, O. Epp, K. Miki, R. Huber, H. Michel, J. Mol. Biol. **180**, 385 (1984)
93. H. Michel, J. Mol. Biol. **158**, 567 (1982)
94. E.W. Knapp, P.O.J. Scherer, S.F. Fischer, BBA **852**, 295 (1986)
95. P.O.J. Scherer, S.F. Fischer, in *Chlorophylls*, ed. by H. Scheer (CRC, Boca Raton, 1991), pp. 1079–1093
96. R.J. Cogdell, A. Gall, J. Koehler, Q. Rev. Biophys. **39**, 227 (2006)
97. M. Ketelaars et al., Biophys. J. **80**, 1591 (2001)
98. M. Matsushita et al., Biophys. J. **80**, 1604 (2001)
99. K. Sauer, R.J. Cogdell, S.M. Prince, A. Freer, N.W. Isaacs, H. Scheer, Photochem. Photobiol. **64**, 564 (1996)
100. A. Freer, S. Prince, K. Sauer, M. Papitz, A. Hawthorntwaite-Lawless, G. McDermott, R. Cogdell, N.W. Isaacs, Structure **4**, 449 (1996)
101. M.Z. Papiz, S.M. Prince, A. Hawthorntwaite-Lawless, G. McDermott, A. Freer, N.W. Isaacs, R.J. Cogdell, Trends Plant Sci. **1**, 198 (1996)
102. G. McDermott, S.M. Prince, A. Freer, A. Hawthorntwaite-Lawless, M. Papitz, R. Cogdell, Nature **374**, 517 (1995)
103. N.W. Isaacs, R.J. Cogdell, A. Freer, S.M. Prince, Curr. Opin. Struct. Biol. **5**, 794 (1995)
104. E.E. Abola, F.C. Bernstein, S.H. Bryant, T.F. Koetzle, J. Weng, in *Crystallographic Databases – Information Content, Software Systems, Scientific Applications*, ed. by F.H. Allen, G. Bergerhoff, R. Sievers (Data Commission of the International Union of Crystallography, Cambridge, 1987), p. 107

105. F.C. Bernstein, T.F. Koetzle, G.J.B. Williams, E.F. Meyer Jr., M.D. Brice, J.R. Rodgers, O. Kennard, T. Shimanouchi, M. Tasumi, J. Mol. Biol. **112**, 535 (1977)
106. R. Sayle, E.J. Milner-White, Trends Biochem. Sci. **20**, 374 (1995)
107. Y. Zhao, M.-F. Ng, G.H. Chen, Phys. Rev. E **69**, 032902 (2004)
108. A.M. van Oijen, M. Ketelaars, J. Köhler, T.J. Aartsma, J. Schmidt, Science **285**, 400 (1999)
109. C. Hofmann, T.J. Aartsma, J. Köhler, Chem. Phys. Lett. **395**, 373 (2004)
110. S.E. Dempster, S. Jang, R.J. Silbey, J. Chem. Phys. **114**, 10015 (2001)
111. S. Jang, R.J. Silbey, J. Chem. Phys. **118**, 9324 (2003)
112. K. Mukai, S. Abe, Chem. Phys. Lett. **336**, 445 (2001)
113. R.G. Alden, E. Johnson, V. Nagarajan, W.W. Parson, C.J. Law, R.G. Cogdell, J. Phys. Chem. B **101**, 4667 (1997)
114. V. Novoderezhkin, R. Monshouwer, R. van Grondelle, Biophys. J. **77**, 666 (1999)
115. M.K. Sener, K. Schulten, Phys. Rev. E **65**, 31916 (2002)
116. A. Warshel, S. Creighton, W.W. Parson, J. Phys. Chem. **92**, 2696 (1988)
117. M. Plato, C.J. Winscom, in *The Photosynthetic Bacterial Reaction Center*, ed. by J. Breton, A. Vermeglio (Plenum, New York, 1988), p. 421
118. P.O.J. Scherer, S.F. Fischer, Chem. Phys. **131**, 115 (1989)
119. L.Y. Zhang, R.A. Friesner, Proc. Natl. Acad. Sci. USA **95**, 13603 (1998)
120. M. Gutman, Structure **12**, 1123 (2004)
121. P. Mitchell, Biol. Rev. Camb. Philos. Soc. **41**, 445 (1966)
122. H. Luecke, H.-T. Richter, J.K. Lanyi, Science **280**, 1934 (1998)
123. R. Neutze et al., BBA **1565**, 144 (2002)
124. D. Borgis, J.T. Hynes, J. Chem. Phys. **94**, 3619 (1991)
125. F. Juelicher, in *Transport and Structure: Their Competitive Roles in Biophysics and Chemistry*, ed. by S.C. Müller, J. Parisi, W. Zimmermann. Lecture Notes in Physics (Springer, Berlin, 1999)
126. A. Parmeggiani, F. Juelicher, A. Ajdari, J. Prost, Phys. Rev. E **60**, 2127 (1999)
127. F. Jülicher, A. Ajdari, J. Prost, Rev. Mod. Phys. **69**, 1269 (1997)
128. F. Jülicher, J. Prost, Progr. Theor. Phys. Suppl. **130**, 9 (1998)
129. H. Qian, J. Math. Chem. **27**, 219 (2000)
130. M.E. Fisher, A.B. Kolomeisky, arXiv:cond-mat/9903308v1
131. N.D. Mermin, J. Math. Phys. **7**, 1038 (1966)
132. M.W. Schmidt, K.K. Baldridge, J.A. Boatz, S.T. Elbert, M.S. Gordon, J.J. Jensen, S. Koseki, N. Matsunaga, K.A. Nguyen, S. Su, T.L. Windus, M. Dupuis, J.A. Montgomery J. Comput. Chem. **14**, 1347 (1993)

Index

Fitzhugh–Nagumo, 149
Flory, 19
Flory parameter, 21, 26
Fokker–Planck, 87
force–extension relation, 8, 10, 15
Franck–Condon, 189
Franck-Condon factor, 202, 207
free electron model, 249
freely jointed chain, 3
Förster, 267

Gaussian, 213
Goüy, 52

heat of mixing, 21
Henderson–Hasselbalch, 64
Huggins, 19
Hückel, 45, 251

ideal gas, 30
implicit model, 37
interaction, 11
interaction energy, 20, 25
ion pair, 40
isotope effect, 166

Juelicher, 311

Klein–Kramers equation, 97
Kramers, 101
Kramers–Moyal expansion, 96

ladder model, 233, 237
Langevin function, 8
lattice model, 19
LCAO, 250
level shift, 232
LH2, 280
Lorentzian, 213

Marcus, 173
master equation, 98, 107
maximum term, 7, 15
Maxwell, 94, 159
mean force, 37, 38
membrane potential , 140
metastable, 28
Michaelis Menten, 82
mixing entropy, 19, 25

molecular aggregates, 274
molecular motors, 311
motional narrowing, 213
multipole expansion, 41, 262

Nagumo, 150
Nernst equation, 144
Nernst potential, 140, 142
Nernst–Planck equation, 135
nonadiabatic interaction, 196, 197
nonequilibrium, 125

octatetraene, 252
Onsager, 130
optical transitions, 201

pK_a, 66
parallel mode approximation, 199
phase diagram, 30, 33
point, 222
Poisson, 114
Poisson–Boltzmann equation, 46
polarization, 177
polyenes, 249
polymer solutions, 19
power time law, 119
Prost, 311
protonation equilibria, 61

ratchet, 311
reaction order, 76
reaction potential, 38
reaction rate, 75
reaction variable, 75
reorganization energy, 186, 189, 208

saddle point, 339
Schiff base, 302
Slater determinant, 248, 259
Smoluchowski, 98, 311
solvation energy, 43, 49
spectral diffusion, 209
spinodal, 31
stability criterion, 26, 28
state crossing, 219
stationary, 130
Stern, 57

time correlation function, 202, 205
titration, 62, 63, 65, 70